Building Design, Construction and Performance in Tropical Climates

The design, construction and use of buildings in tropical climates pose specific challenges to built environment professionals. This text seeks to capture some of the key issues of technology and practice in the areas of building design, refurbishment, construction and facilities management in tropical regions.

Using a consistent chapter structure throughout, and incorporating the latest research findings, this book outlines:

- the functional requirements of buildings in tropical climates;
- the challenges associated with the sustainability of the built environment, building form and whole life performance in the context of a tropical setting;
- the impact of potentially hostile tropical conditions upon building pathology and the durability of components, structure and fabric;
- the tasks which face those responsible for appraising the design, condition, maintenance and conservation of built heritage in tropical regions;
- the facilities management issues faced in tropical climates; and
- the refurbishment, upgrade and renewal of the tropical built environment.

The book is ideal as a course text for students of Architecture, Construction, Surveying and FM as well as providing a sound reference for practitioners working in these regions.

Mike Riley is Director of the School of the Built Environment, Liverpool John Moores University, UK.

Alison Cotgrave is Deputy Director of the School of the Built Environment, Liverpool John Moores University, UK.

Michael Farragher is Senior Lecturer in Architectural Technology, Liverpool John Moores University, UK.

Building Design, Construction and Performance in Tropical Climates

Edited by
Mike Riley, Alison Cotgrave
and Michael Farragher

LONDON AND NEW YORK

First published 2018
by Routledge
2 Park Square, Milton Park, Abingdon, Oxon OX14 4RN

and by Routledge
711 Third Avenue, New York, NY 10017

Routledge is an imprint of the Taylor & Francis Group, an informa business

British Library Cataloguing in Publication Data
A catalogue record for this book is available from the British Library

Library of Congress Cataloging in Publication Data
A catalog record for this book has been requested

ISBN: 978-1-138-20387-7 (hbk)
ISBN: 978-1-138-20388-4 (pbk)
ISBN: 978-1-315-47053-5 (ebk)

Typeset in Sabon
by Wearset Ltd, Boldon, Tyne and Wear

Contents

Notes on contributors viii
Acknowledgements xiii

1 Introduction 1
MIKE RILEY AND PAYAM SHAFIGH

Introduction 1
Overview 2
Building design 3
Building construction 7
Building performance 11

2 Environment and sustainability in tropical regions 13
NOOR SUZAINI MOHAMED ZAID AND BRIT ANAK KAYAN

Introduction 13
Overview 13
Impact of climate change 14
Sustainability in the construction industry 22
Life cycle assessment and eco-labelling 29
Embodied energy in buildings 29

3 Functional requirements of buildings: tropical context 37
NORHAYATI MAHYUDDIN, FARID WAJDI AKASHAH AND RAHA SULAIMAN

Introduction 37
Overview 38
Physical performance requirements of buildings in tropical climates 38
Macro, mesa and micro climate in tropical regions 43
Building science in tropical climates 49

4 Historical evolution of buildings in tropical regions 68
NOR HANIZA ISHAK, NUR FARHANA AZMI AND NOOR SUZAINI MOHAMED ZAID

Introduction 68
Overview 68
General characteristics of buildings in response to tropical climate 68
Building in context 82
Evolution of tropical buildings typology and morphology 90

5 Construction technology for tropical regions 103
MICHAEL FARRAGHER

Introduction 103
Overview 104
Building form: the humid tropics 107
Building form: the hot and dry tropics 117
Urban morphology 131
Climate analyses 136
Domestic building elements in the tropical regions 137
Industrial and commercial building in the tropical regions 158
Environmental services 176
Sustainable construction materials in the tropical regions 180

6 Building pathology, maintenance and refurbishment 188
ZAHIRUDDIN FITRI ABU HASSAN, AZLAN SHAH ALI, SHIRLEY JIN LIN CHUA AND MOHD RIZAL BAHARUM

Introduction 188
Overview 189
Common building pathology 190
Remediation of building defects 203
Maintenance consideration and techniques 203
Building refurbishment 208

7 Operational building performance in tropical climates 218
NIK ELYNA MYEDA, SYAHRUL NIZAM KAMARUZZAMAN AND CHEONG PENG AU-YONG

Introduction 218
Overview 218
Assessment trends 220
Building information modelling 225
FM practice in tropical zones 231
Performance planning and control 234
Optimising building performance 235

8 Case studies 240
MICHAEL FARRAGHER AND MIKE RILEY

Africa: Uganda 240
South America: Brazil 247
Australasia: Australia 252

Index 267

Contributors

Alison Cotgrave is Subject Leader in the Department of Built Environment at Liverpool John Moores University, UK and Associate Dean (Quality Assurance and Enhancement) in the faculty of Engineering and Technology. She has worked in academia for 23 years and was previously employed in the construction industry as a site manager. Alison is a Chartered Builder and Chartered Surveyor and has significant experience in the design and validation of construction programmes both in the UK and internationally. She is the co-author of a number of books on Construction Technology and has also published extensively in her area of research which is sustainability in construction education. She is currently on the supervision teams for five PhD students whose topics range from the psychological aspects of changing attitudes to sustainability via the curriculum to the design of a decision making aid for the selection of renewable/sustainable energy systems for buildings.

Mike Riley is the Head of the Department of Built Environment and Professor of Building Surveying at Liverpool John Moores University, UK. He is also Visiting Professor in Building Surveying at University of Malaya and has more than 25 years' experience in Building Surveying practice and Construction and Property education. Mike is a consultant to several national organisations in the UK and internationally, including Malaysia and Ghana. Mike is the joint author of several text books in the field of Construction Technology and Sustainability and he has published numerous academic papers in the areas of building pathology, property management and building performance appraisal. In addition he has presented papers and keynote addresses to numerous international conferences and is a regular reviewer for several internationally renowned academic journals and conferences. He obtained his first degree from Salford University (Building Surveying), followed by a Master of Science from Heriot-Watt University (Building Services Engineering) and PhD from Liverpool John Moores University (Building Performance). He is a Fellow of the Royal Institution of Chartered Surveyors and Fellow of the Royal Institution of Surveyors Malaysia, Senior Fellow of the Higher Education Academy and is a Chartered Environmentalist. He is also a Chairman of Assessors for the RICS Assessment of Professional Competence and has assessed candidates in Hong Kong and Malaysia as well as the UK.

Payam Shafigh received his doctoral degree in Structural Engineering and Materials from the University of Malaya as well as an MSc (Civil Engineering-Structure) and a BSc (Civil Engineering) from the University of Mazandaran, Iran. He is currently Senior Lecturer in the Department of Building Surveying, University of Malaya. He has received a number of national and international awards. He has taught courses of Concrete Technology, Theory of Structure, Structural Design and Strength of Material at the undergraduate level. His research interests include Concrete Technology, Lightweight Concrete, Concrete Pavement, Concrete Composites and Strengthening of Concrete Structures.

Noor Suzaini Mohamed Zaid teaches at the Department of Building Surveying, University of Malaya. Suzaini holds a PhD in Planning and Urban Development from the Faculty of the Built Environment, University of New South Wales. She has teaching experience in the fields of building surveying, urban planning and sustainable development. Her current research interest focuses on energy efficiency, zero-carbon development, climate change mitigation and adaptation in built environment. Suzaini was part of the testing of the United Nations Environment Programme's Sustainable Building and Climate Initiative (UNEP-SBCI)'s Common Carbon Metric and Protocol tool Pilot Test Phase 1 and Phase 2 in the Malaysian context, through her PhD research. Her research was conducted in collaboration of the Ministry of Higher Education Malaysia, University of New South Wales, UNEP-SBCI and the City Hall of Kuala Lumpur. Her current research project includes SULED-BIM: Sustainability Led Design Through Building Information Modelling in collaboration with the University of Manchester under the 2015 UK Newton Fund. Other projects she is involved in are within areas of affordability, acoustic properties of residential buildings, and housing and rehabilitation.

Brit Anak Kayan is a Senior Lecturer at Department of Building Surveying, University of Malaya and graduated with Bachelor of Science (Building Surveying) (Hons) in 1999. He received Master of Science (Building) in 2003, from University of Malaya and Doctor of Philosophy (PhD) (Construction Management) from Heriot-Watt University, UK, in 2013. Brit has over 16 years of experience as an academician and two years in industrial practice, latterly specialising in sustainable materials and repair in the context of 'Green Maintenance' modelling, Life-Cycle Assessment (LCA) approaches, Environmental Maintenance Impact (EMI) and low-carbon building materials, both in new built construction and heritage buildings conservation. Professionally, Brit is a full Member of the Royal Institution of Surveyors, Malaysia (MRISM) and a Registered Conservator of Department of National Heritage, Ministry of Tourism and Culture (MOTAC), Malaysia.

Norhayati Mahyuddin holds an honours degree from the University of Science Malaysia, a Master of Science degree from Universiti Teknologi MARA and a PhD degree from the University of Reading, UK. She is also a Registered Building Surveyor and a member of the Royal Institution of Chartered Surveyors (MRISM). She is currently the Head of the Building Surveying Department, Faculty of Built Environment, University of Malaya. Norhayati has been involved in Environmental Sustainability since early 2002. She is currently concerned about the long-term future of Indoor Environmental Quality. Her major research efforts include studies on ventilation effectiveness in built environment and building simulation through the application of Computational Fluid Dynamics (CFD), thermal comfort and Indoor Air Quality. Her research articles are published in many of the world's top ranked journals including *Building and Environment*, *Indoor and Built Environment*, *Renewable and Sustainable Energy Reviews*, *Building Performance and Simulation*, and a few others.

Farid Wajdi Akashah is a Senior Lecturer at the Faculty of Built Environment, University of Malaya and a registered building surveyor. He holds a Doctorate in Fire Safety Engineering from the Institute for Fire Safety Engineering Research and Technology (FireSERT), University of Ulster. He is actively involved in the development of standards related to fire safety, being one of the committee members for Technical Committee on Passive Fire Protection System with Engineering Standards Management & Consulting Services Section, SIRIM Berhad. His areas of interest are in the field of Fire Safety Engineering relating to Quantitative Fire Risk Assessment and Human Behaviour in Fire. He is a member of the Royal Institution of Surveyors Malaysia and the International Association of Fire Safety Science.

Raha Sulaiman is a Senior Lecturer in Building Surveying at the University of Malaya and specialises in indoor environmental quality. She holds the awards of MSc from Heriot-Watt University, UK and BSc in Building Surveying from the University of Malaya.

Nor Haniza Ishak is a Senior Lecturer in the Building Surveying programme in the Faculty of Built Environment, University of Malaya. Her qualifications include a Diploma in Architecture, a Bachelor of Building Surveying, an MSc in Project Management from Universiti Teknologi MARA and a PhD in Building Pathology from Universiti Kebangsaan Malaysia. Her research interests are in Building Performance, Project Management and Sustainability.

Nur Farhana Azmi is currently a Senior Lecturer in the Department of Building Surveying, University of Malaya. She holds a PhD degree in Building Control and a Performance and Bachelor degree in Building Surveying. Her research interests are in Building Control and Urban Conservation.

Michael Farragher graduated from John Moores University, UK in 1991 with RIBA Part II Diploma in Architectural Studies (Distinction) and RIBA Part I Bachelor of Arts (Hons) Architecture (1:1). He was elected onto the membership of the RIBA in 1997 after passing the RIBA Part III examination externally through Liverpool John Moores University. Qualified as a Chartered Architect and member of the RIBA, Mike has also won a government award for Integrated Renewable Technology use within a building project in Kent. Following some 18 years in private practice, Mike moved into the education sector and has recently spent 18 months teaching in central China at Henan University. He has worked in the commercial field of architecture and his built work in Merseyside includes social housing, healthcare and part of the Queen's Square development as well as collaboratively on the Speke Airport Hotel. His teaching experience within the Built Environment Profession has been within both Architecture (LJMU) and Architectural Technology (LJMU and Henan University, China). Mike has travelled extensively in Asia and Africa and is interested in how tropical forms of construction can inform passive design in temperate climates. He has published work on zero-carbon projects in the UK and is currently researching the development of sanitation systems in sub-Saharan Africa.

Zahiruddin Fitri Abu Hassan is a Senior Lecturer in Building Surveying at the University of Malaya. He holds a PhD from the University of Dundee, UK as well as an MSc in Building Surveying from the University of the West of England and a BS (Hons) in Building Surveying from the University of Malaya. His expertise and research is in building pathology, concrete durability and construction technology. An avid e-Learning advocate, he loves tinkering with computers and exploring new softwares and web 2.0 tools for education. His passion led the university to appoint him as the Head of e-Learning at the Academic Enhancement & Leadership Development Centre (ADeC).

Azlan Shah Ali is a professor in the Department of Building Surveying, Faculty of Built Environment, University of Malaya, Kuala Lumpur, where he has been a faculty member since 2004. He is also currently a Visiting Professor at School of Built Environment, Liverpool John Moores University, UK. He completed his PhD at Universiti Teknologi MARA, Malaysia specialising in building refurbishment. His research interests lie in the area of building maintenance and refurbishment with the focus to improve management strategies. He has collaborated actively with researchers in several other disciplines of built environment, particularly surveying and architecture local and internationally. Dr Azlan has been involved in a number of academic researches with a total amount of more than RM 1 million and has published over a hundred technical publications in journals, proceedings and books. He is a Chartered Building Surveyor, UK; a Registered Building Surveyor, Malaysia; a Fellow of The Royal Institution of Surveyors Malaysia (FRISM) and The Royal Institution of Chartered Surveyors, UK (FRICS).

Shirley Jin Lin Chua holds a PhD degree in Asset and Facilities Management and a Bachelor degree in Building Surveying from the Faculty of Built Environment, University of Malaya, Malaysia. Her research interests are in Facilities Management and Analytic Hierarchy Process. She has also been involved in various research grants and has published several ISI and Scopus-indexed technical publications in journals and proceedings pertaining to building maintenance and facilities management. She has also received a number of awards throughout her studies and career path which include the Building Surveying Excellence Award, University of Malaya Book Prize and University of Malaya Excellence Award 2015.

Mohd Rizal Baharum is a consultant at the University of Malaya Consultancy Unit, and also administrative member at the Faculty of Built Environment, University of Malaya. He has been involved in teaching, programme development, research and consultancy projects for more than ten years. He is also a professional member of the International Facility Management Association and the Malaysian Association of Facility Management. He received his MSc in Facilities Management and Asset Maintenance from Heriot-Watt University, UK, and a doctoral degree from Liverpool John Moores University, UK. The study was granted by the University of Malaya–Liverpool John Moores University bursary scheme. Rizal also carries out external advisory roles for the Royal Malaysian Navy; Facility Management programmes both at the Malaysian Polytechnic and Ministry of Higher Education; and Open University Malaysia.

Nik Elyna Myeda is a Senior Lecturer at the Department of Building Surveying, University of Malaya and holds a doctoral degree in Facilities Management from The

Bartlett School of Graduate Studies, University College London. Elyna's main teaching and research interests focus on Strategic Facilities Management and Performance Measurement and Sustainable Environment.

Syahrul Nizam Kamaruzzaman is Associate Professor in the Department of Building Surveying at the University of Malaya. He obtained his PhD from the School of Mechanical, Aerospace and Civil Engineering, University of Manchester, UK. His research interest is in Building Technology, Construction Engineering, Energy Efficiency in Buildings and Facilities Management. Dr Syahrul is a certified Building Surveyor and also a registered member of the Royal Institution of Surveyors Malaysia (MRISM).

Cheong Peng Au-Yong holds a PhD degree in Facilities Management from the Faculty of Built Environment, University of Malaya. His research interest is in facilities management and building maintenance. He has published several articles in reputable journals. His recent publications are 'Improving Occupants' Satisfaction with Effective Maintenance Management of HVAC System in Office Buildings' in *Automation in Construction*; 'Prediction Cost Maintenance Model of Office Building Based on Condition-Based Maintenance' in *Eksploatacja i Niezawodnosc – Maintenance and Reliability*; 'Office Building Maintenance: Cost Prediction Model' in *Gradevinar*; and 'Significant Characteristics of Scheduled and Condition-Based Maintenance in Office Building' and 'Participative Mechanisms to Improve Office Maintenance Performance and Customers' Satisfaction' in *Journal of Performance of Constructed Facilities*.

Acknowledgements

The authors would like to express their thanks to the following people for their support and contribution to the book:

David Tinker for his time and effort in creating some of the illustrations contained within the book.

Robin Hughes for his advice and commentary on the preliminary drafts.

Naomi Jayne Wilson for her general support with the draft and assistance in sourcing images and permissions to publish.

Julie, Steve and Sam for their support and understanding during the writing of this text.

Becky Randles for assistance with indexing.

In addition, the following is a list of people and companies to whom we would like to extend our thanks for the use of original material, case studies and reproduced BIM drawings:

Pilkington Glass, St Helen's for use of Pilkington Spectrum version 7.0.1.

Autodesk for the use of Revit software in producing original drawings.

Chartered Institute of Building for using original graphical data.

World Wildlife Fund, Germany for using original graphical data in Tables 5.8, 5.9 and 5.10. © WWF – World Wide Fund For Nature, 'Illegal wood for the European market: An analysis of the EU import and export of illegal wood and related products', July 2008.

Alex Yong Kwang Tan and Nyuk Hien Wong, Department of Building, School of Design and Environment, National University of Singapore, for use of original data.

Shane Thompson Architects, Queensland, AU.

EMF Griffiths Engineers, Australia.

Thalita Gorban Ferreira Giglio and Ercilia Hitomi Hirota, at Construction Department, State University of Londrina, Brazil.

Sara Wilkinson, University of Technology Sydney.

1 Introduction

Mike Riley and Payam Shafigh

After reading this chapter you should:

- recognize the geographical positioning of areas considered to be 'tropical';
- appreciate the characteristics of the tropical climate;
- be aware of the factors that impact upon the design, construction and use of buildings in tropical climates; and
- understand the importance of considering buildings in tropical climates discretely from those in other climatic zones.

Introduction

The tropical regions of the World are typically characterized by climatic conditions that are hot, humid and lacking in great thermal variability between seasons. The areas that fall within what are referred to as 'the tropics' are located close to the equator, between the Tropic of Cancer to the North and the Tropic of Capricorn to the south. The warm, moist conditions of these areas create a distinctive environment for the creation and use of buildings. By nature the environment would be lush and potentially dense with vegetation as a result of the favourable conditions for growth of flora and fauna. These same conditions provide potential challenges to the creators and users of buildings as the control of unwanted natural agencies can require high levels of consideration. The belt of land and sea that forms the tropics includes many variants of local climate and geography. Similarly it provides home to a multitude of nations and communities. As such it is impossible to generalize about the design, construction and use of buildings in such regions. Similarly, in a text such as this it is impossible to make specific reference to all. However, across these areas there are similarities that are consequent to their positioning within a tropical context. Whether considering sub-Saharan Africa, Central and South America, India or South East Asia, all share common elements in the ways in which creators and users of buildings have responded to the challenges and the opportunities created by their tropical positioning. This text seeks to highlight some of these broad themes with reference to the context of tropical buildings. The case studies within the final chapter provide some specific examples of buildings and the use of materials in some of the regions that we include in our consideration of the tropics.

Historically the buildings in such regions would have been developed using local materials and vernacular design and construction principles. However, as industrial, commercial and residential building forms have become more homogenous across World regions, the specificity of design and construction features has been altered.

Overview

The creation and use of buildings in tropical climates pose specific challenges to designers, surveyors, constructors, facilities managers and occupiers. Traditional approaches to dealing with the challenges of tropical climates exploited natural materials and vernacular design approaches. However, as the nature of buildings and their construction form have become more internationally uniform, different approaches and design features have evolved to allow buildings to perform in different climatic conditions. Designers must recognize the specific functional and environmental requirements of buildings in such situations. Similarly, the impact of potentially hostile tropical conditions upon building pathology and the durability of components, structure and fabric can be significant. Building maintenance and management must take into account these issues in order to diagnose and develop solutions to building defects. At the same time the use and occupation of these buildings must be approached in a manner that ensures appropriate functionality and achievement of performance expectations on the part of developers and occupiers.

Challenges associated with the sustainability of the built environment have become widely recognized in terms of urban development, building form and whole life performance. The context of these issues in a tropical setting presents unique demands on built environment professionals as they seek to respond to these challenges. Alongside the increasingly rapid pace of development in many tropical countries there is the will to conserve the heritage that is manifested within the existing and historic built environment. The technology of such buildings ranges immensely and the task which faces those responsible for appraising their design, condition, maintenance, conservation and use differs fundamentally from that faced by those developing contemporary urban environments capable of delivering the needs of modern commerce and society. Innovative approaches to surveying, building pathology, design, refurbishment, project management, property management, and maintenance and facilities management afford the opportunity to conserve and enhance existing buildings and to create new buildings and facilities that are fit for purpose and sustainable in a tropical setting.

It is important to recognize that the nature of property as a global asset has the potential to result in a degree of homogeneity in the form and function of modern real estate. It is essential that buildings are considered in terms of their local context with reference to the impact of challenging climates. Many of the features that are expected by building occupiers in terms of comfort and internal environmental quality are directly opposed to the natural principles of sustainable design and construction. Evaluative mechanisms such as post-occupancy evaluation need to be cautious of this. The perspectives and skills sets of professionals seeking to recognize and address such challenges must be informed by in-depth understanding of the underlying principles of building pathology and performance. Approaches to meeting these challenges need to be balanced against the economic constraints, competing priorities, cultural issues and aesthetic tastes of people and society in tropical environments.

This text seeks to capture some of the key issues of technology and practice in the areas of building design, refurbishment, construction and facilities management in tropical regions. It is aimed at students of Architecture, Construction, Surveying and Facilities Management as well as providing a sound reference for practitioners working in these regions. Based upon cutting-edge research the text makes a unique and original contribution to knowledge and understanding of buildings, the built environment and built environment practice in a tropical context. The chapters are written in language that is easy for students to understand and each chapter includes information under a series of themed, main headings. Each chapter is suitably referenced so that students can easily determine where further reading can be found on the particular aspect of the content.

Building design

The building sector, worldwide, is responsible for around a quarter of the total greenhouse gas (GHG) emissions associated with the consumption of fossil fuels. At the same time the embodied energy of construction materials has a very considerable impact upon their overall GHG emissions and, if included, result in a far higher effect. As a consequence the construction sector is often considered as one of the primary global generators of greenhouse gasses (UNSBCI 2010).

Hence, the design, construction and use of buildings has a massive effect upon the ability of nations across the globe to achieve the targets for GHG reduction required to arrest the issue of climate change. Many of the countries in the tropical regions are developing countries and their potential impact upon the sustainability agenda is great. Much of the energy consumed in such areas is currently derived from biomass and the overall level of energy consumed per capita is lower than in developed countries. However, there is expected to be significant improvement in living conditions in the future, with the associated increase in levels of energy consumption, which is anticipated to lead to a shift from biomass to fossil fuels. Such a shift will increase CO_2 emissions quite dramatically. The design, construction and use of buildings will play an important part in the overall position of developing countries, many of which are in tropical climates, in the ongoing sustainability agenda. As more advanced building forms are adopted and the nature of development and urbanization shifts there is likely to be an increase in the use of air-conditioning and growth in the number of domestic appliances, allied to an overall increase in the number of buildings constructed. In Europe it is estimated that, by 2050, around 25 per cent of the total building stock will be made up of new buildings; in developing countries it is estimated that the figure will be closer to 75 per cent. The energy demands of buildings in tropical regions, if utilizing air-conditioning and modern building services and amenities, will be higher than traditional, vernacular forms.

If these new buildings are as energy hungry as those currently existing the impact upon GHG production and the ability to achieve targets associated with reduction in CO_2 emissions will be challenging.

Hence, the design, construction and operation of buildings in tropical regions must recognize the long-term goal of reducing energy consumption from the creation and use of buildings. Unlike buildings in temperate areas, which derive much of their energy demand from space heating, buildings in tropical areas derive much of theirs from cooling. Increases in the extent of commercial, retail and high quality residential real estate in developing nations have a significant impact upon overall energy demands. Design approaches for new buildings that aim to use natural ventilation and to minimize solar gain can assist in reducing energy demands to some extent.

There is a degree of variability in the tropical climatic zones but the general tendency is for them to display a fairly consistent set of environmental conditions with respect to heat, humidity and seasonal variability. Most tropical regions will tend to have hot, humid climates with high rainfall and relatively little variance in day and night temperatures or seasonal temperatures. There are, of course, several sub-sets of tropical climatic types, which are described in detail in later chapters.

With such climatic characteristics in mind there is a set of generally accepted design principles that might logically be applied for both vernacular and more modern buildings and for both residential and commercial forms. These include the following:

- Promotion of air movement using large openings in external enclosure and minimal internal sub-division to promote natural cross-ventilation. Such an approach makes the maximum use of the slight cooling breezes that may exist in the absence of a significant difference in internal and external temperatures. To this end buildings in these regions will often adopt a long, thin floor plan to maximize the potential for fenestration and exploit cross ventilation as effectively as possible.
- Minimization of thermal mass of the building by using lightweight materials for the external fabric. This approach is aimed at ensuring that the building fabric does not act as a heat store and that it allows rapid response to temperature changes in the ambient environment. This should assist in shedding heat from the internal enclosure quickly when cooling breezes do exist. This is seen in traditional building forms, using local, lightweight materials such as woven rattan, as well as more modern buildings using lightweight cladding rather than brick and masonry exterior walls. If such an approach is not adopted there is the danger that external fabric, which heats up during the day, will radiate heat to the building interior at night.
- Insulation of building fabric on surfaces and elevations that are exposed to solar radiation. This assists in preventing heat gain in the building interior and can be combined with the adoption of light-coloured, reflective surface treatments to inhibit solar gain.
- Control of solar gain. The use of reduced sizes and number of windows on elevations exposed to solar radiation is useful in reducing the amount of heat generated within the building. This might be in tension with aspiration to maximize cross ventilation but the use of solar shading and tactical vegetation can assist greatly with this aspect, whilst, at the same time, providing protection from high rainfall and possibly monsoon conditions. The potential use of vegetation as solar shading in both winter and summer conditions is shown in Figure 1.1. However, it is most often the case in tropical regions that there is minimal difference between the winter and summer state of vegetation. This is different from moderate climates where there is significant difference in both extent of vegetation and solar aspect between seasons.
- High ceilings within interior spaces allow for increased air movement and are often combined with approaches to window design such as louvred claddings as illustrated in Figure 1.2. In this example, taken from Colombo in Sri Lanka, the contemporary building adopts a bioclimatic approach to design which maximizes passive features to manage the hot, humid conditions. Such an approach is in contrast to the transition of international designs to tropical regions that is typified by the air-conditioned office block.

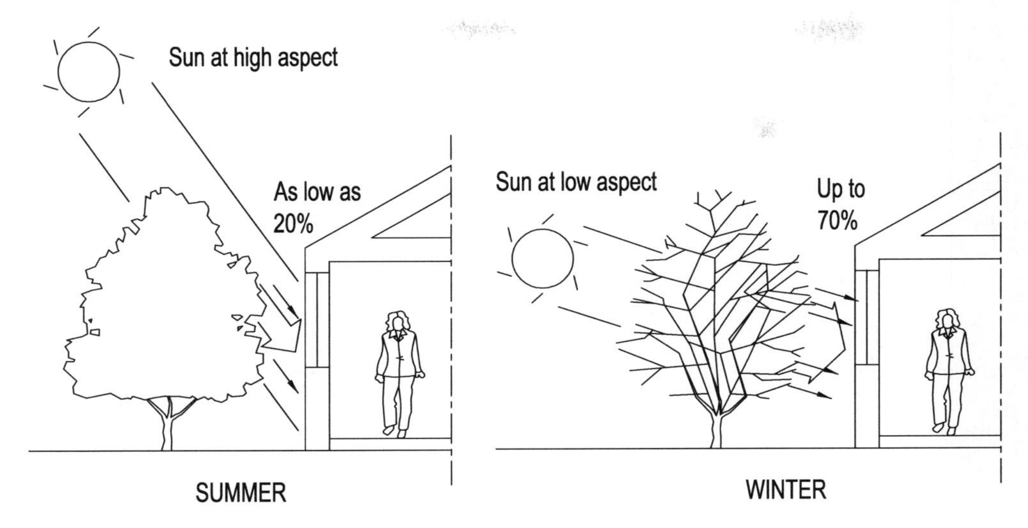

Figure 1.1 Vegetation as solar control.

The design features of buildings in tropical regions are covered in detail in later chapters of this book. Chapter 5 considers the nature of the overall design and the design features of buildings in tropical regions.

The way in which buildings are designed in response to the tropical context reflects the recognized climatic parameters of solar radiation, air temperature, relative humidity and wind. Many areas in the tropics must also cope with the potential dangers of earthquake and tsunami. However, this is not a consistent threat across all tropical areas, and much of central Africa and central South America are free from such issues. The local variation in conditions results in geographically specific responses to the various demands of the tropical climate and specific site conditions. Hence there is tremendous variety of built form in response across the international distribution of countries in the tropical regions. Each reflects the local design influences, materials availability, climatic and environmental demands, and cultural variance.

In many developed areas the techniques of construction are the same as those in non-tropical regions. The use of reinforced concrete, steel and the major materials groups used in international commercial and industrial buildings are reflected in tropical areas also. Whilst the vernacular forms of traditional buildings in localized tropical regions reflect local climate, materials and design approaches, more modern buildings adopt generic international features. Construction has evolved to integrate international materials and construction details and techniques. The development of bioclimatic design, which seeks to maximize the synergies between climate, locality and construction form, has driven an evolution in approach to built form.

The adoption of passive design features, which maximize the environmental and physical performance of buildings in tropical zones results in demonstration of distinctive construction features. Natural ventilation, control of solar gain and the matching of thermal mass with environmental conditions are concepts familiar to vernacular buildings. They are now featured more prominently in modern commercial and industrial built form. Figure 1.2 illustrates the utilization of louvred cladding, high internal ceilings and light external fabric in the modern convention centre in Colombo, Sri Lanka. Such features are also manifest in traditional tropical construction forms such as the Thai or Malay house, which matches thermally light envelope, large openings to promote ventilation and high ceiling voids. This is illustrated in Figure 1.3, the image of which is used several times within this text as an illustration of multiple design and performance concepts.

The increased adoption of bioclimatic design, which is discussed in detail in a later chapter, infers the adoption of natural ventilation, passive solar gain reduction, and non-active design and construction features to match the building performance to the local environment. There are some excellent examples of such a design approach in many tropical areas, as illustrated in the case studies at the end of this text. However, commercial, residential and other forms of modern building still experience a natural tension between such an approach and the more generic adoption of air-conditioned spaces within internationally derived building designs.

The distinguishing typologies of vernacular buildings in tropical climates are far more visible than they are in more modern, commercial forms. The principles of their design and construction, based upon the use of locally sourced materials and appropriate design features and technologies, has not been lost in more modern forms. However, in many tropical countries there has been an overlaying of colonial styles and approaches that draws heavily upon materials and techniques from quite different climatic regions. Many parts of the World display the design heritage of British, Dutch, Portuguese and Spanish colonialization amongst others. Indeed some locations display the combined heritage of multiple different colonial periods in the same place. This is certainly the case in towns such as Malacca.

Figure 1.2 Colombo Convention Centre, Sri Lanka.

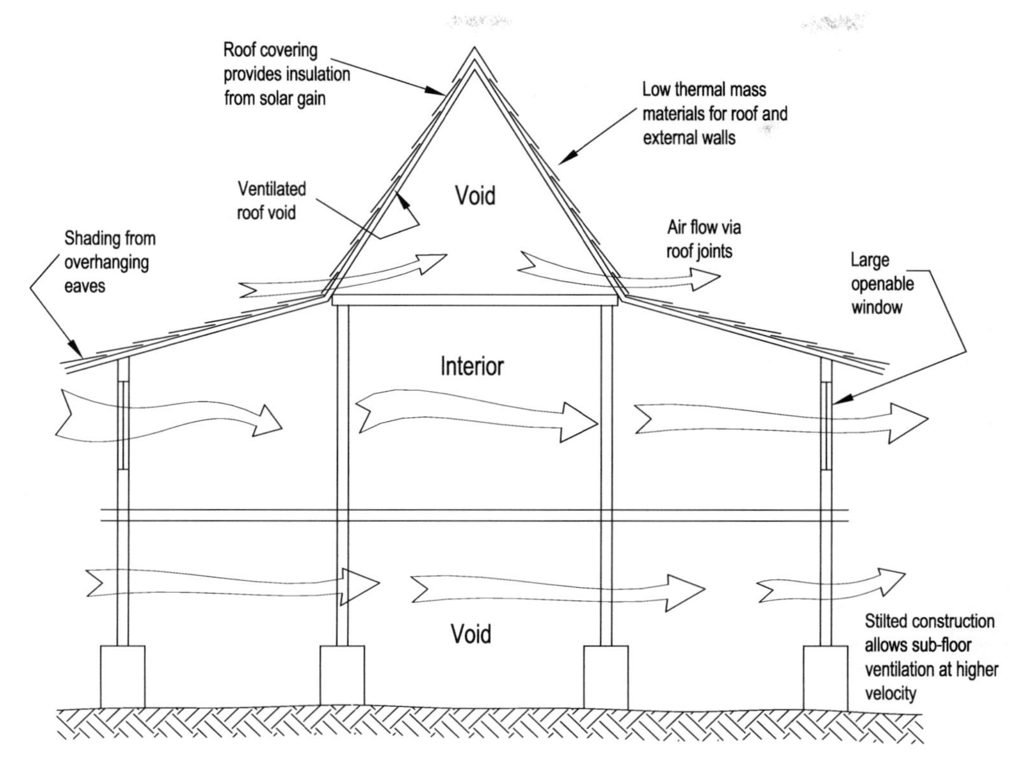

Figure 1.3 Features of a traditional Malay house.

Building construction

The techniques of constructing buildings in tropical regions share much with those adopted in the rest of the World. Vernacular forms have traditionally adopted local craft skills to fashion locally derived materials into traditional typologies. The advent of more significant urban and rural development has seen a shift to the use of internationally adopted materials and techniques. Steel and concrete structural frames, clad with a range of external claddings and finishing, typify commercial construction across the globe.

In that sense the variance in construction practice between tropical and non-tropical regions might be considered to be very limited. However, the manner in which construction materials behave is different in warm, humid environs from that of cooler, dryer conditions. Essentially, however, the principles are the same and the process of construction varies in terms of context rather than principle. The use of cementitious materials needs to be carefully considered and managed, given the potential for high temperatures to cause premature drying, with the subsequent implications for strength, durability and surface appearance.

The behaviour of building elements and components depends primarily upon three factors; the environmental conditions, material characteristics and geometrical form. The prevailing environmental conditions influence the nature of loading on the structural elements. As such the climatic context of the tropics may take on a significant influence upon built form and performance in use.

Changes in temperature result in changes to dimensions of construction materials. The fact that tropical buildings are likely to be subject to a very limited diurnal temperature range means that components are less likely to suffer from defects derived from repeated expansion and contraction. However, rapid cooling of elements such as roof coverings that have been heated by solar gain can occur in the event of the high rainfall that is common in tropical areas. In such circumstances temperature drops of more than 40 degrees in a few minutes are not unusual. This creates potential for rapid contraction of elements such as roof membranes, relative to the more stable underlying structure, which retains heat due to its thermal mass. Such performance characteristics need to be reflected in both design and construction practice. The issue of defects in tropical areas is dealt with specifically in a later chapter of this book.

Thermal variation also derives from solar gain on specific elevations of some buildings. If a structure such as a high-rise building is heated from one side only, by the sun shining on it, its temperature will not rise uniformly; the sunny side will heat up more quickly and there will be a – potentially high – temperature 'drop' across the width of the building. This means that the side of the building facing the sun will get longer than that facing away from the sun. As such the building will seek to bend along its height, inducing strain in connections, components and fabric. In addition the detailing of movement joints to cope with expansion and contraction takes on a huge significance.

Changes in average temperature and temperature differentials are the key thermal loading parameters in tropical climates. High temperature can induce stress in the structural elements beyond allowable limits.

The primary characteristics of the various tropical areas are warm and dry air and dry ground, or warm and moist air together with high rainfall. In warm, dry areas the temperature remains high during the day time and humidity fluctuates between modest to low. Outdoor climate is highly influential on the interiors of the buildings located in such tropical areas. Arrangements are needed in these dry areas to save the exterior and interior of the building from solar radiation and the hot, dusty winds. Consideration should be given to using heavy materials with high thermal resistance to construct walls and roofs. The outer wall, if constructed with a material with increased thermal capacity, will capture a large amount of temperature during the day time. In contrast, the warm-humid climate is characterized by hot, sweaty and sticky conditions as well as continual presence of dampness. Air temperatures remain moderately high with little variation between day and night. Such a scenario is best catered for by the use of lightweight building enclosures with low thermal mass.

Historically, the use of local materials that reflect the performance requirements has resulted in very recognizable vernacular forms that reflect climatic conditions. As buildings of greater scale have developed the nature of their construction has aligned more with generic, international approaches to construction. It is still essential, however, to reflect the particular design requirements of such regions. In tropical climates, the construction of the whole building with the same material is not usually possible. Since the serviceability requirements have differences, it is essential to choose different materials based on their use. As far as the main structural elements (columns, beams, slabs and connections) are concerned, the material which has higher resistance to temperature with lesser moisture content should be preferred.

Tropical and equatorial construction is subject to its own unique set of defect characteristics. This is, in part, because the intensity of the sun, which sits at a higher azimuth from the horizon and which is constant throughout the year, influences the weather and humidity of this area. The tropical and equatorial belt has high humidity due to the downdraft of the jet stream pushing warm and humid air downwards. This climatic feature is one of the main distinguishing characteristics of tropical countries and it is significant in the manner in which building defects derive and propagate. Humidity is perhaps the main factor in the tropical regions that make them different from the climate in other parts of the world. As such it is a factor that also forces distinction in the nature and extent of defects in building structure, fabric and components. In terms of vegetation growth, the warm, humid environs provide an excellent environment for the development of flora. Even well developed, modern cities give the impression that if left untended the forests would consume them rapidly given the chance. These fertile conditions can result in rapid growth of vegetation mould and fungi in and around buildings, which must be controlled. The warm, moist conditions are ideal for vegetation growth but also ideal for the development of organic agencies that damage buildings.

Unlike temperate regions, decay mechanisms influenced by water and water vapour work consistently throughout the year with minimal seasonal variation. In addition the exposure to these consistently elevated temperatures accelerate chemical reactions resulting in degradation of materials and components. The tropical regions are also distinguished, in part, by high levels of rainfall and in some instances monsoon climates. In such conditions even minor defects in the external building enclosure that might result in a modest leakage in a temperate climate might be the cause of catastrophic water ingress in a monsoon climate. This is particularly the case with modern commercial buildings where the external enclosure is likely to be based on some form of non-porous rainscreen cladding. In such circumstances, even minor fissure can be responsible for major ingress of water.

Local building practice has responded and has been largely successful in reflecting the nuances of the climate, with specific adaptation to deal with the threat of defects appropriate to the locality and the micro-climatic conditions.

However, as buildings and construction practice have homogenized internationally to reflect a 'global' construction form for modern commercial buildings, the approach to dealing with pathology and defects has changed. Construction processes and materials selection have shifted to less climatically specific forms using steel, concrete, bricks, mortar and plaster. As such the typography of defects in tropical areas has changed with these materials. Although the construction of buildings does take into account, to some extent, the vagaries of the tropical climate, the defects that occur in these modern buildings are different to the types of defect that befall similar buildings in other climates.

Sustainable construction materials in tropical regions

One of the key characteristics of tropical regions is the extent to which the warm and wet climate supports rapid growth of vegetation. This has traditionally provided a ready source of natural materials for vernacular and traditional construction.

However, as more modern construction techniques have become widespread, such materials have fallen out of mainstream use. More recently, in the strive towards sustainable construction approaches there have been some major innovations which seek to utilize locally available materials and waste products to support modern, mainstream construction technologies. An example of this is the use of oil palm shell (OPS).

An average 10–20 million tonnes of OPS is produced annually in the tropical countries, such as Malaysia, Indonesia and Thailand. Historically the oil palm waste including OPS and palm oil fly ash (POFA) was disposed of by uncontrolled dumping which eventually caused environmental pollution. However, the utilization of these waste materials into potential materials in the production of concrete helps to reduce pollution as well as reducing the requirements to monitor the dumping sites. Abdullah (1984) pioneered the incorporation of OPS as aggregate in concrete and, by using this, lightweight aggregate concrete (LWAC) can be produced.

Similar approaches have been taken with other readily available materials such as coconut shell. The cultivation of the coconut is predominant in East Africa and the tropical regime countries of Asia. This material can be used as a construction material in the concrete industry due to its varying shape, weight, size and colour. Similar to OPSs, coconut shells are produced in very large quantities in the tropical regions of the World. Gunasekaran and Kumar (2008) reported that the coconut shells are one of the most auspicious agro-waste materials and they can easily be used as coarse aggregate in the production of structural and sustainable LWAC. Coconut shell can successfully be used as a lightweight aggregate to produce green and sustainable LWAC.

Corn cob has also been utilized as an aggregate and was first introduced by Pinto *et al.* (2012) to produce lightweight concrete for non-structural applications. It was found that by using corn cob as an alternative aggregate to the natural and artificial materials several economic and environmental benefits could be realized.

Rice husk ash has also been considered as a possible cement replacement for the production of concrete. Concrete containing rice husks can be used as a material with hybrid characteristics between insulating concrete and structural LWAC.

The by-products produced from the incineration and the cogeneration process of sugar cane bagasse at certain temperatures is called bagasse ash (BA). Currently a huge amount of BA is being produced in tropical countries such as Thailand, Pakistan, Brazil, India, the Philippines, Colombia, Malaysia and Indonesia. It has been found that this material is a very promising pozzolanic material which can be effectively utilized as a supplementary binding material with ordinary Portland cement for the production of mortars and concretes.

In many instances tropical countries are developing countries that have, at some point in their history, experienced the impact of colonalization of foreign powers. The colonial powers brought with them their building materials, construction methods and styles. They were extremely successful in adapting these imported materials to the local climate through detail changes to take into account the conditions and context of the local climate. These colonial approaches worked in harmony with vernacular building forms, local materials and cultural context in many instances, although perhaps less successfully in others. More recently, however, such local adaptation has been superseded by more generic and international construction trends imported from more temperate regions. These modern commercial styles characterizing modern international architecture are often overlaid with local contextual or culturally relevant detailing. For example, broadly similar high-rise building forms in Asia, Africa and India are often distinguished only by cosmetic locally specific adornment.

Building performance

In the tropical regions, there has been a tendency for few designers to take into account the climate of the surroundings in the base form of their buildings. A variety of building forms and construction methodologies based on those usually employed in Europe and America have become increasingly prevalent for commercial building in Africa, Asia and the Indian sub-continent. As a result, thermal discomfort afflicts these types of buildings and necessitates the occupants using artificial ventilation and air-conditioning systems (HVAC). The high relative humidity in these regions can lead to problems with Internal Environmental Quality (IEQ) that have been dealt with through mechanical ventilation and cooling rather than natural or passive approaches. Clearly this impacts upon the sustainability of the building development in such areas. As the trend for redevelopment and restoration of existing buildings has spread across the tropical regions, so the issues of rectifying IEQ failings have taken a higher prominence. In newly renovated buildings in central and eastern Cameroon, the high indoor relative humidity is a persistent problem. Consequently, the indoor air almost always contains a higher percentage of water vapour than the outdoor air, which results in condensation risk and potential occupant health issues.

Chapter 6 of this text considers the main forms of defect associated with buildings in tropical regions. One of the key characteristics of such issues is the rapid pace with which organic agencies can develop to cause deterioration to building fabric. The lush tropical climate is a great advantage for developing flora and fauna. However, this very advantage can wreak havoc in even relatively new buildings through the action of mould, fungus, insect attack and vegetation growth. Whilst there is huge variability in the sub-categories of tropical climate, the broad principles and effects of these agencies are applicable throughout.

High levels of humidity and occasional high rainfall, or even monsoon conditions, result in the aggressive weathering of external fabric. In many tropical countries buildings that have been built very recently appear dilapidated due to this effect. The monsoonal climate also brings with it the potential for structural damage, flooding and for the building to be the target of flying debris in high winds. This imposes upon building occupiers a unique set of maintenance and whole life performance issues that are unknown to occupiers in moderate climatic areas. For instance, with a tropical climate and average rainfall of 250 centimetres in a year, waterproofing systems for roofs and external fabric are crucial for buildings. The implications of even minor failure can be significant in terms of water penetration to the interior. However, the performance of waterproofing systems depends on many factors, including the choice and quality of materials, the skill of workers, application methods, substrate condition, prevailing weather conditions, building orientation and the whole life approach to maintenance.

The culture of building maintenance and effective Facilities Management is far less advanced in many developing countries than in Europe or America. As many of the countries in tropical regions fall into this category, it is implicit that the development and application of effective property maintenance is still underdeveloped in these areas.

Poor building maintenance will lead to unchecked building defects and resultant damage and costly repair work. In Malaysia and Singapore, for example, buildings are built broadly in accordance with British Standard and under strict supervision. However, the maintenance approach to buildings is still weak, and developers and occupiers do not yet value ongoing maintenance in the same way that they consider new building developments.

Chapter 7 of this book considers the issues around building performance in use and building management or FM. Generic discussion of the principles and benefits of various established approaches is provided with reflection on the specific context of tropical countries. As noted above the application of these approaches is far less developed in tropical areas than it is in Europe and America. Hence, this chapter focuses to some extent on what is possible and what is needed rather than what is currently being undertaken. As the developing World accelerates urban development, the issues around establishing robust philosophies regarding the whole life performance of existing buildings is in danger of taking even less prominence than currently exists.

The final chapter presents a series of case studies from Uganda, Brazil and Australasia that illustrate the ways in which designers are approaching response to the challenges of the tropical climate. In addition there is illustration of the nature of materials performance and the increasing tendency to apply sympathetic bioclimatic approaches rather than simply transposing international design approaches into tropical climates. This is a welcome shift away from the homogenized application of generic built form which typified the 1970s, 1980s and 1990s. Such approaches relied upon the construction of generic building types with a retro-engineered climate control system based on mechanical ventilation and comfort cooling. More contemporary approaches seek to maximize the benefit of natural, passive approaches to promoting natural ventilation, solar shading and complementary action of the building with the local ambient environment.

References

Abdullah, A. A. (1984). *Basic strength properties of lightweight concrete using agricultural wastes as aggregates*. Paper presented at the Proceedings of the international conference on low-cost housing for developing countries, Roorkee, India.

Gunasekaran, K. and Kumar, P. S. (2008). *Lightweight concrete using coconut shell as aggregate*. Paper presented at the Proceedings of the ICACC-2008. International conference on advances in concrete and construction, Hyderabad, India.

Pinto, Jorge, Vieira, Barbosa, Pereira, Hélder, Jacinto, Carlos, Vilela, Paulo, Paiva, Anabela, … Varum, Humberto (2012). Corn cob lightweight concrete for non-structural applications. *Construction and Building Materials*, 34, 346–351.

United Nations Sustainable Buildings and Climate Initiative (UNSBCI) (2010). Common carbon metric: protocol for measuring energy use and reporting greenhouse gas emissions from building operations. *Draft for Pilot Testing*. Paris, France: UNSBCI.

2 Environment and sustainability in tropical regions

Noor Suzaini Mohamed Zaid and Brit Anak Kayan

After studying this chapter you should be able to:

- understand the concept of sustainability and climate change in the tropical context;
- appreciate strategies of adaptation and mitigation of climate change;
- recognize the drivers and barriers, together with the benchmark and policy and guidelines relating to sustainable construction techniques;
- discuss the link between life cycle assessment (LCA) and environmental consideration; and
- appreciate energy and carbon management to reduce environmental impact.

Introduction

The most commonly accepted definition of the term sustainable development derives from the Brundtland Commission (1987) that defines sustainable development as "development that meets the needs of the present without compromising the ability of future generations to meet their own needs" (Brundtland 1987, p. 43). Sustainable development as a concept was developed alongside acute awareness of the ecological destruction and social concerns manifested as poverty, deprivation and urban dereliction that blight many parts of the world.

In the last few decades, the concept of sustainable development has been gaining attention and becoming a cornerstone to development policies in the built and natural environments. The term 'sustainability' itself has become a contested concept by politicians, economists, academics, theorists, businesses and corporations alike. Such tensions were openly acknowledged by the President of the United Nations General Assembly (Razali Ismail), not more than five years after the Rio summit at which the concept of sustainable development was officially adopted and operationalized.

He mooted that the vision of Agenda 21 is not being achieved and that industry was moving away from, not towards, sustainable development.

Overview

In some ways, progress of sustainable development has been normalized into the mainstream medium of development. There has been a debate about the contested meaning and contradictions of the term 'sustainable development' itself, in which sustainable brings meaning to maintain and sustain; and development is often associated with growth. The Brundtland definition also raised questions of what are the 'needs', what should be sustained, and which direction for development? These issues must be in accordance with specific context, in order to meet the needs of current and future generations (CIB and UNEP-IETC, 2002). There is an underlying tension between the associated aspects of sustainability, as well as the wide interpretation of the concept that has led to a variety of urban form being described as 'sustainable'.

The CIB report on Agenda 21 for Sustainable Construction in Developing Countries (CIB and UNEP-IETC, 2002) defines 'sustainable development' as "the kind of development we need to pursue in order to achieve the state of sustainability" (CIB and UNEP-IETC, 2002, p. 6); and 'sustainability' as "the condition or state which would allow the continued existence of homo sapiens" (CIB and UNEP-IETC, 2002, p. 6).

The report also stresses that sustainable development is not a goal, but merely the process of maintaining a balance between the demands of human beings for equity, prosperity and quality of life in the most ecological way possible. Other debates on sustainable development also surround the tension between the 'North' and 'South', what time frame responses are planned for, and the question of whether the problems that sustainable development attempts to address can be solved by the same consciousness that created them in the first place. The terminology 'the North and the South' that is increasingly being used in development circles contains geographical inaccuracy when many of the countries termed as 'the South' are located in the northern hemisphere or equatorial, tropical zones.

There is an increasing need for the developing countries – many of which are in tropical zones – to address sustainability if real global change is to be achieved. According to the World Bank glossary, developed countries are known as industrial countries, industrially advanced countries, or high-income countries with Gross National Income (GNI) of more than $11,905 per capita (measured by quantity of various goods and services consumed) (World Bank, 2004, 2010). The developing countries are those with GNI below $11,906 per capita (including five high-income developing economies that are Hong Kong (China), Israel, Kuwait, Singapore and United Arab Emirates (UAE)) (World Bank 2004, 2010). It is also worth noting that 80 per cent of the world's population lives in the developing world (World Bank, 2004) and most of these regions are predominantly urban, with an average of 5 million growth every month (UN-HABITAT 2008). The developed world's urban population has largely remained unchanged and is experiencing a decline due to low rates of natural population increase and declining fertility rates. This phenomenon is however not explicit, as some countries in the developing world have shown decline in urban population (UN-HABITAT, 2008, p. xiii). Urban populations are increasing in countries that can least afford to provide appropriate infrastructure. Providing basic services such as affordable housing, educational facilities and sanitation and transportation systems are therefore pressing concerns. It is for these reasons that sustainable practices need to be adopted at a much greater rate than in developed countries.

Impact of climate change

The United Nations Framework Convention on Climate Change (UNFCC) defines climate change as "a change in climate which is attributed directly or indirectly to human activity that alters the composition of the global atmosphere and which is in addition to natural climate variability observed over comparable time period" (UNFCC, 1992, p. 3). Human activities are the main drivers of climate change, either directly or indirectly. Climate change is attributed to changes in atmospheric concentration, land cover and solar radiation that alter the energy balance of the climate system (IPCC, 2007a). The earth receives energy from the sun and the flow of energy is scattered back into space through infrared radiation. The growing intensity of greenhouse gases (GHGs) in the atmosphere blocks the earth's infrared radiation from escaping directly to space, subsequently warming the earth's surface and increasing global temperature, i.e. global warming.

Concentrations of GHGs such as carbon dioxide (CO_2), methane (CH_4), nitrous oxide (N_20) and halocarbons which are gases containing fluorine, chlorine or bromine, in the atmosphere have exceeded the natural range over the last 650,000 years, as a result of human activities since industrial revolution in the eighteenth century (IPCC, 2007a). The biggest contributor to GHGs is CO_2 emitted from using fossil fuels, at 56.6 per cent, while the largest sector contributing to CO_2 is energy supply at 25.9 per cent, and industry at 19.4 per cent (IPCC, 2007a, p. 36).

It is estimated that global temperatures will rise by about 3°C over the next century (IPCC, 2007b). The severity of climate change and environmental degradation will be felt globally, even when the action that causes it, happens in a different continent (Table 2.1). Such catastrophic events as sea-level rise, draught, biodiversity loss and decreased rainfall will be a direct impact on temperature increase anticipated from 0.5 to 6°C. Subsequent to the sea-level rise, there is a growing concern as to how it will impact human health, especially for small islands that are located in the tropical regions, where the weather and climate are already conducive to many

Table 2.1 Tipping elements in the Earth's climate system

Tipping element	*Global warming parameter (°Celsius)*	*Key impacts*
Disappearance of Arctic summer sea ice	0.5–2	Amplified warming, ecosystem change
Greenland ice-sheet meltdown	1–2	2–7 m rise in sea-level
West Antarctic ice-sheet collapse	3–5	5 m sea-level rise
Reorganization of the Atlantic thermohaline circulation impacts on inter-tropical convergence	3–5	Regional cooling, sea-level affected
El Nino Southern Oscillation (ENSO): increased amplitude elsewhere	3–6	Drought in South East Asia
Indian summer monsoon: change in variability	Temperature not concerned	Drought
Sahara/Sahel and West African monsoon: collapse	3–5	Increased rainfall in the Sahel
Amazon rainforest: dieback	3–4	Biodiversity loss, decreased rainfall
Boreal forest: dieback	3–5	Biome switch

Source: Lenton *et al.* in IUCN (2008).

transmitted diseases such as malaria, dengue, filariasis, schistosomiasis, and other food- and water-borne diseases (Mimura *et al.*, 2007).

With growing innovations in technology, urbanization, population growth and consumerism behaviour, the Earth's resources are being consumed faster than ever, polluting the environment and emitting toxins without proper consideration for the planet's future. The rate of consumption and depletion of the Earth's resources are now faster than the pace of natural reproduction and nations are going into ecological debts. The Living Planet Report (WWF, 2008) records the growing rate of environmental degradation and introduces the notion of ecological debt, which estimates national consumption that exceeds their country's bio-capacity. It stresses the consequences of living beyond our means, using resources at a quicker rate than they can be replenished.

There is a need for transitional economies to deal with issues of environmental impact and sustainable development, especially when cities are competing against each other to reach world-class status and achieve advanced development. The current world footprint is at a global overshoot – humanity currently uses the equivalent of 1.6 planets to supply the demand on natural resources, and that exceeds the biosphere's supply or regenerative capacity. The condition of generating waste faster than waste can be turned back into resources is called an ecological overshoot, and it is estimated now that the Earth takes one year and six months to regenerate what we use in a year (Global Footprint Network, 2015). Figure 2.1 illustrates the overshoot of 1.6 planet earths needed to cope with humanity's demand on resources and absorb waste, and calculated that the Earth Overshoot Day for 2016 was on 8 August, where humanity had exhausted the ecological supply and for the rest of the year we consume earth's resources on an ecological deficit (Global Footprint Network, 2015).

With reference to tropical regions the global trend of rising CO_2 emission from fuel combustion has seen a particularly drastic increase from Asia. The current culture of consumption, which is perpetuated by economic growth, is a major challenge in achieving sustainability. The implication of current consumption levels is that we might have to change some of our fundamental beliefs about how we will behave and how the future will unfold.

According to the Fifth Assessment Report from the IPCC, adaptation and mitigation are only complementary strategies to reduce and manage risks of climate change impact. The amount of CO_2 in the atmosphere has caused the Earth's temperature to rise to warmer temperatures than ever recorded before any preceding decade since 1850 (IPCC, 2014), breaking records of temperature anticipated by scientists.

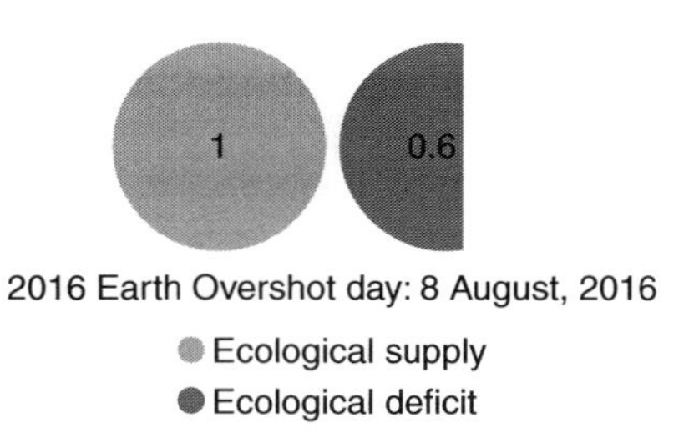

Figure 2.1 World Ecological Footprint,

Source: Global Footprint Network (2015).

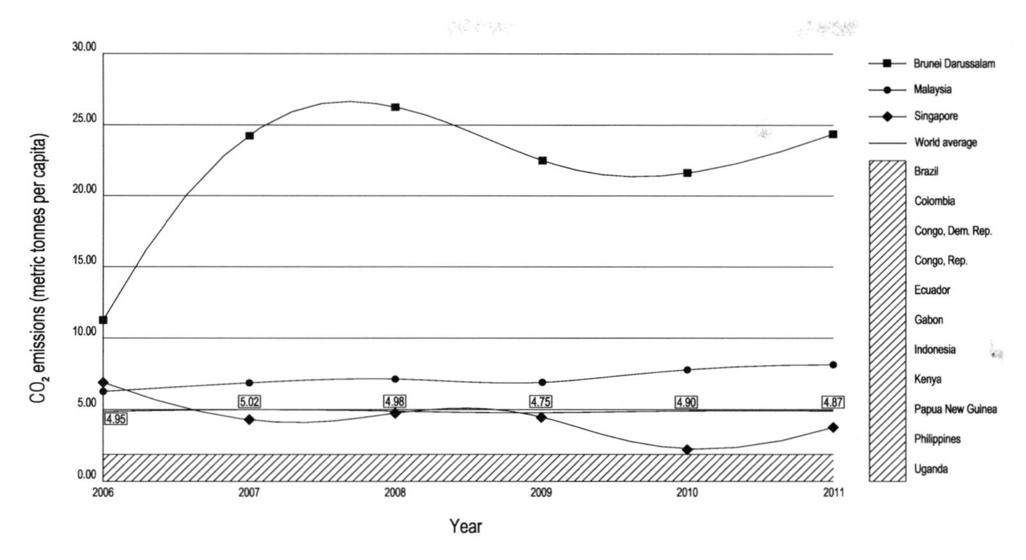

Figure 2.2 CO_2 emissions (metric tons per capita) for tropical countries.

Source: adapted from World Bank (2015).

Sustainability and climate change in tropical climates

Sustainable development does not only deal with environmental issues, it also entails economic and social features that are more commonly known as the three pillars of sustainable development or the 'triple bottom line': economic, environmental and social. One of the remaining principal challenges of sustainability is to make the Brundtland definition operational, that is, use it to guide direction, and developing sustainability criteria or indicators for the design of policies must be clearly defined in order to make the definition operational. Some have used sustainable development as a political strategy and resource to pursue differing sets of interests, and it is becoming a politicized battleground that struggles over value, meaning and knowledge.

For developing world cities, sustainability is becoming an increasingly elusive object, because of the impacts by forces that are beyond their own borders. Some argue that the cost incurred in the interests of sustainable development falls unevenly on different social groups, resulting in the poor being disproportionately affected. This supports the hypothesis that sustainable development's benefits are being enjoyed by those privileged few. Therefore, it is important to explore how the tropical regions have dealt with sustainability, and more importantly climate change issues.

According to the Koppen-Geiger Climate Classification, the tropical region is a climatic region that is specific to the equatorial line and categorized as Tropical Moist Climate (A). The Tropical Moist Climate can be further classified according to the distribution of precipitation patterns such as Tropical Fully Humid (Af), Tropical Monsoonal (Am) and Tropical Wet and Dry (Aw) (Institute for Veterinary Public Health, 2011; Pidwirny, 2011). The locations of these classifications are listed in Table 2.2.

Table 2.2 Locations of tropical moist climate classification

Tropical Fully Humid or Tropical Wet (Af)	• Amazon River Basin (South America) • Congo River Basin (Africa) • Eastern coast of Central America • Eastern coast of Brazil • The Philippines • Coast of Madagascar • Eastern India • Southern Bangladesh • Malaysia • Indonesia
Tropical Monsoonal (Am)	• Coast of south-western India • Sri Lanka • Bangladesh • Myanmar • South-western Africa • Guyana • Surinam • French Guiana • Parts of north-east and south-east Brazil
Tropical Wet and Dry (Aw)	• Northern and eastern India • Central Myanmar • The Indo-Chinese Peninsular • Northern Australia • Region around the Congo River Basin • South-central Africa • Western Central America • Parts of Venezuela • Parts of Brazil • Southern tip of Florida • The Caribbean Islands

Source: Pidwirny (2011).

World Bank metadata, relating to climate change indicators such as the amount of CO_2 emissions (metric tons per capita), electric power consumption (kWh per capita), energy use (kg of oil equivalent per capita), forest area (per cent of land area or sq. km), and so forth are available publically. Based on the World Bank metadata, the combined figures for individual countries listed in the Tropical Fully Humid (Af) climate classification are taken to display the amount of CO_2 emissions (metric tons per capita) for the tropical region against the World Average in Table 2.3.

From Table 2.3, it can be seen that the highest emitter per capita is Brunei Darussalam (24.39 metric tons per capita), well beyond the world average CO_2 emissions (4.87 metric tons per capita) by approximately 400 per cent in 2011 but has successfully declined its emission in the consequent years to 18.9 metric tons per capita in 2013 (World Bank, 2015). Another South East Asian country that has emitted more than the world average is Malaysia, by approximately 61 per cent more in 2013 (8.03 metric tons per capita) (World Bank, 2015). Figure 2.2 illustrates this worrying trend of CO_2 emissions for two rapidly developing South East Asian countries, Brunei and Malaysia.

Table 2.3 CO_2 emissions (metric tons per capita) for the tropical region

Country name	*2006*	*2007*	*2008*	*2009*	*2010*	*2011*
Brazil	1.82	1.88	1.99	1.87	2.11	2.19
Brunei Darussalam	11.60	24.18	26.34	22.15	21.87	24.39
Colombia	1.44	1.41	1.45	1.48	1.71	1.56
Congo, Dem. Rep.	0.05	0.05	0.05	0.05	0.05	0.05
Congo, Rep.	0.31	0.33	0.34	0.43	0.47	0.54
Ecuador	2.07	2.20	2.06	2.36	2.29	2.35
Gabon	1.48	1.38	1.40	1.42	1.46	1.42
Indonesia	1.51	1.62	1.75	1.90	1.81	2.30
Kenya	0.26	0.26	0.27	0.31	0.31	0.33
Malaysia	6.50	7.04	7.66	7.37	7.99	7.90
Maldives	2.66	2.63	2.79	2.96	2.93	2.93
Papua New Guinea	0.69	0.96	0.73	0.76	0.68	0.75
Philippines	0.77	0.78	0.84	0.82	0.88	0.87
Singapore	7.00	3.97	4.93	4.78	2.66	4.32
Uganda	0.09	0.10	0.10	0.11	0.11	0.11
World Average	4.95	5.02	4.98	4.75	4.90	4.87

Source: World Bank (2015).

Other developing countries located in the tropical region in South America (Brazil, Colombia, Ecuador), Africa (Gabon, Congo, Democratic Republic of Congo, Uganda, Kenya, Maldives), Asia (Indonesia, Philippines) and Australasia (Papua New Guinea) were emitting less than the world average with the exception of Singapore. Singapore has managed to steadily reduce its CO_2 emissions lower than world average since 2007.

One of the key factors that affects global climate change is the change in our forests. The world's forest has a significant role in the Earth's carbon cycle, where it absorbs and stores carbon in its trees and soil – known as carbon sequestration. When deforestation happens, the carbon that is stored is released as carbon dioxide (CO_2) and other GHGs, which contribute to approximately a fifth of global GHGs (WWF-UK, 2016). Climate change mitigation options through forest mitigation include reducing emissions from deforestation, enhancing sequestration rate in existing and new forests, providing wood fuels to substitute fossil fuels and providing wood products for energy-intensive materials.

The estimated global forest cover on Earth is 3,952 million hectares, which is approximately 30 per cent of the global land area. Between 2000 and 2010, gross deforestation was at 13 million hectares per year, due to conversion of forest to agricultural land, expansion of settlements, infrastructure and unsustainable logging practices (Nabuurs *et al.*, 2007; WWF, 2016). Deforestation threatens species and biodiversity loss, jeopardizes people's livelihoods and intensifies global warming, and most of the global deforestation took place in tropical countries such as Malaysia, Indonesia and the Philippines.

Environmental degradation of tropical forests

Tropical forest and its soil provide one of the world's largest terrestrial (land) natural carbon capture and storage functions, sequestering an equivalent of 1.5 Gt of global carbon emissions annually (or 17 per cent) (IAP, 2010). However, when the forest is cut down, the same amount of carbon is released into the atmosphere, which consequently makes the deforestation impact on climate change bigger than emissions from transportation (Dattaro, 2014).

The global deforestation rate annually is at an average of 8–15 million hectares per year (IAP, 2010) and Malaysia saw a 115 per cent increase in deforestation during the first three months of 2013 according to National Aeronautics and Space Administration (NASA) researchers, and noted Sarawak with the most forest cover change since 2011 (Sukumaran, 2013). The alarming rate of deforestation has been linked to illegal logging activities in both Peninsular and East Malaysia, and Malaysia's deforestation was the highest among the top five countries with massive deforestation detected in 2013: Nepal (114 per cent), Mexico (92 per cent), Argentina (72 per cent) and Madagascar (51 per cent) (Butler, 2013; Sukumaran, 2013). Malaysia's forest loss from 2000 to 2012 accounted for 14.4 per cent or 47,278 square kilometres (km^2), a total area larger than the country of Denmark. Realistically, land that was once forest and has been deforested will never become forest land again unless it is replanted. If the land is used for development then this will never occur.

In 2014, NASA researchers found severe forest loss occurred in Bolivia, Malaysia and Cambodia using a satellite called MODIS that orbits and captures images of the Earth, orbiting the North and South poles (Dattaro, 2014). The data image in a given location is compared with an image taken on the same day in previous years to detect whether there is loss of over 40 per cent of green vegetation (Dattaro, 2014). The green vegetation loss can be from natural causes such as hurricanes or wildfires, or can be the result of human activities such as cutting down or burning trees for lumber or clearing of land for grazing, crops or other developments.

Climate change and ecological impact in tropical South East Asia region

All of South East Asia (SEA) falls under the definition of tropical and subtropical climate zone, and stretches to some 13,000,000 km^2 of land and sea (Encyclopædia Britannica, 2016). According to the Association of Southeast Asian Nations (ASEAN), there are ten member states that comprise Brunei Darussalam, Cambodia, Indonesia, Lao People's Democratic Republic (PDR), Malaysia, Myanmar, Philippines, Singapore, Thailand and Vietnam (ASEAN, 2009).

SEA is annually affected by climate change impacts and climate extreme conditions such as flooding, drought and tropical cyclones, which severely threaten the livelihood of poor population in rural areas with limited adaptive capacity and resources (IFAD, 2009). Climate change is affecting many sectors in SEA such as agriculture, coastal systems, ecosystems and water resource (IFAD, 2009; IPCC, 2007a). Coastal hazards such as large tidal variations, tropical cyclones and heavy regional rainfall is predicted to increase with climate change impact due to the geology and geography of SEA's regional coastal zone, which puts the growing coastal population and infrastructure at risk (IFAD, 2009). Studies have projected the displacement of several million people in coastal areas with only a one metre rise in sea level, and mitigation measures to reduce the sea-level rise (to between 30 and 50 centimetres) could cost millions of dollars annually (IFAD, 2009; IPCC, 2007a).

In recent years, NASA researchers have not been able to link the deforestation in Bolivia, Malaysia and Cambodia to any natural causes; it would be most likely lost through human activities. Based on data gathered by Global Forest Watch, in 2014 Malaysia accounted for 200,715 hectare (ha) of forest loss outside of plantation purposes, and a total tree cover loss of 40.68 per cent (Hansen *et al.*, 2013), which causes an ecological deficit.

An ecological deficit occurs when the ecological footprint of a population exceeds the biocapacity of the area available to that population. A national ecological deficit means that the nation is importing biocapacity through trade, liquidating national ecological assets or emitting carbon dioxide waste into the atmosphere. An ecological reserve exists when the biocapacity of a region exceeds its population's ecological footprint. Biocapacity is the capacity of biosphere to regenerate and provide for life (Carbon Footprint Network, 2016).

Table 2.4 further highlights the state of biocapacity deficit or biocapacity reserve for countries located in the SEA region, in terms of global hectare (biologically productive hectare with the world average productivity) per capita. Singapore has an biocapacity deficit of 5.9 global hectares per capita and is one of the highest biocapacity deficits in the world (Carbon Footprint Network, 2016). There are only two countries in the SEA region that have a biocapacity reserve, which are Lao PDR and Myanmar at 0.3 and 0.4 global hectares per capita respectively. This deficit will continue to grow if the region continues to allow deforestation at alarming rates, which in turn will affect its climate change contribution in reduction of carbon sequestration capacity.

Therefore, according to the facts presented earlier in terms of CO_2 emissions rate, the climate change and ecological impact occurring in the tropical region and alarming rate of some SEA countries, we need to find strategies for adaptation and mitigation. One of the biggest areas of potential to reduce these impacts lies within the tropical forests in increasing carbon sequestration capacity, and in the building sector to reduce CO_2 emissions in its operation for the existing building stock.

Table 2.4 South-east Asian ecological deficit/reserve

Country	*Ecological deficit*	*Ecological reserve*	*Biocapacity (deficit) or reserve (gha per capita)*
	Total ecological footprint (gha per capita)	*Total biocapacity (gha per capita)*	
Brunei	Not available	Not available	Not available
Cambodia	1.1	1.1	(0.0)
Indonesia	1.3	1.2	(0.1)
Lao PDR	1.2	1.5	0.3
Malaysia	2.9	2.3	(0.6)
Myanmar	1.5	1.9	0.4
Philippines	1.0	0.5	(0.5)
Singapore	5.9	0	(5.9)
Thailand	1.9	1.2	(0.7)
Vietnam	1.4	1.0	(0.4)

Source: Carbon Footprint Network (2016).

Adaptation and mitigation of climate change

The building sector, comprising residential and commercial buildings, contributes 7.9 per cent to global CO_2 emissions (IPCC, 2007a, p. 36). The building sector primarily contributes to climate change through the causal effect of using fossil fuels in building operations, more than its construction (UNEP, 2009). According to the Asian Business Council, Asian cities account for more than half of the world's total new construction, with China and India leading Asia in levels of new construction and the expansion of built-up areas (ABC, 2007). China alone accounts for 18.7 per cent of total CO_2 global emissions (IPCC, 2007b), and this increasing trend is expected to continue with the construction of two new coal-fired plants every week to meet high electricity demands (ABC, 2007, p. 2). The Indian gross built-up area is fast growing with a rate of 10 per cent per annum since the last decade (UNEP, 2010). The rising trend of CO_2 emissions from Asia is replicated in the striking rate of economic development and expansion of international market forces (IEA, 2009).

Sustainability in the construction industry

The construction industry has not made significant efforts in addressing the sustainability and efficiency issues within the built environment (ABC, 2007), and an urgent call for a more sustainable construction industry is noted. Cities and buildings are in themselves not sustainable, as they are causally linked to the global ecological degradation, but they also play a major role in contributing to global sustainability (Burnett, 2007; Rees and Wackernagel, 1996). Buildings consume approximately 40 per cent of energy and 60 per cent of electricity in most countries, producing 30 per cent of annual GHG emissions and contributing 10–40 per cent of urban solid wastes (Graham, 2003; WBCSD, 2008, 2009). Higher levels of consumption are predicted to increase with the expansion of the construction industry. The majority of energy consumption during building occupation is for indoor space heating (10–30 per cent) and cooling purposes (0–50 per cent) (IPCC, 2007b; WBCSD, 2008, 2009) but varies according to culture, climate, building design and choice of building envelope (WBCSD, 2009). In tropical areas by far the most significant issue is cooling.

However, globally the construction industry generates employment for millions of people, contributes approximately 11 per cent to a nation's GDP and monetary growth, and provides society with housing and services (CIB, 1997; CIB and UNEP-IETC, 2002; UNCED, 1992; WBCSD, 2009). Therefore construction, it can be argued, contributes towards two of the pillars of sustainability: economic strength and social improvement. Residential properties cover the vast majority of the existing building stock in most countries, with approximately 30 billion square miles of existing residential covering China's built landscape and an estimated 2 billion square meters being added annually (WBCSD, 2008, p. 22). The construction industry plays a fundamental role in shaping a sustainable future as it impacts substantially on building human settlements and infrastructure that are needed for development.

At the same time, the built environment and the construction industry have the largest potential to address these issues head on (UNEP, 2009; WBCSD, 2008). The construction industry's direction has slowly shifted from traditional resource consumption with little regard to the environment to a more environmentally integrated agenda. As more awareness of environmental protection policies, climate change and environmental degradation issues are introduced to developers, contractors, policy makers and the general public, a more sustainable construction industry is presented. Identified measures for buildings to reduce GHG emissions can be addressed within three categories: "reducing energy consumption and embodied energy in buildings, switching to low-carbon fuels including a higher share of renewable energy, and controlling the emissions of non-CO_2 GHG gases" (IPCC, 2007b, p. 389).

According to the OECD's Sustainable Building Project, governmental policies should address environmental impacts of the building sector through reduction of CO_2 emissions, minimization of construction and demolition waste, and prevention of indoor air pollution (OECD, 2001). Actions taken now in the building sector will continue to affect its contribution of GHG emissions. This is due to the relatively long lifespan of buildings and by using the life-cycle approach as it is estimated that over 80 per cent of GHG emissions are during building operation for heating, cooling, lighting and other applications (UNEP, 2009, p. 6). Much of building's GHG emissions and energy consumption is wasted through three significant, but changeable factors (WBCSD, 2009, p. 10):

1 Poor building design
2 Inadequate technology
3 Inappropriate behaviour

Based on the evidence available, the direct link between the built environment and contribution to climate change is undeniable. Cities and the built environment play a pivotal role in mitigating climate change and addressing environmental degradation (Burnett, 2007).

Sustainable construction is seen as a subset to the ideology of sustainable development, and has the potential to contribute positively and proactively towards environmental protection (CIB, 1997; CIB and UNEP-IETC, 2002; Zainul Abidin, 2009). According to the first International Conference on Sustainable Construction (1994), sustainable construction is defined as "the creation and responsible maintenance of a healthy built environment based on resource efficient and ecological principles" (Kibert, 1994 cited in Bourdeau, 1999, p. 354). This broad definition plays only a starting point role (Bourdeau, 1999) as the construction industry in itself is multidimensional and with diversified professionals. In order to achieve the goal of sustainable construction, all sectors of the construction industry, the civil society and voluntary organizations must work together in finding the best possible solutions (Gill, 2000).

Sustainable construction can be identified as a process that, with time, sustainability is achieved (Zainul Abidin, 2009) in terms of methods of production, design and existing stock of the built environment. The linear method of the construction industry's production, operation, maintenance and disposal has caused separation of responsibilities, which resulted in failure of feedback loops among stakeholders involved within the creation and operation of the built environment (Kibert, 2000). Sustainable construction should entail a closed loop cycle between production, use and demolition/refurbishment. The need for a more 'cradle to cradle' approach, which recycles the waste at the end of a building's lifespan for production of new construction has been recognized (Braungart *et al.*, 2007; Graham, 2003).

The construction industry is an active player in consumption of extensive amounts of global resources and waste emissions, but also plays a role in socio-economic development and quality of life (du Plessis, 2002). Sustainable construction must then incorporate the triple-bottom-line concept that integrates economic prosperity, social well-being and environmental protection within the construction industry (Zainul Abidin, 2009). It has also been recognized for an international agreement on what constitutes sustainable construction. The International Council for Research and Innovation in Building and Construction (CIB) in 1999 prepared a report specifically for this reason, the Agenda 21 on Sustainable Construction Report (CIB, 1997; du Plessis, 2002).

Sustainable construction as defined by CIB is "a holistic process aiming to restore and maintain harmony between the natural and built environments, and create settlements that affirm human dignity and encourage economic equity" (du Plessis, 2002, p. 8). Sustainable construction is also recognized as "a way for the building industry to respond towards achieving sustainable development on the various environmental, socio-economic and cultural facets" (CIB, 1997, pp. 18–19). Sustainable buildings are then the most significant and logical manifestation of the many aspects to sustainable construction (Gill, 2000). Constructing more sustainable buildings will have an impact on the larger built and natural environment.

However, the more mainstream sustainable development concept is built upon market-driven technologies that promise to steer sustainability without dramatic changes, or "upsetting the comfortable, the rich or the powerful" (IUCN, 2008, p. 12). There is a huge potential for new buildings to be more energy efficient with appropriate building design approaches, but one of the biggest challenges to overcome is a shift of behaviour barriers (Graham, 2003, 2008; IPCC, 2007b; Watson, 2004; WBCSD, 2008).

The combination of Western Europe and North America's way of life, which represents 10 per cent of the world's population and consumes approximately 50 per cent of the world's resources, has inspired the standards and direction for global development (Hartkopf and Loftness, 1999). The American built landmarks of supermarkets, restaurant chains and highways have changed the built environment's landscape with stocks of McDonalds, Starbucks and K-Marts (Smith *et al.*, 1998). This wave of change is not only seen in America but also in many developing world cities, as with globalization and economic growth (Sassen, 2009). As many cities develop tropical regions there is a tendency to mimic these international characteristics of affluence. Affluence and wealth, which are influenced by economic, social and psychological factors, have empowered the behaviour of consumption of non-renewable resources at distressing rates (Hester, 1995; WBCSD, 2008). This changing and sprawling landscape increases travelling distance as suburbs are being built and connected through layers of highway (Smith *et al.*, 1998). It is detrimental that changes are needed in the way we build, especially in the developing world where the majority of the world's new construction is taking place (ABC, 2007). It has been recognized that the developing world requires a different approach to the direction of development, as they have different needs and priorities, and their problem, scale and capacity are very much different from the developed world (du Plessis, 2002).

According to the World Bank, in China alone, it is estimated that every year lost in failure to build efficient buildings locks in approximately 800 million square meters of urban built space of inefficient energy use for decades in the future (ABC, 2007). China's construction industry consumes almost 50 per cent of total energy and produces 650 million tonnes of waste per annum (Wang *et al.*, 2010). It is also projected that by the year 2020 China will have constructed a total of 70 billion square meters of buildings (Wang *et al.*, 2010). This raises the important issue of locking the future of the built environment to be 'unsustainable', through constructing more new buildings, which are inefficient in the way they consume energy and emit CO_2 gas.

The developing world needs to make a decision. To follow the example of the developed world in contributing to an unsustainable world via its choice of building design and development, is to fail to learn from the developed world's mistakes.

CIB produced a report following this concern which is directed specifically to the developing world's construction industry and the Agenda 21 for Sustainable Construction in Developing Countries (du Plessis, 2002). Sustainability depends on an institution of consensus of contribution made by each sector within the economy, e.g. buildings, and specific targets need to be identified to make an appropriate contribution to sustainability (Zimmermann *et al.*, 2005). The construction industry is, by nature, fragmented (Spence and Mulligan, 1995), hence reaching a consensus between all the stakeholders involved is essential to improve the current conditions. Without consensus of the identified target of each sector, it is difficult to determine a standard within the construction industry, and to identify which direction that contributes to sustainability. As Zimmermann *et al.* (2005) stress, "without binding subtargets for the different sectors, it will be almost impossible to move systematically towards a sustainable society" (Zimmermann *et al.*, 2005, p. 1148).

A study conducted in 2009 to investigate awareness levels of the sustainable construction concept amongst developers in Malaysia concluded that little efforts are made to implement it, despite the rising awareness (Zainul Abidin, 2009). The study also deduced that developers, as a majority, perceived sustainability was only about environmental protection, with social and economic considerations being out of the remit of the construction industry (Zainul Abidin, 2009). As low-cost housing is seen as a government effort to provide affordable housing for the lower income population, and not as a means to generate capital return, it is unlikely for developers to consider quality and performance towards low-cost housing projects. Most of the initiatives and demonstration projects for sustainable construction were implemented in commercial or office buildings, and private houses. Factors recognized as impeding the implementation of sustainable practices are, foremost, lack of legislation and enforcement; lack of knowledge and awareness; and a passive construction industry culture (Zainul Abidin, 2009, 2010).

The building sector holds tremendous potential in addressing the challenge of carbon mitigation through the lifespan of buildings in tropical areas. Mitigation of GHG emissions and assessing actual building performance during its operational phase is more likely to have a positive impact on the environment. Actual building performance depends of the quality construction, design, operation and maintenance systems (Levine *et al.*, 2007). Measuring energy performance of buildings will play a large role in contributing to sustainable building performance, as carbon emission is recognized as the largest building pollutant to climate change. The operational stage in buildings account for 80–90 per cent of total carbon emissions from energy use (United Nations Sustainable Buildings and Climate Initiative, 2010), which justifies the need to measure actual operating building performance. Monitoring the performance level of buildings will also facilitate its operating efficiency (Levine *et al.*, 2007). The construction industry can play a major contributor role towards sustainable development, locally and globally, and in accordance with the 20 per cent CO_2 emissions reduction by the year 2050 as set by the Intergovernmental Panel on Climate Change (IPCC) (UNDP Communications Office, 2007).

Sustainable assessment tools

A sustainability assessment must measure the performance and interactions of ecological, economic and social quality, to identify possible directions of development (Draaijers *et al.*, 2003). The assessment is addressed through three main aspects of sustainable development; ecological, economical, and social and cultural (Draaijers *et al.*, 2003, p. 4). Sustainability assessment does not necessarily include a quantitative assessment, but should however strive to provide an assessment or guidance for all domains of sustainability (Draaijers *et al.*, 2003).

The growing concern on improving the construction industry's detrimental impact on the natural environment has also consequently given birth to many eco-labelling, green rating schemes and environmental building performance assessment amongst professionals in the building sector (Cole, 1999; Ding, 2008; George, 2009; Watson, 2009). An abundance of assessment and rating tools has been created in an effort to lessen the built environment's impact and provide a more sustainable construction future.

International collaboration to develop a building environmental assessment tool started with the Green Building Challenge by the International Initiative for a Sustainable Built Environment (iiSBE) in 1998, and since 2008 has evolved as the Sustainable Building Challenge (Cole and Larsson, 2000; iiSBE, 2009). The Green/Sustainable Building Challenge is a series of ongoing conferences to explore and refine the development and testing of a comprehensive assessment method, which focuses on the controversial aspects of wholesome building performance (Cole and Larsson, 2000; iiSBE, 2009). Through this international collaborative process, the GBTool was developed as a generic software that can be locally modified to suit national conditions (Cole, 2005; iiSBE, 2005; Watson, 2004).

Environmental assessment tools, like the Building Environmental Assessment (BEA), have been used as a means to rate the environmental performance of buildings in many countries (Scipioni *et al.*, 2009; Watson, 2004). BEA tools measure environmental performance through methods such as the Environmental Impact Assessment (EIA) method, which predicts the environmental impact of a development; and the life cycle assessment (LCA) method that determines the environmental impacts of products, materials and building components (Watson, 2004). Tools like the Building Research Establishment Environmental Assessment Method (BREEAM) in the UK, and EcoProfile in Norway, are credit assessment methods that rate the post-design performance of whole buildings based on environmental weightings on an overall scale (BRE, 2009; Watson, 2004).

Another rating tool based on a point scoring system is the widely recognized Leadership in Energy and Environmental Design (LEED), developed by the U.S. Green Building Council, and the Malaysian version of the Green Building Index (GBI) that evaluates building on a 100–point scale weighted against its performance areas (GBS, 2012; Graham, 2003; USGBC, 2010; Watson, 2004). The LEED building certification programme promotes 'sustainability by recognizing performances' through nine identified key areas (USGBC, 2010). BEA tools have also been recognized as a contributor to environmental awareness and environmentally responsible design in buildings (Watson, 2004). BEA tools can be conceptualized as a means to bridge the environmental assessment of the built environment together with decision making by stakeholders in building design, construction and operation (Watson, 2004).

Most of these BEA tools have included emissions accounting in their rating systems, such as BREEAM, LEED, CASBEE, Green Mark and Green Star (Ng *et al.*, 2013). However, these emission assessments vary from tool to tool, due to different environmental regulations, accounting methodology and building life-cycle phase. Table 2.5 (Managan and Araya, 2012; Ng *et al.*, 2013) compares the different assessment methodologies. These building tools cater mostly for individual buildings, designed to be used in their country and/or region and climatic conditions, and are also implemented on different levels as voluntary or partially mandatory mechanisms, which also have different environmental criteria (Ng *et al.*, 2013; Reed *et al.*, 2009).

Drivers for sustainable construction techniques

Driven by future investment, sustainable construction is the tenet of conservation of energy, water and natural resources. It is attained by reuse, recycling, innovative design and the minimization of waste and pollution. According to Suliman and Omran (2009) sustainable construction delivers built assets and their immediate surroundings that are able to:

- improve the quality of life and offer customer satisfaction;
- offer flexibility and the potential to cater for user changes in the future;
- provide and support desirable natural and social environments; and
- maximize the efficient use of resources.

Moreover, sustainable construction signifies one way of approaching the complex issues of sustainability and their drivers to the construction process: foundation for the

Table 2.5 Building life-cycle phase regarding emissions assessment

Building environmental assessment (BEA) tools	*Stages concerning carbon emissions evaluation*			
	Pre-construction	*Operational*	*Renovation*	*End-of-life*
BREEAM		***		
BEAM Plus	*	***		
LEED	**	***		
CASBEE	**	**	**	**
Green Mark		***		
Green Star	**	***		

Sources: Managan and Araya (2012); Ng *et al.* (2013).

Notes
* Qualitative assessment of CO_2 emissions performance.
** Quantitative assessment of CO_2 emissions performance.
*** Both qualitative and quantitative assessment of CO_2 emissions performance.

whole process in achieving balance in financial, environmental and operational considerations; assessment and remediation of contaminated land and consideration of materials, energy, design, construction and community. Philosophically, sustainable construction was expounded and promoted under different terms such as 'Intelligent Buildings', 'Energy Efficient Buildings' and 'Green Buildings'. As a whole, sustainable construction promotes environmental, social and economic beneficial impacts – now and for the future by creating a better quality of life at the present day and for generations to come, regardless of regional climates.

In tropical climates, reuse of crushed aggregates on site, harvesting of rainwater for flushing toilets, reuse of road planning in asphalt production, natural ventilation and schemes are the common approach adopted in working towards sustainable construction.

Benchmark, policies and guidelines

Philosophically, significant benefits of sustainable construction can be obtained by innovative design or consideration of alternative methods of project delivery. From a Malaysian perspective, the adoption of environmental quality standards such as the Building Research Establishment Environmental Assessment Method (BBREEAM), Civil Engineering Environmental Quality Assessment (CEEQUAL) or the National Health Service Environmental Assessment Tool (NEAT) are useful. Darus and Hashim (2012) expounded that these standards significantly stimulate the consideration of sustainable issues at the early stages of project development.

In addition, significant numbers of environmental methodologies and methods for evaluating environmental performance of buildings are being developed in tropical climate countries, based on international policy and guidelines. Commonly adopted policy and guidelines adopted include Canada's SB (Sustainable Building Tool, formerly known as Green Building Tool), USA's LEED (Leadership in Energy and Environmental Design) and Japan's CASBEE (Comprehensive Assessment System for Building Environmental Efficiency) (Suliman and Omran, 2009). Also, European policy and guidelines were commonly adopted in Malaysian construction sectors including United Kingdom's BREEAM (Building Research Establishment Environmental Assessment Method), France's HQE (High Environmental Quality) and Spain's VERDE method (resulting certification scheme).

Barriers

a The lack of awareness

In order to achieve sustainable construction, capacity, technologies and tools, total and ardent commitment by all players in the construction sector, including the governments and the public at large, is essential. This will increase demand for sustainable building. In tropical climate countries such as Malaysia, more buildings will try to be certified with the Sustainable Building Rating System (SBRS), mainly to compete in the niche market. In Malaysia particularly, recycling and reuse are promoted everywhere by the government and non-government organizations, and energy saving is the main driver. Darus and Hashim (2012) highlighted that awareness of sustainable development by the Malaysian public has created industry drive and demand for sustainable building design. In addition, demand for energy saving equipment has also increased recently because the equipment lowers the cost of energy bills. Conversely, however, there is often the perception among the Malaysian public and construction sector that 'green' or sustainable features will increase the cost of projects.

b Lack of training and education

Nowadays, there is an overarching need for developing countries (including countries of tropical climate including Malaysia) to be able to assess the sustainability of their infrastructure projects using economic, societal and environmental impact. The three parameters need to be incorporated with specific indicators for sustainable development. San-Jose *et al.* (2007) suggested that consideration of these parameters will help the projects' decision-making process.

Life cycle assessment and eco-labelling

A good example of an initiative to address this in a tropical context was the LCA Malaysia that was created in 2008, in connection with the National LCA Project for the establishment of the national life cycle assessment (LCA) related database. This aims to support a cleaner production approach in the country's industrial activities. The ultimate goal of this organization is to bring together the different players namely practitioners, software developers, users, decision-makers, students and the public at large in LCA activities for sustainability development. As a voluntary initiative LCA Malaysia is working with the SIRIM (SIRIM Berhad, formerly known as the Scientific and Industrial Research Institute of Malaysia) LCA Team with the aim of promoting the development and application of LCA activities in Malaysia with the following objectives:

- to create an LCA dissemination platform in Malaysia;
- to promote the life cycle thinking approach in sustainable consumption and production; and
- to provide linkages to LCA related tools and applications available locally and internationally.

However, unless these approaches become policy, and adherence mandatory, little impact will be seen.

Eco-labelling involves undertaking an evaluation of the inputs, outputs and the potential environmental impacts of a product system or process by considering the whole life cycle chain. This can be of benefit when considering which materials and systems to use based on ecological good practice.

Embodied energy in buildings

Previously, several LCA studies have been undertaken to evaluate embodied carbon in different types of buildings. The focus of these works are centred largely on embodied energy figures (rather than embodied carbon expenditure) for limited types of buildings, such as new residential and commercial buildings.

Embodied energy is defined as the energy consumed by all of the processes associated with the production of a building, from the mining and processing of natural resources to manufacturing, transport and product or materials delivery. Calculations on embodied energy in buildings is commonly based on the relevant data as follows:

a All direct embodied carbon use from fuels and electricity at raw material extraction (embodied carbon coefficient for quarrying, mining, manufacturing and processing); and
b Off-site embodied carbon (CO_2 emissions) used related to materials transportation.

It must be noted that the mode of transportation used for materials delivery from their respective resourcing locations to building sites was commonly based on type of mode of transportation. The average gross weight in tonnes of mode of transportation, their height laden (percentage), size, body type (rigid and articulated), distance travelled (in kilometres), number of deliveries, delivery weight (in tonnes) and what is carried on the return journey (on percentage part load) were also to be considered in embodied calculation. Also, an estimate of the tonnage from each delivery was also included in the calculation of embodied energy.

For instance, stainless steel has a high embodied carbon coefficient value. This is mainly due to the high energy that was expended in production. Conversely, a preference for using low embodied carbon materials for building maintenance and repair (such as locally sourced lime and sand) is commonly accepted wisdom. However, the benefits potentially derived from the use of locally available materials of similar durability for buildings needs to be considered. The use of low carbon locally sourced materials commonly contribute to lower embodied energy.

Studies undertaken show that developing countries have more buildings with high embodied energy. Jones (1991) asserts that industrialization and urbanization in developing countries obviously affect embodied energy consumption. Industrialization and urbanization are commonly known to accompany each other during economic development of these countries, but urbanization exerts influences on embodied energy use, for example in hospital buildings.

A considerable amount of research has been undertaken in developing countries with tropical climates concerning the analysis of the Energy Efficiency Index (EEI) for hospital buildings. Commonly, worldwide organizations implemented EEI analysis on their buildings by using the kWh/m^2 approach, where energy consumption evaluations are based on per unit floor area. For example, Bakar *et al.* (2015) compared the end-use energy in a large-scale hospital building in Malaysia by using this index. The case study was conducted at Universiti Kebangsaan Malaysia Medical Centre (UKMMC). In this research, Bakar *et al.* (2015) calculated the EEI of this hospital then compared it to the EEI of other hospitals to measure the level of energy usage in this hospital. In Malaysia, the EEI standard value for hospital buildings is estimated at 200 kWh/m^2/year. This is the baseline rating to achieve energy efficiency in the building. However, the EEI comparison of this research reveals that this hospital, had an EEI value of approximately 384 kWh/m^2/year, which is significantly higher than the Malaysian recommended rating standard. Thus, UKMMC requires some strategies for reducing the EEI. Moreover, various factors influence the energy consumption of a building system, such as the types of activities carried out in the building, weather conditions, building materials, HVAC system and occupancy.

Studies undertaken by the Australian CSIRO (Scientific and Industrial Research Organization) have concluded that there is a relationship between CO_2 emissions and the embodied carbon expended in the building materials manufacturing process (an average of 0.098 tonnes of CO_2 per GJ of embodied energy) (Holtzhausen, 2007).

Existing buildings have an important role in reducing carbon emissions and energy consumption, to meet global targets for 2050. Maintenance interventions clearly expend energy, with some leading to higher CO_2 expenditure than others, e.g. the significance of transport.

The development of tools to select embodied energy construction systems is considered timely as it may help less experienced designers with limited energy efficiency knowledge to make decisions in the same way as experts. Common construction systems for buildings can be grouped into several main groupings such as Structural Frame Systems, Wall Systems and Slab Systems. In construction sectors, the main criteria that need to be considered by the designer in order to evaluate embodied energy of these construction systems that have been identified include among others recycling of materials, CO_2 emission and transportation.

In the case of steel, its embodied energy values are calculated based on its main materials, namely 50 per cent recycled, market average, predominant recycled, unspecified, virgin and other specification. Hammond and Jones (2011) suggested that the embodied energy value for steel is the total of the energy consumption mixture in its production. However, this total value does not include the final cutting of the steel product to the specified dimensions (such as an estimate from World Steel Association (Worldsteel) data. With regard to embodied energy, a breakdown of fuel use or carbon emissions was not commonly included as this generally was not possible. This is because the steel industry is complicated by the production of by-products (which may be allocated energy or carbon credits), excess electricity production (producers generate some of their own electricity) and non-fuel related emissions (from the calcination of lime during the production process).

Material selection

Building maintenance forms a large component of the construction sector. It is therefore clear that maintenance has a substantial potential capacity to reduce carbon emissions through repair intervention and the selection of materials and techniques of building in tropical climates. Building repair is an integral part of the maintenance sector and appropriate techniques can reduce carbon expenditure, whilst inappropriate techniques can increase carbon.

In order to achieve a good environmental outcome (with low embodied carbon expenditure and fewer CO_2 emissions), and in order to fulfil building conservation philosophical defensibility, appropriate LCA could be adopted to evaluate embodied carbon expenditure for repair. The selected LCA boundary and the associated inputs, and maintenance periods and longevity are essential in determining the embodied carbon expenditure or how 'green' the interventions are. The concept of 'green' maintenance provides benefits for those involved in the building maintenance decision-making process, enabling rational selection of repair, not solely based on cost.

Low carbon impact

Generally, different types of materials used in repair are derived from different locations. Different resourcing locations for materials used contribute significantly towards variation in transportation distances to building site. The greater the distance of resourcing location of materials, the greater the CO_2 that is emitted during the transportation process, i.e. within the 'gate-to-site' boundary. Therefore, the use of locally sourced materials for repair of buildings in tropical climates is to be encouraged when evaluating the selection of materials.

It also has been found that the resourcing location for materials is determined based on where they are being produced, processed and manufactured. It has been ascertained that each repair material has a different resourcing location as they have a different nature of procurement: quarry and processing plant.

References

ABC. 2007. *Building Energy Efficiency: Why Green Buildings Are Key to Asia's Future*. Hong Kong: Asia Business Council (ABC).

ASEAN. 2009. *ASEAN Member States*. Association of Southeast Asian Nations (ASEAN). Available at: http://asean.org/asean/asean-member-states/ [accessed 10 July 2016].

Bakar, N. N. A., Hassan, M. Y., Abdullah, H., Rahman, H. A., Abdullah, M. P., Hussin, F. and Bandi, M. 2015. Energy efficiency index as an indicator for measuring building energy performance: a review. *Renewable and Sustainable Energy Reviews*, 44, 1–11.

Bourdeau, L. 1999. Sustainable development and the future of construction: a comparison of visions from various countries. *Building Research & Information*, 27, 354–366.

Braungart, M., McDonough, W. and Bollinger, A. 2007. Cradle-to-cradle design: creating healthy emissions – a strategy for eco-effective product and system design. *Journal of Cleaner Production*, 15, 1337–1348.

BRE. 2009. *What is BREEAM*. BRE Global Ltd. Available at: www.breeam.org/page.jsp?id=66 [accessed 28 May 2010].

Brundtland, G. 1987. *Our Common Future*. Oxford: World Commission on Environment and Development.

Burnett, J. 2007. City buildings – eco-labels and shades of green! *Landscape and Urban Planning*, 83, 29–38.

Butler, R. A. 2013. Malaysia has the world's highest deforestation rate, reveals Google forest map. 15 November 2013. Available at: http://news.mongabay.com/2013/1115-worlds-highest-deforestation-rate.html [accessed 19 December 2016].

Carbon Footprint Network. 2016. *Ecological Wealth of Nations: Earth's Biocapacity as a New Framework for International Cooperation*. Oakland, CA: Carbon Footprint Network.

CIB. 1997. Agenda 21 on Sustainable Construction. *CIB Report Publication 237*. Rotterdam, the Netherlands: International Council for Research and Innovation in Building and Construction.

CIB and UNEP-IETC. 2002. *Agenda 21 for Sustainable Construction in Developing Countries*. Plessis, C. D. and WSSD (eds). South Africa: CIB, International Council for Research and Innovation in Building and Construction, UNEP-IETC, United Nations Environment Programme, International Environmental Technology Centre.

Cole, R. J. 1999. Building environmental assessment methods: clarifying intentions. *Building Research and Information*, 27, 230–246.

Cole, R. J. 2005. Building environmental assessment methods: redefining intentions and roles. *Building Research & Information*, 33, 455–467.

Cole, R. J. and Larsson, N. 2000. Green building challenge: lessons learned from GBC '98 and GBC 2000. *In:* Boonstra, C., Rovers, R. and Pauwels, S. (eds) International Conference

Sustainable Building 2000, Proceedings 22–25 October 2000. Maastricht, the Netherlands: Eneas, Technical Publishers.

Darus, Z. M. and Hashim, N. A. 2012. Sustainable building in Malaysia: the development of sustainable building rating system. *In:* Sustainable Development-Education, Business and Management-Architecture and Building Construction-Agriculture and Food Security, InTech Europe, Rijeka: Croatia.

Dattaro, L. 2014. NASA catches deforestation in the act. 24 April 2014, ed. The Weather Channel.

Ding, G. K. C. 2008. Sustainable construction: the role of environmental assessment tools. *Journal of Environmental Management*, 86, 451–464.

Draaijers, G. P. J., Verheem, R. A. A. and Morel, S. A. A. 2003. *Developing a General Framework for Sustainability Assessment.* Utrecht: Netherlands Commission for Environmental Impact Assessment.

Du Plessis, C. 2002. *Agenda 21 for Sustainable Construction in Developing Countries. A discussion document.* WSSD (ed.). South Africa: International Council for Research and Innovation in Building and Construction (CIB); United Nations Environment Programme, International Environmental Technology Centre (UNEP-IETC).

Encyclopædia Britannica. 2016. *Southeast Asia.* Encyclopædia Britannica, Inc.

GBS. 2012. *Greening Malaysia.* Kuala Lumpur: Greenbuildingindex Sdn. Bhd (GBS).

George, O. B. 2009. Taking the LEED. *BusinessWest*, 26, 47.

Gill, C. C. 2000. The impacts of facilities/maintenance management on the long term sustainability of the built environment: performance indicators at post-construction stage (operation + maintenance phases) of building life cycle. *In:* Boonstra, C., Rovers, R. and Pauwels, S. (eds) International Conference Sustainable Building 2000, Proceedings 22–25 October 2000. Maastricht, the Netherlands: Eneas, Technical Publishers.

Global Footprint Network. 2015. *World Footprint.* 20 November 2015, ed. Global Footprint Network.

Graham, P. 2003. *Building Ecology: First Principles for a Sustainable Built Environment.* United Kingdom: Blackwell Science Ltd.

Graham, P. 2008. *Sustainability and the Struggle for Hegemony in Australian Architectural Education.* Doctoral (PhD), University of New South Wales.

Hammond, G. P. and Jones, C. I. 2011. *Inventory of Carbon and Energy (ICE), Beta Version V2.0.* Department of Mechanical Engineering, University of Bath. Available at: www.bath.ac.uk/mech-eng/sert/embodied/ [accessed 21 December 2016].

Hansen, M. C., Potapov, P. V., Moore, R., Hancher, M., Turubanova, S. A., Tyukavina, A., Thau, D., Stehman, S. V., Goetz, S. J., Loveland, T. R., Kommareddy, A., Egorov, A., Chini, L., Justice, C. O. and Townshend, J. R. G. 2013. Hansen/UMD/Google/USGS/NASA Tree Cover and Tree Cover Loss and Gain, Country Profiles. Global Forest Watch – www.globalforestwatch.org: University of Maryland, Google, USGS and NASA.

Hartkopf, V. and Loftness, V. 1999. Global relevance of total building performance. *Automation in Construction*, 8, 377–393.

Hester, R. T. 1995. Life, liberty and the pursuit of sustainable happiness. *Places*, 9.

Holtzhausen, H. J. 2007. Embodied energy and its impact on architectural decisions. *WIT Transactions on Ecology and the Environment*, 102, 377–385.

IAP. 2010. *IAP Statement on Tropical Forests and Climate Change.* Trieste, Italy: The Inter Academy Partnership (IAP).

IEA. 2009. *CO_2 Emissions from Fuel Combustion Highlights.* Paris: International Energy Agency (IEA).

IFAD. 2009. *Climate Change Impacts – South East Asia.* International Fund for Agricultural Development (IFAD).

iiSBE. 2005. *GBTool.* International Initiative for a Sustainable Built Environment (iiSBE).

iiSBE. 2009. *GB/SB Challenge.* International Initiative for a Sustainable Built Environment (iiSBE).

Institute for Veterinary Public Health. 2011. Observed and projected climate shifts 1901–2100. Depicted by world maps of the Köppen-Geiger climate classification. *World Maps of Köppen-Geiger Climate Classification*. Institute for Veterinary Public Health.

IPCC. 2007a. Climate change 2007: synthesis report. *In:* Allali, A., Bojariu, R., Diaz, S., Elgizouli, I., Griggs, D., Hawkins, D., Hohmeyer, O., Pateh Jallow, B., Kajfez-Bogataj, L., Leary, N., Lee, H. and Wratt, D. (eds) *An Assessment of the Intergovernmental Panel on Climate Change*. Valencia, Spain: Intergovernmental Panel on Climate Change (IPCC).

IPCC. 2007b. Climate change 2007: working group III: mitigation of climate change. *In:* Metz, B., Davidson, O. R., Bosch, P. R., Dave, R. and Mayer, L. A. (eds) *Contribution of Working Group III to the Fourth Assessment Report of the Intergovermental Panel on Climate Change*. Cambridge, UK; New York: Cambridge University Press/Intergovernmental Panel on Climate Change (IPCC).

IPCC. 2014. Climate change 2014: synthesis report. *In:* Pachauri, R. K. and Meyer, L. A. (eds) *Contribution of Working Groups I, II and III of the Fifth Assessment Report of the Intergovernmental Panel on Climate Change*. Geneva, Switzerland: Intergovernmental Panel on Climate Change (IPCC).

IUCN. 2008. Transition to sustainability: towards a humane and diverse world. *In:* Adams, W. M. and Jeanrenaud, S. J. (eds) *IUCN Future of Sustainability Initiative 2008*. Geneva, Switzerland: International Union for Conservation and Nature (IUCN).

Jones, D. W. 1991. How urbanization affects energy use in developing countries. *Energy Policy*, 19(7), 621–630.

Kibert, C. J. 2000. Construction ecology and metabolism. *In:* Boonstra, C., Rovers, R. and Pauwels, S. (eds) International Conference Sustainable Building 2000, Proceedings 22–25 October 2000. Maastricht, the Netherlands: Eneas, Technical Publishers.

Levine, M., Urge-Vorsatz, D., Blok, K., Geng, L., Harvey, D., Lang, S., Levermore, G., Mehlwana, A. M., Mirasgedis, S., Novikova, A., Rilling, J. and Yoshino, H. 2007. Residential and commercial buildings. *In:* Metz, B., Davidson, O. R., Bosch, P. R., Dave, R. and Meyers, L. A. (eds) *Contribution of Working Group III to the Fourth Assessment Report of the Intergovernmental Panel on Climate Change*. Cambridge, UK; New York: Cambridge University Press/Intergovernmental Panel on Climate Change (IPCC).

Managan, K. and Araya, M. 2012. *Driving Transformation to Energy Efficient Buildings*. Washington, DC: Institute for Building Efficiency.

Mimura, N., Nurse, L., McLean, R. F., Agard, J., Briguglio, L., Lefale, P., Payet, R. and Sem, G. 2007. Small islands. *In:* Parry, M. L., Canziani, O. F., Palutikof, J. P., Linden, P. J. V. D. and Hanson, C. E. (eds) *Climate Change 2007: Impacts, Adaptation and Vulnerability. Contribution of Working Group II to the Fourth Assessment Report of the Intergovernmental Panel on Climate Change*. Cambridge, UK: Intergovernmental Panel on Climate Change (IPCC).

Nabuurs, G. J., Masera, O., Andrasko, K., Benitez-Ponce, P., Boer, R., Dutschke, M., Elsiddig, E., Ford-Robertson, J., Frumhoff, P., Karjalainen, T., Krankina, O., Kurz, W. A., Matsumoto, M., Oyhantcabal, W., Ravindranath, N. H., Sanz Sanchez, M. J. and Zhang, X. 2007. Forestry. *In:* Metz, B., Davidson, O. R., Bosch, P. R., Dave, R. and Meyers, L. A. (eds) *Climate Change 2007: Mitigation. Contribution of Working Group III to the Fourth Assessment Report of the Intergovernmental Panel on Climate Change*. Cambridge, UK; New York: Cambridge University Press/Intergovernmental Panel on Climate Change (IPCC).

Ng, S. T., Chen, Y. and Wong, J. M. W. 2013. Variability of building environmental assessment tools on evaluating carbon emissions. *Environmental Impact Assessment Review*, 38, 131–141.

OECD. 2001. Workshop on the Design of Sustainable Building Policies. *The OECD Environment Programme*. Organisation for Economic Co-operation and Development (OECD).

Pidwirny, M. 2011. Tropical moist climates – a climate type. *The Encyclopedia of Earth* (EoE). Available at: www.eoearth.org/view/article/162264/ [accessed 10 July 2016].

Reed, R., Bilos, A., Wilkinson, S. and Schulte, K.-W. 2009. International comparison of sustainable rating tools. *Journal of Sustainable Real Estate*, 1, 1–22.
Rees, W. and Wackernagel, M. 1996. Urban ecological footprints: why cities cannot be sustainable – and why they are a key to sustainability. *Environmental Impact Assessment Review*, 16, 223–248.
San-Jose, J. T., Losada, R., Cuadrado, J. and Garrucho, I. 2007. Approach to the quantification of the sustainable value in industrial buildings. *Building and Environment*, 42, 3916–3923.
Sassen, S. 2009. Cities today: a new frontier for major developments. *Annals of the American Academy of Political and Social Science*, 626, 53–71.
Scipioni, A., Mazzi, A., Mason, M. and Manzardo, A. 2009. The Dashboard of Sustainability to measure the local urban sustainable development: the case study of Padua Municipality. *Ecological Indicators*, 9, 364–380.
Smith, M., Whitelegg, J. and Williams, N. 1998. *Greening the Built Environment*. UK: Earthscan Publications Ltd.
Spence, R. and Mulligan, H. 1995. Sustainable development and the construction industry. *Habitat International*, 19, 279–292.
Sukumaran, T. 2013. NASA: Malaysia sees 115% jump in deforestation. *The Star Online*.
Suliman, L. Kh. M. and Omran, A. 2009. Sustainable development and construction industry in Malaysia. *Manager*, 10, 76–85.
UN-HABITAT. 2008. *State of the World's Cities 2008–2009: Harmonious Cities*. UK/USA: United Nations Human Settlements Programme (UN-HABITAT).
UNCED. 1992. *Agenda 21. Earth Summit: The United Nations Programme of Action from Rio*. Rio de Janeiro: United Nations Department of Economic and Social Affairs; United Nations Conference on Environment and Development (UNCED).
UNDP Communications Office. 2007. *UNDP Report: Multi-Billion Facility Crucial to Spur Climate Mitigation Initiatives*. UNDP Communications Office: Press Release. Kuala Lumpur: United Nations Development Programme (UNDP).
UNEP. 2009. *Buildings and Climate Change: Summary for Decision-makers*. Yamamoto, J. and Graham, P. (eds). Paris: United Nations Environment Programme – Sustainable Buildings and Climate Change Initiative (UNEP-SBCI).
UNEP. 2010. *The 'State of Play' of Sustainable Buildings in India*. UNEP DTIE Sustainable Consumption & Production Branch.
UNFCC. 1992. United Nations Framework Convention on Climate Change (UNFCC).
United Nations Sustainable Buildings and Climate Initiative. 2010. *Common Carbon Metric: Protocol for Measuring Energy Use and Reporting Greenhouse Gas Emissions from Building Operations*. Draft for Pilot Testing. Paris: United Nations Sustainable Buildings and Climate Initiative (UNEPSBCI).
USGBC. 2010. *Introduction – What LEED is* [Online]. U.S Green Building Council. Available at: www.usgbc.org/DisplayPage.aspx?CMSPageID=1988 [accessed 29 May 2010].
Wang, N., Chang, Y.-C. and Nunn, C. 2010. Lifecycle assessment for sustainable design options of a commercial building in Shanghai. *Building and Environment*, 45, 1415–1421.
Watson, P. 2009. Addressing the need for common definitions of building performance in respect of the sustainability agenda. *Journal of Building Appraisal*, 5, 67–74.
Watson, S. 2004. *Improving the Implementation of Environmental Strategies in the Design of Buildings: Towards a Life-Cycle Based, Front-Loaded, Framework for Building Environmental Assessment During Design*. Doctoral (PhD), University of Queensland.
WBCSD. 2008. *Energy Efficiency in Buildings: Business Realities and Opportunities*. Geneva, Switzerland: World Business Council for Sustainable Development (WBCSD).
WBCSD. 2009. *Energy Efficiency in Buildings: Transforming the Market*. Geneva, Switzerland: World Business Council for Sustainable Development (WBCSD).
World Bank. 2004. *Beyond Economic Growth*. The World Bank Group. Available at: www.worldbank.org/depweb/english/beyond/global/glossary.html [accessed 22 April 2010].

World Bank. 2010. *Country Classification*. The World Bank Group. Available at: http://data.worldbank.org/about/country-classifications [accessed 22 April 2010].
World Bank. 2015. CO_2 emissions (metric tons per capita). *Data*. World Bank Group.
WWF. 2008. *The Living Planet Report*. World Wide Fund For Nature (WWF).
WWF. 2016. *Deforestation*. World Wide Fund For Nature (WWF).
WWF-UK. 2016. *Deforestation and Climate Change*. World Wide Fund for Nature-UK (WWF-UK).
Zainul Abidin, N. 2009. Sustainable construction in Malaysia – developers' awareness. *World Academy of Science, Engineering and Technology*, 53, 807–814.
Zainul Abidin, N. 2010. Investigating the awareness and application of sustainable construction concept by Malaysian developers. *Habitat International*, 34, 421–426.
Zimmermann, M., Althaus, H. J. and Haas, A. 2005. Benchmarks for sustainable construction: a contribution to develop a standard. *Energy and Buildings*, 37, 1147–1157.

3 Functional requirements of buildings

Tropical context

Norhayati Mahyuddin, Farid Wajdi Akashah and Raha Sulaiman

After reading this chapter you should be able to:

- appreciate the main physical functions of buildings in tropical climates;
- describe the factors that must be considered in creating an acceptable living and working environment;
- discuss links between these factors and the design of buildings in tropical climates;
- recognize the sources and nature of loadings applied to building elements in tropical climates and the ways in which they affect those elements; and
- appreciate the influence of the choices of materials and the selection of design features on building performance.

Introduction

Functional requirements of buildings, in a broad context, will be very much related to where the buildings are located. In addition, the location greatly influences the diversity in building design from one place to another throughout the world. What lies behind such significant variance in the physical and environment requirements of buildings is mainly derived from the weather and climatic conditions of a place. An analysis of the climate of a particular region can help in assessing adverse climatic effects, while simultaneously identifying those that are beneficial in determining the performance of a building. Climates throughout the world are described as tropical, dry, moderate, continental and polar and are mainly determined by the latitude of the zones. Accurate information about climate is required to provide suitable design for the building as well as the selection of building materials.

Designers strive for long-term sustainability when designing and constructing a building. Building design, materials and their response to the climate if being implemented successfully will, to a great extent, protect the building and the occupants from the occasionally hostile environment. For the end users, the building must provide an environment that does not harm their health. Moreover, it should provide living and working conditions which are comfortable.

Building materials and the construction systems should take into consideration their individual and collective properties, which should be suitable to the local climatic conditions.

Hence, physical and environment requirements of buildings in tropical climates are potentially quite different from those in other climatic zones. The nature, form and use of such buildings have a uniqueness that is driven by their tropical context. This in turn drives the construction form and the manner in which they perform throughout their lives.

Overview

After reading this chapter you should be familiar with the nature of buildings as environmental envelopes, providing protection from outdoor effects and variations of a tropical climate. The fundamental building science within this chapter will be focused on the tropical climate and its effect on buildings. The nature of physical building design involves response and adaptation of the building design, elements and materials. Understanding the fundamental requirements, physically and environmentally, is an essential part of understanding the buildings at an early stage. There is an important relationship between environmental factors and building form and function which in future use affect the internal environment. It is important to appreciate and understand the need to moderate the environment. This helps in making the right decisions when designing a building that is climate responsive, particularly towards tropical climate.

Factors such as temperature, humidity, atmospheric pressure, wind, precipitation and atmospheric particle count in a tropical climate region need to be taken into consideration as early as possible. Ideally this would be during conceptualization of design. This is important as these factors affect the indoor climate of buildings greatly.

By doing this, it will help to ensure that the design intended delivers a functional building that is comfortable to the end users.

The first part of this chapter focuses on the physical performance requirements of buildings in tropical climates. This is followed by discussion on macro, mesa and micro climate in tropical regions. The final part of this chapter deliberates on the aspect of building science in tropical climates.

Physical performance requirements of buildings in tropical climates

The structural behaviour of elements

The structural behaviour of building elements depends mainly upon three factors: the environmental conditions, material characteristics and geometrical features. The environmental conditions influence the nature of loading on the structural elements. For instance, a structural element behaves differently at elevated temperatures as compared to moderate ambient temperature. This behaviour varies further when cyclic or seismic loading is applied. Therefore, while designing a structural member, it should be considered that in addition to the mechanical loading, environmental types of loading may also be applied to the structural elements of the building. The material characteristics determine the strength of the structural element. For instance, plain concrete is strong under compression but weak under tensile loading. However, when it becomes reinforced with mild steel or fibre reinforced polymers, it behaves as a very strong material in any type of climatic conditions. Similarly, the material characteristics of cold formed steel make it stronger in tension permitting a ductile mode of failure whereas the compressive resistance is relatively smaller. The geometrical features determine the stiffness of the structural element. For instance, increasing the thickness, depth and width of the structural member increases its stiffness at a higher rate compared to its strength. It should be noted that on most occasions, all these three factors are inter-connected. However, in order to achieve an errorless design of structural elements, it is mandatory to clearly understand which of the factors (loading, material or geometry) has supremacy over the others.

Design considerations for buildings in tropical climates

The design of building structures in tropical climates is a complex issue and demands experience and sufficient knowledge of the subject. An incorrect estimation of resources and the nature of loading on the elements and/or selection of material may result in collapse of the structure or the service life of the building may be reduced. Thus, in order to obtain successful design of structural elements of a building in tropical climates, a few considerations are essential.

Types of environmental loading

In addition to the conventional mechanical loading on structural elements, there are two main environmental loading parameters affecting the physical performance of the structural elements of buildings located in tropical climates. These parameters are: high temperature and increased moisture content.

Effect of thermal loading

It is a universal fact that changes in temperatures will eventually lead to changes to dimension of materials in a way where they either expand or contract. In the event that temperature increases, materials expand. Should the opposite happen, where the temperature decreases, materials contract. It is important to note that the degree of the temperature movement varies based on the material characteristics of the structural element. Temperature changes can have another effect: if a structure such as a super high-rise building is heated from one side only, for instance by the sun shining on it, its temperature will not rise uniformly; the sunny side will heat up more quickly and there will be a temperature 'drop' or 'gradient' across the width of the super high-rise building. This means that the side of the building facing the sun will elongate relative to that facing away from the sun.

Changes in average temperature and temperature differentials are the key thermal loading parameters in tropical climates. High temperature can induce stress in the structural elements beyond allowable limits, which initiates the deformation of the element and both the elastic modulus and strength of the elements decrease. The geometry of the element changes from its original position which affects the service life of the building. However, it is not necessarily the case that all the elements have the same type of deformation. Also, the thermal expansion and contraction and change in elastic modulus of the structural element depends upon the thermal coefficient of the material employed in the construction of the building. This shows that an efficient design of the structural elements must take into account the effects of temperature on both material and physical properties of the material.

The magnitude of the temperature distribution in a single structural element and its transmission from one element to the other depends upon a number of factors, which should be determined properly. However, in most cases, it is difficult to interpret all those factors numerically. The widely recognized considerations are the average climate and maximum and minimum temperatures during the year of the geographical location of the construction site. Then there are highly aleatory parameters, such as the presence of perturbations, which influence air temperatures and solar radiation, often with fluctuations on a daily scale or, in any event, over relatively short periods of time. Lastly, there are parameters strictly linked to the conditions of the particular building in question: the presence of other nearby structures that act as solar radiation screens, the building orientation, its total mass (and consequent thermal inertia), the properties of its finishing (i.e. the degree of the solar energy absorption and thermal isolation) and the characteristics of the interior heating, air conditioning and ventilation.

Effect of temperature on major structural components

In building construction, the members which are considered critical under thermal loading are columns, beams and connections. In concrete construction, slabs are also subjected to thermal loading and the joints in it experience the effects of temperature variation. However, the connections or joints are the most critical region under temperature. In a case of high temperature, the moment transfers to the connections from the adjacent members. Thus, the serviceability degradation of a building structure situated in a tropical region often initiates with failure of the connections. The behaviour of columns, beams and slabs is dependent upon the material properties or in other words the value of thermal co-efficient of the material.

Effect of moisture content

The changes in moisture content will eventually lead to changes in the dimensions of materials in a way where it either swells or shrinks. Moisture can be stored in porous materials. The amount of moisture stored in the pores of these materials is dependent on the relative humidity (RH) in the atmosphere. In the event where RH increases, moisture will be absorbed by the materials. High moisture content will lead to the material experiencing changes to its dimension where it swells. If the opposite is to happen, where RH decreases, moisture will be released. Low moisture content will lead to the material drying and henceforth experiencing changes to its dimensions where it shrinks.

Moisture content is a key design parameter when a building structure is required to be constructed from concrete. The concrete strength is affected by alterations in the moisture content throughout the structural element and/or at the surface of the structural element exposed to open air.

When the aggregates of concrete include biomass materials, such as oil palm shell (lightweight aggregate concrete) as is increasingly the case in tropical areas, it is essential that the moisture content should be kept as low as possible to utilize the full efficiency of the material regarding energy saving. The net energy density by mass of a biomass material decreases with an increased moisture content, owing to the weight of the water, but also by volume owing to the energy required to evaporate the water. This creates difficulty in the transportation of biomass since most of the load is of water instead of the original material. Moreover, the effect of moisture content is also considerable for the storage of biomass materials.

Similar to other materials, moisture content influences both the material and geometrical properties of timber. Influencing the material properties, the moisture content highly affects the strength and stiffness when the timber is subjected to bending. The higher the moisture content, the lower the strength and stiffness. The effect on geometrical properties produces dimensional changes. Diminishing moisture content produces shrinkage of the material. Timber shrinks by about 5 per cent of its original volume when moisture content falls from 24 to 6 per cent. The main shrinkage occurs in directions perpendicular to the grain only, whilst the shrinkage in length is often marginal.

The evaluation of detrimental effects of variance in relative humidity (RH) on the building structure is often more difficult than evaluation of effects of environmental temperature. This may be attributed to the fact that a larger number of factors influence moisture equilibration (ME) than thermal equilibration. For example, enclosures or housing situations may act as moisture barriers and thus influence how quickly (or how slowly) the objects are exposed to the new humidity conditions. The temperature will also influence the rate of moisture equilibration. Furthermore, there is more variation in the capacity of the individual objects or materials to control moisture equilibration than thermal equilibration.

The above explanation illustrates how RH, moisture content and dimensional changes are inter-related. This inter-relationship needs to be carefully considered when designing buildings in the tropics together with factors such as ambient temperature, material temperature and thermal movement.

Choice of material for buildings in tropical areas

The primary characteristics of the various tropical areas are warm and dry air and dry ground, or warm and moist air together with high rainfall. In warm, dry areas the temperature remains high during the daytime and humidity fluctuates between modest to low. Outdoor climate is highly influential on the interiors of the buildings located in such tropical areas. Arrangements are needed in these dry areas to save the exterior and interior of the building from solar radiation and the hot, dusty winds. Consideration should be given to using heavy materials with high thermal resistance to construct walls and roofs. The outer wall, if constructed with a material with increased thermal capacity, will capture a large amount of temperature during the daytime. In contrast, the warm-humid climate is characterized by hot, sweaty and sticky conditions as well as continual presence of dampness. Air temperatures remain moderately high with little variation between day and night. Such a scenario is best catered for by the use of lightweight building enclosures with low thermal mass.

Historically, the use of local materials that reflect the performance requirements has resulted in very recognizable vernacular forms that reflect climatic conditions. As buildings of greater scale have developed, the nature of their construction has aligned more with generic, international approaches to construction. It is still essential, however, to reflect the particular design requirements of such regions. In tropical climates, the construction of the whole building with the same material is not usually possible. Since the serviceability requirements have differences, it is essential to choose different materials based on their use. As far as the main structural elements (columns, beams, slabs and connections) are concerned, the material which has higher resistance to temperature with lesser moisture content should be preferred. For example, cold-formed steel (CFS) has minimal temperature resistance and the degradation starts at relatively low temperature. Thus, CFS is usually avoided in those regions. Reinforced cement concrete (RCC) is usually considered as the best solution for the construction of key structural elements for the buildings situated in tropical areas. However, materials with nearly equal tensile and compressive strengths will expand and contract, even when some restraint is imposed. Materials or elements with a tensile strength substantially less than their compressive strength, e.g. stone or masonry, will expand when heated, but cannot contract by the same amount when cooled, and may crack because their tensile strength is not enough to resist the force from the restraint.

The choice of material for the construction of buildings located in tropical areas should be chosen on the basis of the behaviour of their thermal expansion and contraction. A best idea can be achieved through the values of the co-efficient of thermal expansion (alpha [α]), which is different for every material. Table 3.1 shows the values of α for different materials.

α is used to calculate the thermal expansion. When an object is heated or cooled, its length changes by an amount proportional to the original length and the change in temperature. The linear thermal expansion of an object can be expressed as:

$$dl = L_0 \, \alpha \, (t_1 - t_0) \tag{1}$$

where
dl = change in object length (m, inches)
L_0 = initial length of object (m, inches)
α = thermal expansion coefficient (m/m °C, in/in °F)
t_0 = initial temperature (°C, °F)
t_1 = final temperature (°C, °F)

In fact, the structural elements of a building situated in a tropical area should be designed according to the linear expansion calculated with the help of thermal

Table 3.1 Values of alpha, α for different materials

Material	*alpha, α value*
Timber	3×10^{-6} °C^{-1} to 9×10^{-6} °C^{-1}
Clay brick	5×10^{-6} °C^{-1}
Glass	6×10^{-6} °C^{-1} to 9×10^{-6} °C^{-1}
Granite	8×10^{-6} °C^{-1}
Limestone	8×10^{-6} °C^{-1}
Concrete	11×10^{-6} °C^{-1}
Iron and steel	12×10^{-6} °C^{-1}
Copper	17×10^{-6} °C^{-1}
Aluminium	24×10^{-6} °C^{-1}
Lead	29×10^{-6} °C^{-1}
PVC (hard)	55×10^{-6} °C^{-1}

co-efficient. Each of the structural elements has to perform a different function, for instance, the roof and cladding has to act differently from joints. Therefore, the choice of material for the construction of a tropical building should be based on its strength, stiffness, ductility and serviceability characteristics.

Macro, mesa and micro climate in tropical regions

The variations that characterize the climate regions are generated by the five main components of climate. These components, as tabulated in Table 3.2, dynamically interact and are closely interrelated. As can be seen in Figure 3.1, the world is divided into different latitudes, altitudes and terrains as well as featuring nearby water bodies and their currents.

These factors – the components and the coordinates – will be the main reference in classifying the climatic regions into tropical, dry, moderate, continental and polar. In the built environment, we are generally concerned with the different scale of climatic investigation in each climatic region, namely the macroclimate, mesa climate and microclimate.

Macroclimate and mesa climate in tropical region

Macroclimate is a character of a regional climate, described in terms of sun radiation, sunshine, cloud, temperature, wind, humidity and precipitation, within a large geographic area. In built environment, building design may not affect the condition of the macroclimate. However, sound knowledge of the macroclimate affects how the building is designed and constructed. Knowledge of macroclimate gives a general impression of the climate where the building is located and the building design can be planned accordingly.

In general, countries located in the tropical and equatorial regions must adopt approaches to buildings that reflect their climates. Examples of such climatic situations occur *inter alia* in Asia, Africa, Australasia and Central/South America.

Table 3.3 below compiles the majority of tropical countries with their general climate characteristics. It can be noted that they have more or less similar climatic characteristics as long as they are within the Tropics of Cancer and Capricorn. Average temperatures are in the range of 18–38°C with two main general seasons: either cold with heavy rain or hot with less rain; and precipitation in the range of

Table 3.2 Five main components of climate

Components	*Coverage*
Atmosphere	The air that envelops the earth's surface
Hydrosphere	The fresh and salt waters of the earth such as lakes, rivers and oceans
Cyrosphere	The ice on earth's surface including ice sheets, glaciers, snow covers and others
Lithosphere	The solid land that built up the earth which includes soils, plains, mountains and other composition of land and soil
Biosphere	The living plants and animals, flora and fauna and living creatures, including marine and terrestrial organisms

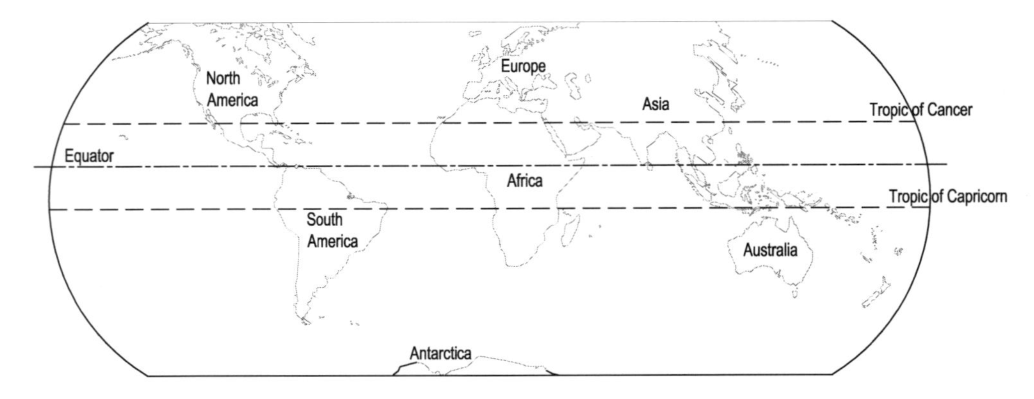

Figure 3.1 Tropical climate between Tropic of Cancer and Tropic of Capricorn.

1,000–4,000 mm a year. Average wind speed up to maximum 10 m/s in maritime regions may relate to their similar severe weather throughout tropical countries – tropical cyclones, heavy rainfall and thunderstorms. Amongst other things, these characteristics will definitely affect how people live in tropical areas and influence the design of their buildings.

They may have similar building attributes due to similar effects from outdoor weather, although there is still regional uniqueness that portrays specific factors other than climate that relate to their individual country. It is necessary to understand the relationship of macro/mesa climate with microclimate. Therefore, the discussion on building science in tropical climates will be covered later in this chapter. First it is appropriate to consider microclimate in the tropics.

Microclimate

Microclimate reflects the deviation in local outdoor environment around a building. The conditions for comfortable living in a building may be directly influenced by the design of the building and those for determining the comfort response of people are local and site-specific. These conditions are generally what we may identify as 'microclimate', which includes wind, radiation, temperature and humidity experienced around a specific building; compared to macroclimate which affects a bigger region.

The location of the building at any place will change the microclimate of the site; for instance the extended wall may obstruct the direction of the wind flow, create shadows on the ground or on other buildings. We have to understand these conditions and to take account of their effect in the design.

Microclimates are the weather conditions that are found in a relatively small, very local geographical area, space or zone that compound within the area of a building. It may be located in rural, urban, coastal or forest area, for example. The microclimate of a site is affected by the following factors:

i Elevation and slope of the site topography.
ii Elements of water nearby the buildings.

Table 3.3 Regional tropical characteristics

Country	*Climate overview*	*General climate*				*Severe weather*
		Temp.	*Wind*	*Rainfall*	*Sunshine*	
Australia (Geoscience Australia & ABARE 2010)	The Northern part of Australia experiences tropical, sub-tropical and equatorial climate	12.9°C–27.7°C	Average wind speed at 7 m/s	504.8 mm	Tropical region receives about 21 MJm^2–24 MJm^2 approximately per day	• Cyclones between November and April • Tropical depression with heavy rainfall between October and April
Bangladesh ("Meet Bangladesh: Climate of Bangladesh" n.d.)	Tropical monsoon-type	26°C–36°C	• 3–6 km/h during cold season • 8–16 km/h during hot season	1,525–5,080 mm **Most rain during June–September and little during November–February*	Decline sunshine duration at 5.3 per cent a decade from 1960 to 2009. Normal sunshine hours in range of 7.18–7.49 hours except in the months of June–September (4.39 hours) (Institute of Water and Flood Management 2012).	• Cyclones between April to May, and September to November • Storm surge, flash flood, thunderstorm
Brazil	Mixture of tropical and equatorial climate • North region is hot and humid tropical forest and equatorial climate • North-east region is a mixture of tropical, semi-arid and semi equatorial climate • South-east region is tropical climate	• 13°C–18°C during cold season • 20°C–40°C during hot season	Wind speed was observed in north-east region with max at 6.16–6.48 m/s and min at 2.78–2.88 m/s depending on height (Lima and Filho 2012)	• Areas in north-east region receive minimum 500 mm to maximum 2,000 mm of rainfall • Other regions on average range between 1,500 mm and 3,000 mm • Some parts in south-east region receive rainfall between 3,600 mm and 4,457 mm	Annual average of daily global horizontal solar irradiation in any region of Brazil is 1,500–2,500 kWh/m^2 (Martins *et al.* 2008)	Mostly affected by rain

continued

Table 3.3 Continued

Country	*Climate overview*	*General climate*				*Severe weather*
		Temp.	*Wind*	*Rainfall*	*Sunshine*	
Brunei Darussalam (Mathew *et al.* 2009)	Hot and wet throughout the year	• The highest temperature may reach 34°C–38°C • The lowest may fall to 18°C–21°C • Average temperature is 27.6°C	Annual average wind velocity is 2.7 m/s in certain parts but most frequent wind velocity varies from 4–8 m/s annually	• Annual rainfall exceeds 2,300 mm • Increases inland to more than 4,000 mm • Average rainfall is 3,147 mm	Annual average daily intensity of 5.43 kWh/m^2	
Indonesia	Tropical rainforest. Tropical monsoon and tropical savannah. Wet and dry all year round.	• Coastal plains average temperature 28°C • Inland and mountain 26°C • Higher mountain 23°C	Average 5 m/s	From 25 mm during dry season to 360 mm during rainy season	Sunny hours from 5–7 h in rainy season and 8–10 h during dry season. Average 4.8 kWh/m^2/day.	Tropical cyclone, rainstorm and heavy rains
Malaysia (Makaremi *et al.* 2012; Ngan and Tan 2012; Noriah and Rakhech 2001)	Hot and humid tropical climatic conditions with overcast cloud cover through the year	Daily air temperature varies from a low of 24°C up to 38°C while the recorded minimum temperature is usually during night	Wind speed ranges from 1.9 m/s to 4 m/s	Annual average rainfall about 2,285 mm with large amounts in the hilly north-east rather than on the plains	• Solar radiation ranges from 4.07 kWh/m^2/day to 5.22 kWh/m^2/day • The annual average of solar irradiance is estimated to be 4.56 kWh/m^2/day	Flash flood
Mexico (Ramírez and Pech 2013)	• Tropical wet and dry in the region of South-east Mexico • Tropical wet in the central region of North-west Mexico	• Minimum average temperature is 22.1°C • Maximum temperature in range of 21.9°C–33.9°C according to different regions	• The average wind speed across Mexico based on several stations and points is in the range of 3.16–8.94 m/s (Hernández-Escobedo *et al.* 2014)	• Maximum average rainfall in 2015 was 2,426.3 mm • Average rainfall 872 mm	North of Mexico city recorded minimum average of 17.04 MJ/m^2 in the year 2002, and a maximum of 19.99 MJ/m^2 in 2011 (Matsumoto *et al.* 2014)	• Tropical cyclone which includes hurricanes, storms • Heat wave in 2015 lasted 1–3 months according to different regions

Philippines (Philippines Atmospheric Geophysical and Astronomical Services Administration 2016)	Tropical monsoon and maritime	• Mean annual temperature is 26.6°C • Mean minimum temperature is 25.5°C in January • Mean maximum temperature is 28.3°C in May	Wind speeds range from 6.4 m/s to 10.1 m/s (Jain *et al.* 2014)	Average rainfall varies from 965 mm in some of the sheltered valleys to 4,064 mm in the mountains annually from one region to another	Country's annual average would be 5.1 kWh/m^2/day (Jain *et al.* 2014)	Typhoons
Singapore	Typical tropical climate, with plenty of rainfall, high temperatures with uniform fluctuation, and high humidity throughout the years	• Daily temperature range is 23°C–25°C during the night and 31°C–33°C during the day • Highest temperature range is 27.7°C–27.8°C • Coolest temperature is 26°C	• Wind speed normally less than 2.5 m/s • During thunderstorm presence and monsoon surge, wind speed can reach 10 m/s–25/35 km/h • Sumatra Squalls bring wind gusts of 40–80 km/h	Annual rainfall is 2,331.2 mm	Daily sunshine hours on average are from 4–5 hours, at least 120 watts/m^2	Heavy rainfall and thunderstorm
Sri Lanka (Emmanuel *et al.* 2007; Goto *et al.* 2013; Zubair 2002)	Tropical climate	Average temperature varies from 20°C to 30°C	• Wind speed usually occurs less than 2.7 m/s • Moderate wind speed at 3.05–4.16 m/s in certain regions • Highest wind speed at 4.4–5.5 m/s at a few places	Average annual precipitation is approximately 1,300–2,400 mm. The highest is approximately 4,000–5,000 mm in the valley (21 m above sea level)	Daily sunshine is 8–9 hours from February to March, 6.5–7 hours for other months	

continued

Table 3.3 Continued

Country	*Climate overview*	*General climate*				*Severe weather*
		Temp.	*Wind*	*Rainfall*	*Sunshine*	
Thailand	Tropical climate	Temperatures range from 23.1°C–29.6°C across the region	Experiences generally very low wind speeds with typically average speeds of not above 3 m/s (Commins 2008)	• General annual rainfall from 1,200–1,600 mm • Windward areas may receive more than 4,000 mm • Leeward areas may receive less than 1,200 mm	Average daily solar exposure at around 18–20 MJ/m^2/day to 18–19 MJ/m^2/day depending on the region (Department of Alternative Energy Development and Efficiency 2016)	• Tropical cyclones usually happen 3–4 times a year • Thunderstorms usually occur in northern and southern parts
Vietnam (Polo *et al.* 2015)	Tropical monsoon type of climate	• Cold season between October and April, temperatures range between 15°C and 20°C • Hot season between June and September, temperatures range between 30°C and 40°C	Average wind speeds of 6–7 m/s	The annual rainfall is above 1,000 mm throughout the country. It may rise to 2,000–2,500 mm in hilly regions particularly facing the sea	Overall solar resources in Vietnam show an average GHI of 4–5 kWh/m^2/day in most regions of southern, central and partially even northern Vietnam (corresponding to 1,460–1,825 Wh/m^2/year)	Tropical cyclones usually affect northern and central regions

iii Elements of vegetation, trees and plants.
iv Elements of buildings which are exposed or sheltered from solar radiation.
v Outdoor elements such as street width, open spaces and others.
vi Last but not least, the building's passive design itself (i.e. orientation, openings for natural ventilation and lighting, materials selection and others).

Further discussion in the next section may in general elaborate this issue.

Building science in tropical climates

During recent years, efforts have been made to provide an international policy framework of environmental improvement leading to sustainable development. Many countries have assigned this policy framework to move towards a better environment. For instance, in Malaysia, the Green Building Index (GBI) is used to increase the productivity and efficiency of buildings in relation to the surrounding environment (Ahmad *et al.* 2011; Chua and Oh 2011). In Singapore, with the densely built-up urban environment, limited land space and few natural resources, the Building Control Authority (BCA) kick-started Green Mark, a rating system to evaluate a building's environmental impact and recognize its sustainability performance, designed specifically for buildings in the tropics. On the other hand, Thailand and Bangladesh applied the Leadership in Energy and Environmental Design (LEED) for green building certification.

Buildings typically account for 40 per cent of primary energy consumption through the energy required to generate, transmit and distribute electricity, as well as energy consumed directly on site. Furthermore, over 60 per cent of the total building energy is consumed for purposes such as cooling, heating and ventilation systems (Chan *et al.* 2010; Masoso and Grobler 2010). Buildings produce substantially more carbon emissions than the transportation sector (WBCSD 2008). It was also predicted that emissions from buildings as a result of energy consumption are estimated to increase by 48 per cent above 1990 levels.

The rapid growth of urban population and increasing income intensify the development of buildings in tropical regions such as South-east Asia. There is likely to be continued and growing demand for buildings in the future. Studies show that the number of buildings, and in particular high-rise residential buildings, has increased by three times in the last six years in Singapore, whilst in Kuala Lumpur the number has increased two times in the same period. The statistics released from the National Property Information Centre (NAPIC) in Malaysia show that the number of apartment buildings grew up to 41 per cent in the last ten years, and the number keeps growing annually (Ta 2009). Similarly, in Brazil the Federal Government Program focused on the low-income population, resulting in the construction of 2.5 million new houses in 2013 and 2014 alone. It is estimated that 75 per cent of the buildings that will exist in India by 2030 have yet to be built. Due to this high demand in the construction sector in such regions, it is argued that new buildings and renovation should be directed more towards utilizing bioclimatic principles to make better use of available natural energy and energy sinks, hence, reducing energy use. The use of bioclimatic principles, strategies and best practice examples will assist in transforming the building industry towards this goal.

Bioclimatic design and its concept

Urban sprawl, without consideration of climate-related issues, can progressively reduce the sustainability of outdoor environments. The concept of bioclimatic design in building was introduced by Olgyay during the 1950s and was widely applied as a design process during the 1960s (Olgyay 2015; Olgyay and Seruto 2010). It brings together multi-disciplines such as climatology and building physics. In the context of buildings, the term bioclimatic refers to spaces with natural ventilation where exposition to the climatic environment can be tackled by passive design measures. Recently, it has been cited as a stepping stone to achieve sustainable building and environment (Szokolay 2014). Bioclimatic studies and related issues have been followed through passive and low energy architecture studies and it is applied worldwide within various fields. The Passive and Low Energy Architecture (PLEA) conference is held annually and many designers and professionals participate to present their latest designs and products (Bowen and Vagner 2013). PLEA is dedicated to the development and documentation of bioclimatic design principles. It also encourages architects and professionals to apply natural and innovative techniques for heating, cooling and lighting.

This section explores the bioclimatic design fundamentals that are applied by design professionals as a first response to design for climate. In general, the application of bioclimatic design is easy for low and medium scale buildings because they are relatively easy to make bioclimatically interactive. To respond to environments climatically, the building's form and fabric should reflect human and climate factors (Hyde 2008, 2013). Low energy building, (net) zero energy building and passivehous are some of the standards that have been introduced within the context of improvements of this strategy. However, these standards vary from region to region in their exact definition and can be similarly applied to bioclimatic building design.

Buildings of a large scale have generally ignored the issue of bioclimatic design due to the complexity of buildings, high-density in an urban context and the availability of cheap energy for providing indoor thermal comfort. The bioclimatic impacts are normally excluded in the design process of complex buildings and they mostly rely on mechanical cooling and ventilation to provide suitable indoor thermal conditions. As a result, the energy consumption of these types of buildings increases rapidly and can lead to an unsustainable built environment. However, there are some exceptions to this methodology, as can be found in the pioneering high-rise buildings in a tropical climate (Santamouris 2006).

Various aspects of bioclimatic design are fully researched and acknowledged in different climates. Although bioclimatic design has been applied on small- and large-scale buildings for a long time, there are limited studies to show that the most effective principles are being followed. Bioclimatic design in large-scale buildings can be linked to energy conservation through using renewable energies like wind and solar energies, occupants' comfort through natural means and design with regard to microclimate. These elements should be connected to the building design strategies in the early stages to provide basics in the context of building science that can be applicable in small- and large-scale buildings.

There are five basic modes of strategies to the design of a building, which provide comfortable conditions for occupants including passive mode, composite mode, full mode, productive mode and mixed mode. Basically, the application of passive mode produces improved internal comfort conditions based on renewable energy resources. The main reason for using this mode before applying any of the other above-mentioned modes is that the passive mode uses up relatively little or zero non-renewable energy and so has the least impact on the environment.

Building design and characteristics

Bioclimatic architecture can make significant contributions to health and well-being, and to healthy economy and ecology. Bioclimatic design in building can be achieved through various design elements and strategies. They include building form and orientation, specific design of the building's façade, application of landscape and vegetation, solar shading, application of natural ventilation through sophisticated elements, roofing systems and building mass and form.

The following building design objectives should be considered as guiding principles in the complete planning process and construction of the building:

1. Create liveable spaces that fulfil functional purpose as well as being physically and psychologically healthy and comfortable in order to promote the optimum development of human beings and their activities.
2. Make effective use of resources and energy that is sufficient to serve the building.
3. Preserve and improve the environment by integrating human activity into a balanced ecosystem.

As an example Figure 3.2 shows effective design strategies that have been applied in Malay traditional houses to respond to climate effectively.

A typical courtyard built form attempts to enhance wind-induced cross-ventilation and functions as an air funnel, promoting maximum air circulation to the interiors. Cross-ventilation occurs through the openings from the entrance door through the central courtyard and out of openings in the building fabric at the leeward side. The entrance veranda on the windward side acts as a wind tunnel, focusing the incident wind into the courtyard that lies on this air funnel, which, in turn, ventilates the indoor spaces surrounding the courtyard. Clearly such features are much more difficult to incorporate into larger commercial and residential buildings of modern design. However, the principles of bioclimatic design can offer much in terms of design approach.

Climate change and the need for bioclimatic design

In tropical climates the temperature sometimes exceeds 32°C, which is too high for human comfort. Comfort 'zones' are defined for various conditions based upon the interaction of temperature and relative humidity. These are utilized to inform design decisions that impact upon ventilation approaches within buildings. The various approaches that are taken to ventilating and cooling building interiors can mitigate discomfort. However, the use of natural ventilation may be insufficient to deal with the issue. In Bangkok for example, even with a 1 m/s indoor air velocity, the conditions may be outside the accepted comfort zone.

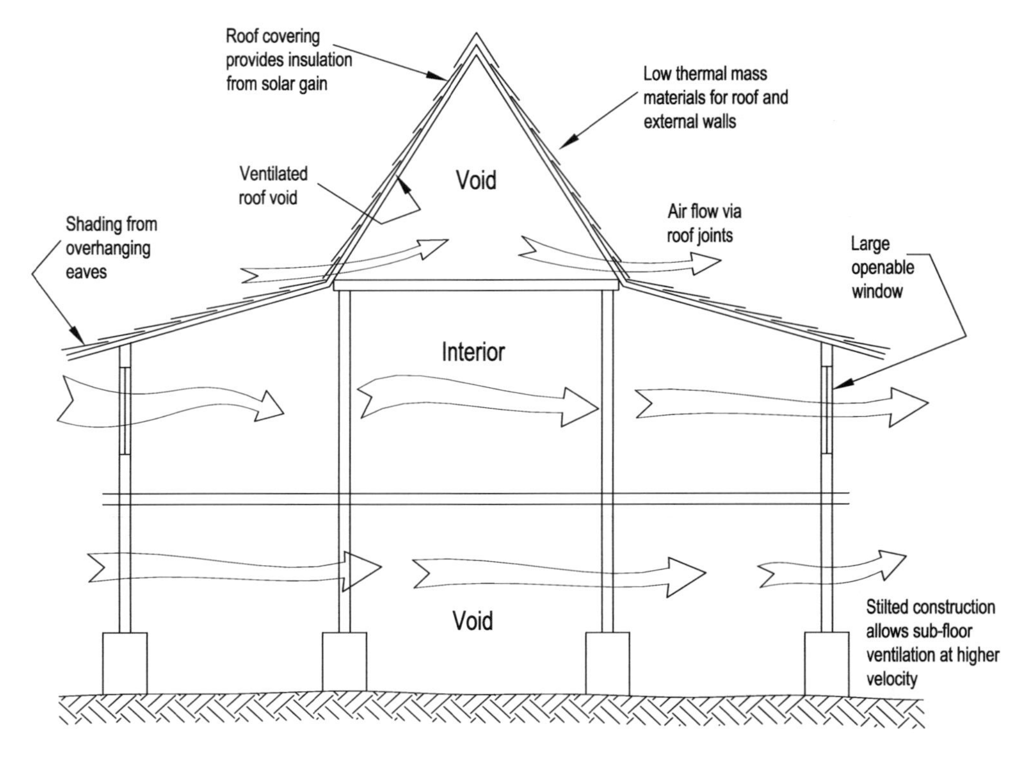

Figure 3.2 Climatic design of Malay house (after Sahabuddin 2012).

Within the example above the internal climate is classified into five groups, as follows:

1 Hot air; natural ventilation does not work.
2 Warm air; high natural ventilation is needed.
3 Comfortable air; moderate ventilation is appropriate.
4 Humid air; natural ventilation is not appropriate.
5 Cool, humid air; minimal ventilation will help.

Considered relative to the American Society of Heating Refrigeration and Air-conditioning Engineers (ASHRAE) human comfort zone, these illustrate the limitations placed upon building designers arising from conditions of temperature and humidity. The example poses differing potential responses based on internal conditions and ventilation levels.

Four main arguments are cited for the need to utilize tools such as this to maximize the utility of naturally ventilated solutions for buildings (Roaf 2003). These are as follows:

1 The rate of change in the level of climate variability and modification is increasing, requiring human adaptation to a rapidly warming world.

2 The fundamental means to this adaptation in the built environment is the adoption of more effective, and widely used, methods for passively cooling buildings.
3 Air-conditioning systems are increasingly seen as a part of the climate change problem, as well as its solution. Rising cost of energy is seen to be a major problem in terms of cost, especially to those who could not afford air conditioning. In addition, the energy used to run these systems is a major contributor to greenhouse gas emissions.
4 It is crucial to create a new 'cool vernacular' or passive design strategies building approach, which matches human and environmental needs.

Landscape and vegetation

Using vegetation in bioclimatic architecture has several advantages, such as air and surface temperature reduction. It provides effective shading on building façades and protects them from direct solar radiation. It can also be used as an intercepting device to control the sun's rays penetration into buildings. As such a bioclimatic design, using vegetation and greenery, can be applied on the east and west walls where they receive a high amount of solar radiation during a day.

Greenery and vegetation can prevent excessive heat gain surrounding the building and decrease unwanted glare. Moreover, the right location for trees is an effective strategy to shade the building during the hottest days of the year and allow the sun to warm the same building during the coldest days. In addition, using vegetation on building façades is more effective for diffusing radiation because of rough textures of plants compared to common materials which are used. Figure 3.3 illustrates how the greenery prevents excessive heat gain absorbed by the building's surface, especially in summer time.

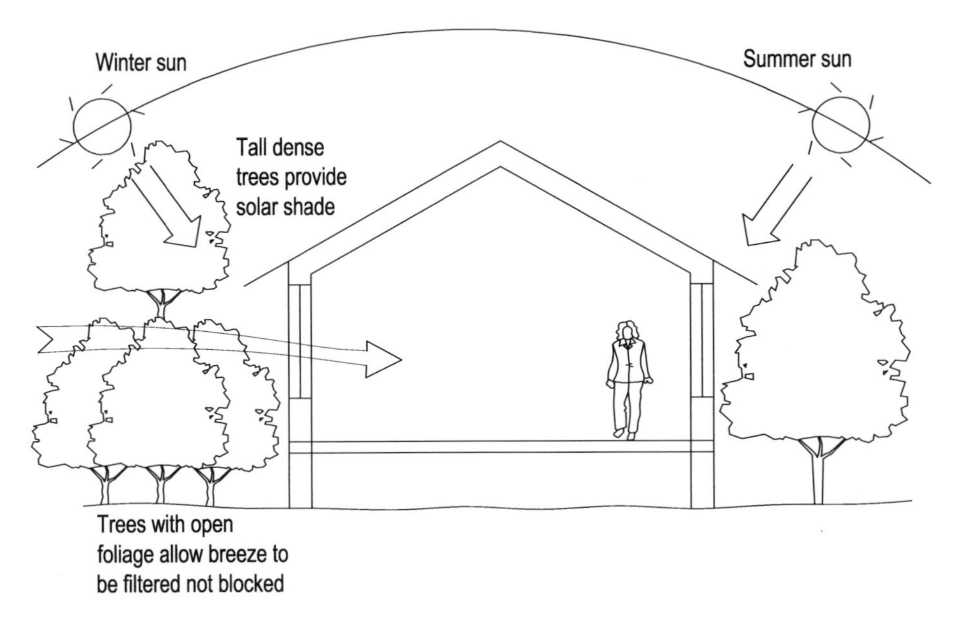

Figure 3.3 Trees as shading device and increasing wind speed.

Building form, orientation and site layout planning

In bioclimatic design the building layout is shaped and planned based on the meteorological data of the locality as a passive response to the environment. The building form should be laid out on the site to consume minimum energy without compromising the site's benefits such as views, privacy and security of occupants. The cooling energy requirement in buildings can be reduced through building form, orientation on the site and volume ratio to surface ratio. Figure 3.4 presents the building layout according to the main compass directions.

The building form for tropical climates near the equatorial zone should have a ratio of 1:3 width:length. Furthermore, the service core placements for buildings in tropical climates should be arranged longitudinally from west to east.

Façade design

Façade design should give controllable permeability to the light, heat and air. Moreover, the building's façade should be able to be modified to respond to changes in the local climate conditions. Appropriate design of a building's façade can significantly help to save energy. In bioclimatic design, the building's façade is required to function as protector, insulator and integrator with the surrounding environment.

In hot and humid climates, it is necessary to insulate the building's façade by using materials with high thermal mass that will increase the time lag. Thermal mass is defined as the ability of a building material to absorb and store heat, before releasing it at a later time when necessary. In other words, if designed and applied correctly, thermal mass provides cooling during warmer months and heating during cooler months (Figure 3.5). A material with the relevant thermal lag delays the heat movement from one surface to the other. As the direct and indirect solar radiations

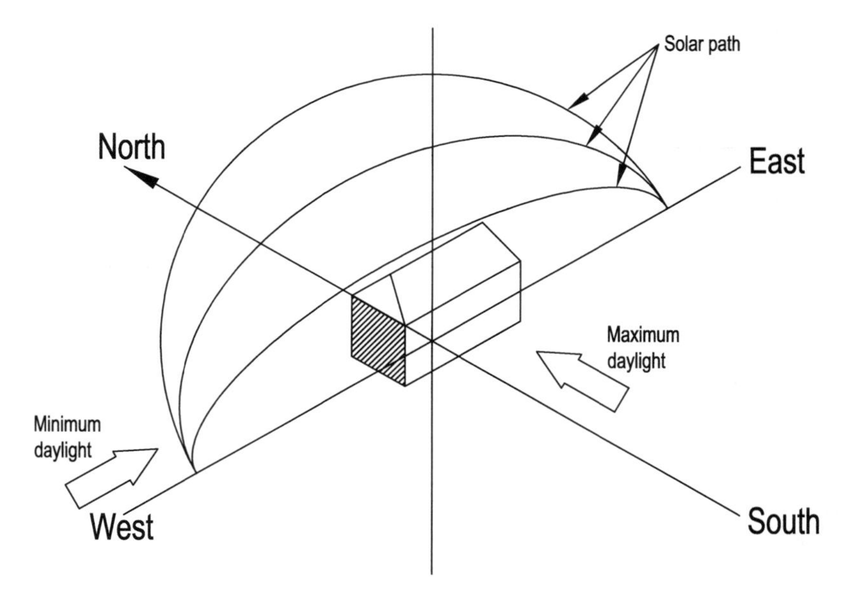

Figure 3.4 The building layout according to the main directions.

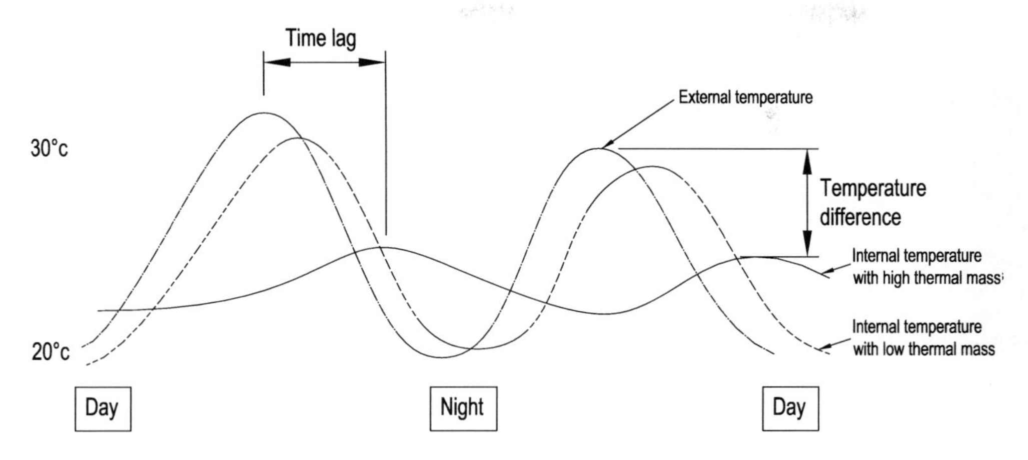

Figure 3.5 Effect of thermal mass on internal temperature.

are intensive in the equatorial region, selection of a suitable material for the building's façade is critical.

The appropriate design criteria for tropical climate must be based on sharing heat gains by high radiation all year and to stimulate the natural ventilation to control internal temperature. Therefore, it is convenient to use natural lighting but needs extra attention to reduce heat loads through radiation.

Solar shading

In hot and humid regions, heat avoidance technique is an effective strategy to provide comfortable indoor conditions for occupants. Solar shading is usually used on outside surfaces to protect the building envelope from sunrays. By using shading elements in bioclimatic architecture, heat transfer through the surfaces will be minimized. As a result, the need for HVAC systems will be reduced and noticeable amounts of energy can be saved (Wines and Jodidio 2000).

There are various types of shading that have been used to protect buildings from the sunrays. Each controls, reflects and transmits solar radiation in different ways. Some of the elements that have been used to protect the buildings from negative impacts of climate are mentioned below:

- fixed shading devices like wing walls, vertical and horizontal overhangs;
- low shading coefficient glasses;
- greenery and landscape;
- interior glare control devices like adjustable louvre windows.

As the most effective shading element, fixed shading devices are usually used to control sunrays in bioclimatic architecture. They are cost effective and durable. Also, they are more convenient and do not restrict daily human activities. However, they cannot be adjusted to various positions to cope with solar movement. Therefore, they should be precisely designed and dimensioned before construction. These types of shading are mostly installed on external walls and windows and they are more efficient than internal fixed shading devices like curtains. The dissipated heat between the fixed internal shading devices and the glazing reduces the efficiency of this type of shading significantly (Datta 2001).

As for fixed solar shading, overhangs help to decrease the degree of radiation reaching in buildings. Furthermore, it protects the buildings from uncomfortable light contrast levels around the apertures. On the other hand, wing walls as fixed shadings are normally used in the east and west of a building's façade where solar radiation is too intensive during the morning and afternoon time in the equatorial regions (Skias and Kolokotsa 2007). Figure 3.6 shows a designed overhang on top of the window based on the azimuth and altitude of the sun to protect the indoor space from direct sunrays.

Natural ventilation applications in buildings

Natural ventilation is normally used for health ventilation, comfort ventilation and structural cooling. Natural ventilation can provide fresh air to a space and significantly reduce the indoor pollution concentration. This type of ventilation is referred to as health ventilation. Comfort ventilation is the simplest strategy for providing comfort when the indoor temperature under still air conditions is too warm. It is easiest to open the apertures to promote comfort by ventilation:

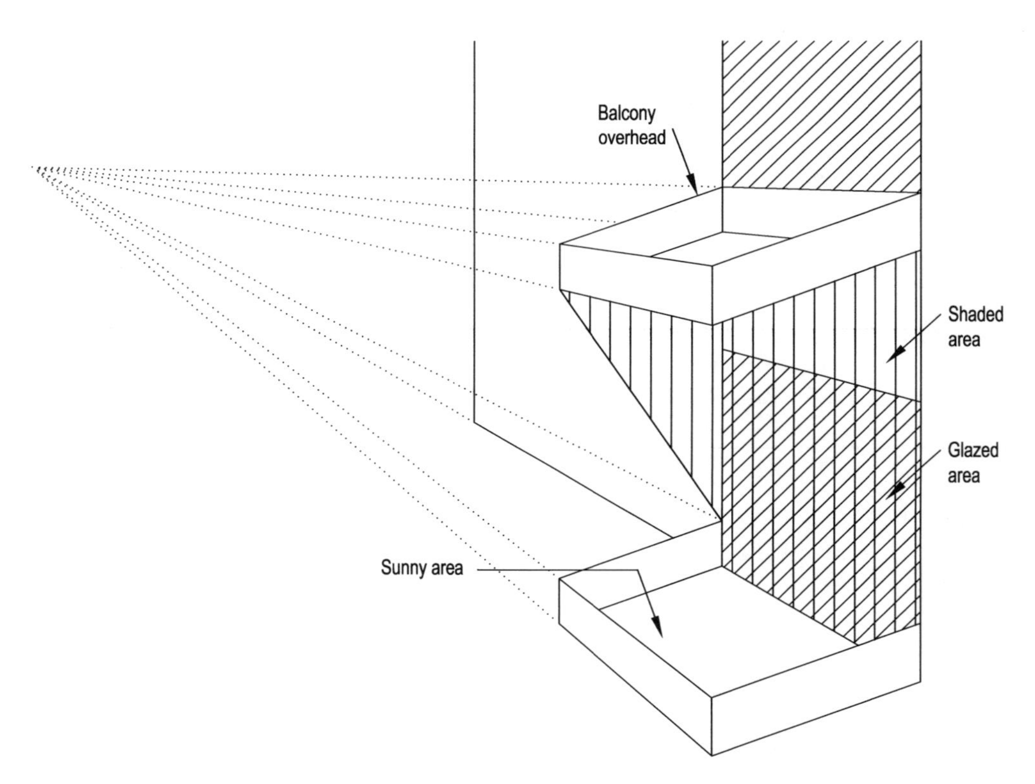

Figure 3.6 A fixed overhang on top of window for sunrays protection.

providing comfort through higher indoor air speeds. Cooling of a building's structure is another main function of natural ventilation. It was termed structural cooling ventilation by Givoni (1998). The effectiveness of this strategy depends on the difference between indoor and outdoor temperature. Structural cooling ventilation is applicable when the indoor air temperature is more than outdoor air temperature. It is a simple mechanism because outdoor air enters the building and mixes with indoor air temperature.

Air movement is the key requirement in the overall ventilation process when integrating and designing the building façade, building form, apertures and building orientation (Almeida *et al.* 2005). Moreover, outdoor air velocities can also affect indoor air movements and temperatures as a result of differences in air pressure applied through building façades in the appropriate location of openings and passive design strategies. Figure 3.7 shows how the elevated floor in the Malay vernacular house allows the wind to circulate effectively through the building. It is a clear example of bioclimatic design in order to respond to microclimate.

In general, natural ventilation inside buildings can be categorized into air pressure ventilation, known as wind-driven ventilation, and stack effect ventilation or buoyancy-driven ventilation (Ghiaus and Allard 2005). Figure 3.8 shows cross and stack ventilation in a building where the cool air removes the heat effectively.

Roof systems

As roofs receive the intensive solar radiation in the building's envelope, they should be thermally insulated to avoid penetration of heat into interior space. In tropical climates, roofs are integrated with overhangs to provide proper shading for the building's envelope as well. In addition, pitched roofs are used to respond to heavy rain during the monsoon seasons.

Green roofs or vegetated roofing systems have been applied, more recently, in buildings to provide suitable thermal insulation. Studies show that green roof

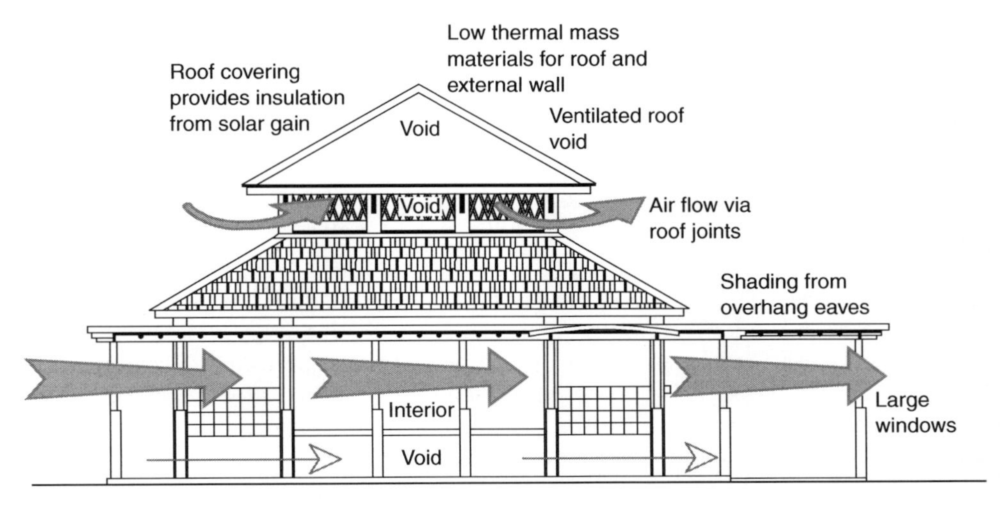

Figure 3.7 Elevated floor for sufficient wind circulation in Malay houses (after Sahabuddin 2012).

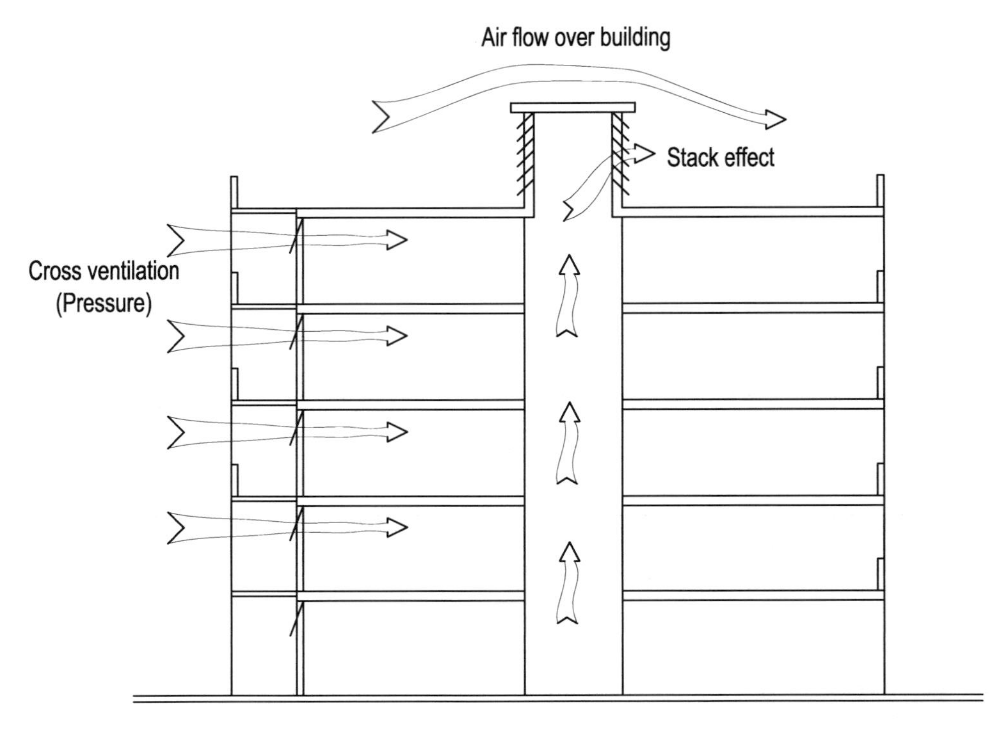

Figure 3.8 Bioclimatic design through effective cross and stack ventilation.

insulation levels are typically higher than the conventional roofing systems (Niachou *et al.* 2001). Green roofs also provide many advantages including extended roof life, increase of water run-off quality and mitigation of urban heat island effects (Oberndorfer *et al.* 2007). As a result, a building equipped with a green roof may be a sustainable development that can properly respond to microclimate.

Green roofs provide additional environmental benefits as they decrease surface and ambient temperature, improve outdoor air quality and balance the water cycle. Also, they have some economic benefits such as recycling water, saving from air purification and increasing property value.

Building mass and form

Building form and elements should be selected to respond to natural daily and seasonal cycles. Also, buildings should be designed to exploit energy sources like solar or wind energy. This can be attained by the connection of temporal and spatial interactions in natural and human-made environments. In order to know which strategies of bioclimatic architecture can be used in the building mass and form, it is necessary to analyse the climate characteristics of the site. Bioclimatic charts help designers to design buildings based on the climate conditions of a specific region.

These charts facilitate the analysis of the climate characteristics of a given location from the viewpoint of human comfort, as they present, on a psychrometric chart, the concurrent combination of temperature and humidity at any given time. They can also specify building design guidelines to increase indoor comfort conditions when the building's interior is ventilated naturally. All such charts are structured around, and refer to, the 'comfort zone'. The comfort zone is defined as the range of climatic conditions. The bioclimatic chart shows the relationship of the four major climate variables that make human comfort. By analysing air temperature and relative humidity, it can be found that the final condition is comfortable which is the comfort zone or it is too hot which is above the top of the comfort zone, or the condition is too cold which is below the bottom of the comfort zone.

Indoor environmental quality (IEQ) in tropical climates

People tend to spend about 80–90 percent of their time indoors. Exposure to any situation of indoor environmental quality (IEQ) may affect people's comfort, health and daily performance. The indoor environment of a building is largely affected by a combination of conditions shown in Figure 3.9.

The combination of these conditions is closely related to the earlier discussions in this chapter around climate and its relationship to building science. Understanding and maintaining each of these individual conditions within recommended benchmarks and standardized ranges will improve the overall IEQ and further increase occupant satisfaction.

Since IEQ is embedded from the early stage of building design to the building operations and maintenance, any changes to the design or operational practices would significantly impact its indoor environment. Comprehensive integration towards good IEQ performance in the building design can lead to healthy, comfortable and satisfied occupants.

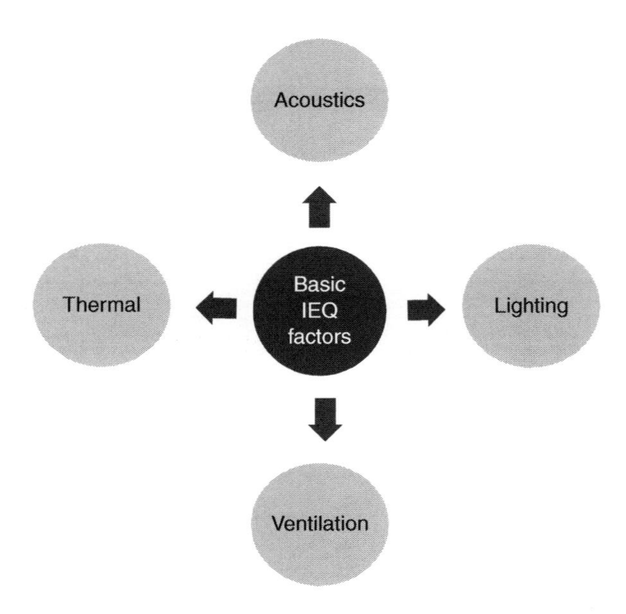

Figure 3.9 Basic environmental factors in an indoor space.

These parameters will be described further by explaining the situation of each environmental factor in the context of tropical climate. The parameters, definition and mechanisms might be the same in other climate regions. However, due to different building design approaches in a different type of climate, tropical buildings may have different measurement and strategies to control these environmental factors.

Thermal comfort

Thermal comfort in indoor climate comprises generic parameters such as moisture, air velocity, air temperatures, mean radiant temperatures, relative humidity, individuals' metabolism rate and clo value. Figure 3.10 shows an example of indoor environment and thermal parameters which may portray the situation in tropical climates – the presence of sun radiation throughout the year with high yearly average outdoor temperatures (often in excess of 30°C).

People in tropical climates may perceive that comfort is influenced by all-year-round high temperatures and humidity. Clothes they wear and personal indoor activities are very much affected by outdoor climate, as are expectations of comfort. Choices of wall thickness and materials with the exposure to sun radiation can make significant contribution to indoor comfort. Means of cooling to provide thermal comfort may include choices of cool surfaces, ceiling fans or mechanical ventilation air conditioning (MVAC), rather than heating ventilation air conditioning (HVAC) as would be found in temperate climates. It is not a surprise that air conditioning accounts for more than 60 per cent of the overall energy consumption in buildings in tropical regions.

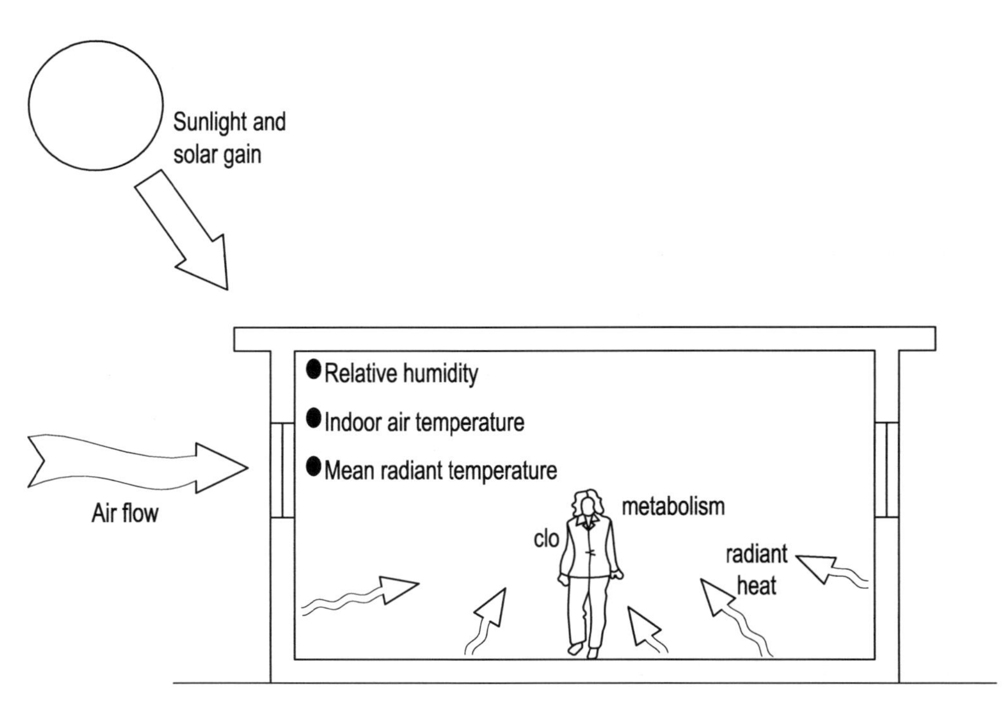

Figure 3.10 Parameters of the thermal comfort factors in tropical countries.

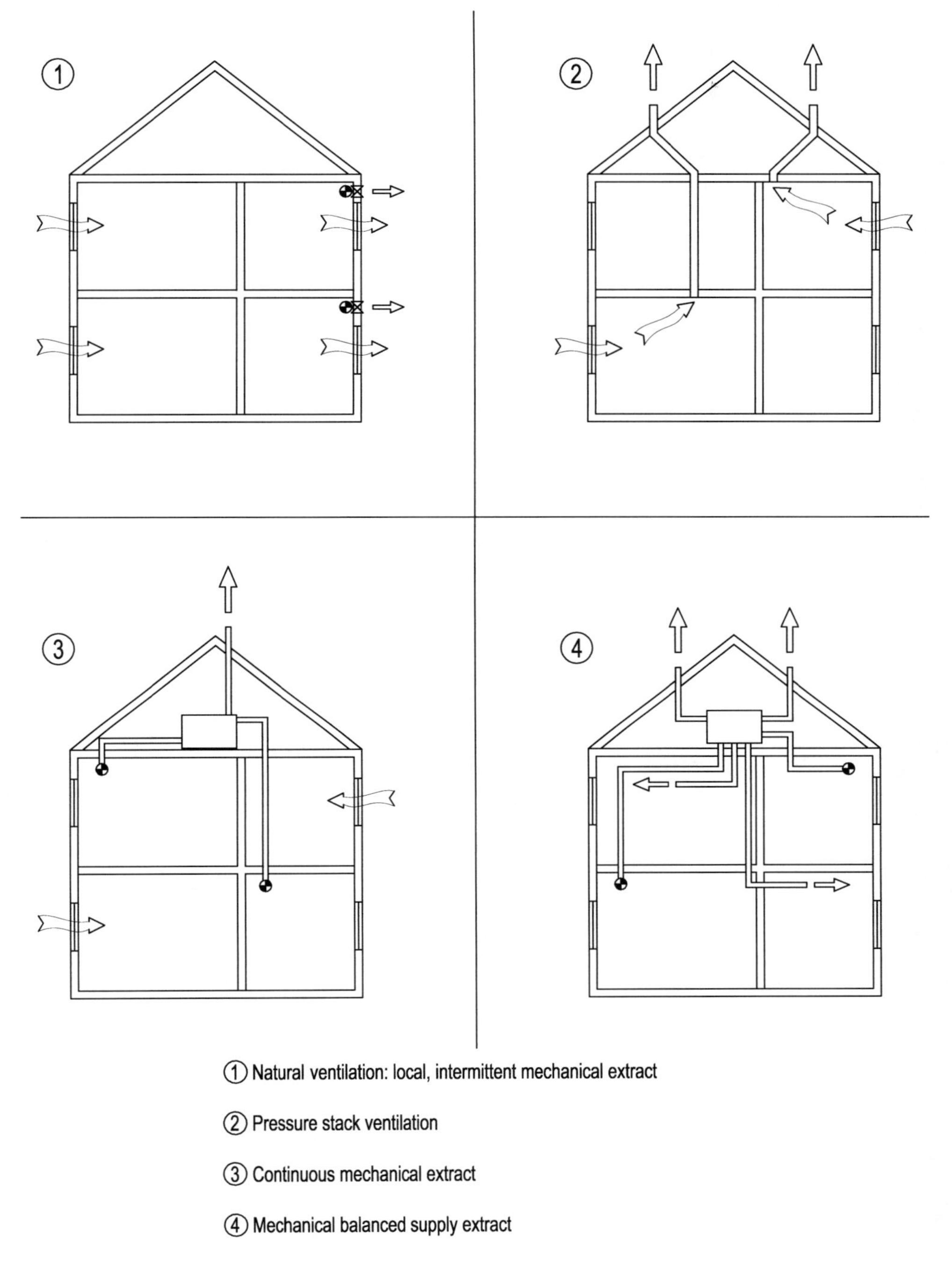

Figure 3.11 Different ventilation situations which can be the factors determining the indoor concentration of an air pollutant (after Vatal 2015).

Earlier discussion on building science has already elaborated topics on bioclimatic design and the building mass and form. Most of them would very much affect occupant thermal comfort.

Ventilation and indoor air quality

Ventilation is the process of supplying the fresh air from the outside into the building and dilutes or removes any indoor pollutants to the outside.

Lighting

Lighting may come from two sources; daylight and artificial light. Both are a non-stop movement of energy in electromagnetic radiation that people can see by their naked eye. The propagation of light may radiate equally to all directions and will diffuse over a larger area. The farther they emit, the weaker their intensity can be. The quality of light in a building at any climate region is determined by the light sources, whether daylight and/or artificial, the light distribution in the building/room, and people's perception of indoor lighting as a whole. Please note that this section may not deeply elaborate on lights spectrum, their mechanism and others. However, basic parameters listed in Figure 3.12 will be explained in brief. Lighting scenarios in tropical countries will follow subsequently.

- *Lumen* is the emission of the light flux (ϕ) from light source. Sources of daylight consist of direct sunlight and sun radiation, whilst artificial light may come from several sources depending on the number of lamps identified in a room.

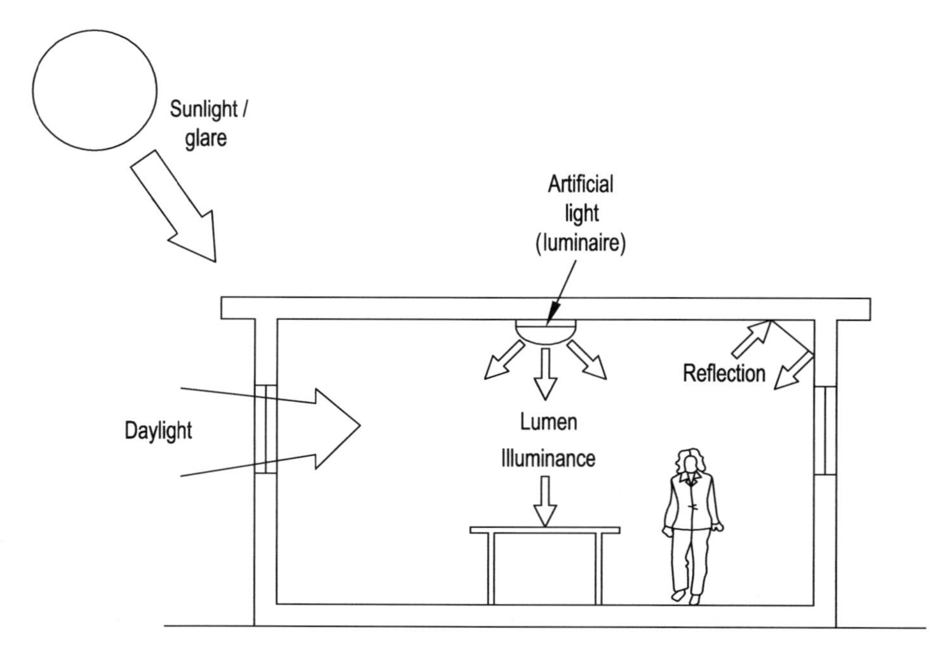

Figure 3.12 Light parameters in an indoor environment.

- *Lighting intensity (I)* is the amount of light flux (lumen) emitted from a source in a solid angle of one direction. It is expressed in lumen/steradian or candela.
- *Illuminance (E)* is the amount of light which has been emitted from the source, and spreads and falls onto specific working areas/surfaces. It is expressed in lumen/m^2 or lux.
- *Reflectance (R)* is the ability of a certain surface to reflect the illuminance to other solid angles and directions. Light may not be fully reflected. Some lights may be absorbed according to the colours and type of surfaces.
- *Luminance (L)* is the amount of reflected light emitted from a surface per unit area in one solid angle of direction.

The penetration of daylight will usually be through the clear glass or windows of the building façade. It has been highlighted that the building orientation, façade design with window and solar shading will greatly influence the performance of daylighting into a building. Therefore, this section will have minimum discussion on the daylighting mechanism and strategies.

It may be difficult to estimate the measurement of lighting in a building as there would be an integration of both sources during the day. Even though the availability of daylight in tropical countries seems to be abundant, with sunshine approximately 6–8 hours daily throughout the year, people still rely on the use of artificial light.

Space further away from a window will receive very weak light intensity as the daylight has been diminished. In order to guarantee that the visual comfort is achieved and to control certain levels of recommended lux artificial lighting is still necessary. However, consideration needs to be taken to ensure that the numbers of luminaires are only meant to increase the lux up to the recommended level. Levels beyond the recommended lux may affect people's visual comfort, especially glare. Besides, the excessive number of electric lights will contribute to higher energy bills. Even though low energy bulbs and lamps are widely available nowadays, such as LED lamps, some people could not afford them, especially in residential and dwelling buildings.

Studies conducted in several apartments in Dhaka, Bangladesh concluded that there were high dependencies on artificial lighting due to the increasing standard of living and ineffective daylighting caused by the small opening size and close proximity of buildings (Ahsan 2009). Research on energy efficient lighting in residential buildings also revealed that lighting contributes the second highest energy consumption of 25 per cent of the total energy consumed (Kamaruzzaman and Zulkifli 2014).

Acoustics

Acoustics is the physics of sound that deals with the reflection, absorption, transmission and incidence which is stimulated by mechanical radiant energy through pressure of waves in the air or other medium. The threshold of acoustic comfort may be subjective from one person to another person, and also depends on the type of room. Generally human comfort is in the range of 0–50 decibel (dB), such as activities of whispering and talking to each other face to face. Comfortable sound for one can be an unwanted sound or noise for others. The higher the dB the more potential sound intensity produces sensation of pain in the human ear. Acoustical design may need to be included from the planning and early designing stage of the building. However, people may always put priority on the thermal performance of a building (Abd Jalil *et al.* 2014) as it may closely affect the building energy performance. This will be discussed further in other sections.

The outdoor environmental noise that may influence the acoustic comfort of the building occupants may come from traffic noise, neighbourhoods, nearby buildings, nearby development and facilities activities (e.g. construction, people assemblies, etc.) among others, be it in urban or rural areas. The façade quality and design (e.g. materials selection, openings sizes, types and locations, orientations from the localization of noise sources, etc.) will contribute to the success of noise propagation before this unwanted sound is attenuated and absorbed into the rooms. Therefore, these building attributes will mostly influence the noise intrusion into the buildings (Ribeiro *et al.* 2001).

The building attributes in tropical countries may not be the same as those in the temperate and cold regions. Buildings in tropical climates may have little thermal inertia with operable windows or louvres for natural ventilation. Fixed openings with glass panes are also used to allow natural daylighting. Attempts to consolidate a comprehensive passive design as the best design solution for the tropical buildings may contribute to improvement of other comfort factors, especially thermal and visual comfort. However, it is often the case that these solutions present some difficulties to acoustic comfort, normally referred to as 'acoustic restrictions' (Slama and Silva, 2000).

Figure 3.13 shows the schematic general situation of a building, as an example in an office environment. Current awareness and development on the practice of green building rating tools is likely to improve the performance of IEQ. The practice of natural ventilation, daylight harvesting, minimum usage of finishes and open plan office layout are thought to be the main factors for a low acoustic performance.

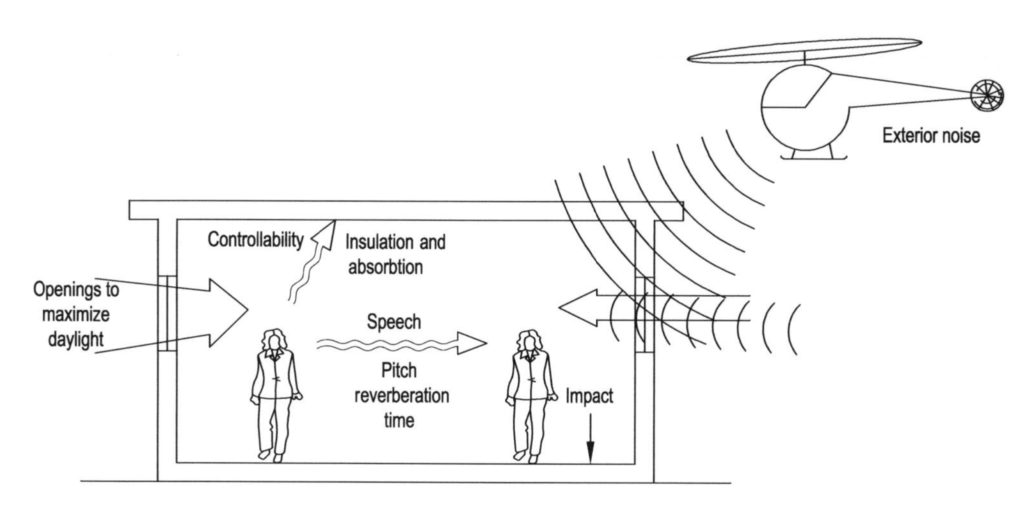

Figure 3.13 Sound parameters in an indoor environment.

References

Abd Jalil, N. A., Nazli, Che Din and Keumala, N. 2014. Assessment on acoustical performance of green office buildings in Malaysia. *Indoor and Built Environment* 25(4), 589–602.

Ahmad, Salsabila, Zainal Abidin Ab Kadir, Mohd and Shafie, Suhaidi. 2011. Current perspective of the renewable energy development in Malaysia. *Renewable and Sustainable Energy Reviews* 15(2), 897–904.

Ahsan, T. 2009. *Passive Design Features for Energy–Efficient Residential Buildings in Tropical Climates: The Context of Dhaka, Bangladesh.* Unpublished MSc thesis. KTH Department of Urban Planning and Environment. Division of Environmental Strategies research – fms. Kungliga Tekniska hogskolan, Stockholm.

Almeida, Manuela, Maldonado, Eduardo, Santamouris, Matheos and Guarracino, Gérard. 2005. The design of optimal openings. *Natural Ventilation in the Urban Environment: Assessment and Design*, 168–194.

Bowen, Arthur and Vagner, Robert. 2013. *Passive and Low Energy Alternatives I: The First International PLEA Conference, Bermuda, September 13–15, 1982.* Burlington, VA: Elsevier.

Chan, Hoy-Yen, Riffat, Saffa B. and Zhu, Jie. 2010. Review of passive solar heating and cooling technologies. *Renewable and Sustainable Energy Reviews* 14(2): 781–789.

Chua, Shing Chyi and Oh, Tick Hui. 2011. Green progress and prospect in Malaysia. *Renewable and Sustainable Energy Reviews* 15(6), 2850–2861.

Commins, T. 2008. Potential of wind power for Thailand: an assessment. *Maejo International Journal of Science and Technology* 2(2), 255–266.

Datta, Gouri. 2001. Effect of fixed horizontal louver shading devices on thermal performance of building by TRNSYS simulation. *Renewable Energy* 23(3), 497–507.

Department of Alternative Energy Development and Efficiency. 2016. *Areas with Solar Power Potential.* Retrieved 29 July 2016 from http://weben.dede.go.th/webmax/content/areas-solar-power-potential.

Emmanuel, R., Rosenlund, H. and Johansson, E. 2007. Urban shading – a design option for the tropics? A study in Colombo, Sri Lanka. *International Journal of Climatology* 27, 1549–1555.

Geoscience Australia and ABARE. 2010. *Australian Energy Resource Assessment.* Canberra. Retrieved 29 July 2016 from http://arena.gov.au/files/2013/08/Australian-Energy-Resource-Assessment.pdf.

Ghiaus, Cristian, and Allard, Francis. 2005. *Natural Ventilation in the Urban Environment: Assessment and Design.* London: Earthscan.

Givoni, Baruch. 1998. *Climate Considerations in Building and Urban Design.* New York: John Wiley & Sons.

Goto, K., Kumarendran, B., Mettananda, S., Gunasekara, D., Fujii, Y. and Kaneko, S. 2013. Analysis of effects of meteorological factors on dengue incidence in Sri Lanka using time series data. *PLoS One* 8(5), 1–8.

Hernández-Escobedo, Q., Saldaña-Flores, R., Rodríguez-García, E. R. and Manzano-Agugliaro, F. 2014. Wind energy resource in Northern Mexico. *Renewable and Sustainable Energy Reviews* 32, 890–914.

Hyde, Richard. 2008. *Bioclimatic Housing: Innovative Designs for Warm Climates.* London: Earthscan.

Hyde, Richard. 2013. *Climate Responsive Design: A Study of Buildings in Moderate and Hot Humid Climates.* London: Taylor & Francis.

Institute of Water and Flood Management. 2012. Ahammad Ali, Syed. Assessing the sustainability of community based fisheries management approaches through beel user group at Sunamganj haor area. Postgraduate thesis, IWFM.

Jain, A., Fichaux, N. and Gianvenuti, A. 2014. *The Philippines: Solar, Wind and Bioenergy Resource Assessment.* International Renewable Energy Agency.

Kamaruzzaman, Syahrul Nizam and Zulkifli, Nursyahida. 2014. A review of the lighting performance in buildings through energy efficiency 2014. 2nd International Conference on Research in Science, Engineering and Technology (ICRSET'2014) 21–22 March. Dubai (UAE).

Lima, L. de A. and Filho, C. R. B. 2012. Wind resource evaluation in São João do Cariri (SJC) – Paraiba, Brazil. *Renewable and Sustainable Energy Reviews* 16(1), 474–480.

Makaremi, N., Salleh, E., Jaafar, M. Z. and Hoseini, A. G. 2012. Thermal comfort conditions of shaded outdoor spaces in hot and humid climate of Malaysia. *Building and Environment* 48, 7–14.

Martins, F. R., Abreu, S. L., Chan, C. S., Rüther, R. and Amarante, O. a. C. 2008. *Solar and Wind Energy Resource Assessment in Brazil. SWERA Report.* (E. B. Pereira and J. H. G. Lima, eds). *Art and Design.* San Paolo, Brazil: Sao Jose dos Campos.

Masoso, O.T. and Grobler, Louis Johannes. 2010. The dark side of occupants' behaviour on building energy use. *Energy and Buildings* 42(2): 173–177.

Mathew, S., Ming, L. C. and Patrick, L. K. C. 2009. Sustainable electricity model for Brunei Darussalam. In *Solar09, the 47th ANZSES Annual Conference* (pp. 1–9). Townsville, Queensland, Australia. Retrieved 21 December 2016 from http://solar.org.au/papers/09papers/182_Mathewsetal.pdf.

Matsumoto, Y., Valdés, M., Urbano, J. A., Kobayashi, T., López, G. and Peña, R. 2014. Global solar irradiation in north Mexico city and some comparisons with the south. *Energy Procedia* 57, 1179–1188.

"Meet Bangladesh: Climate of Bangladesh". n.d. *Discoverybangladesh.com.* Retrieved 21 July 2016 from www.discoverybangladesh.com/meetbangladesh/climate.html.

Ngan, M. S. and Tan, C. W. 2012. Assessment of economic viability for PV/wind/diesel hybrid energy system in southern Peninsular Malaysia. *Renewable and Sustainable Energy Reviews* 16(1), 634–647.

Niachou, A., Papakonstantinou, K., Santamouris, M., Tsangrassoulis, A. and Mihalakakou, G. 2001. Analysis of the green roof thermal properties and investigation of its energy performance. *Energy and Buildings* 33(7), 719–729.

Noriah, M. M. D. A. and Rakhech, P. 2001. Probable maximum precipitation for 24 h duration over southeast Asian monsoon region – Selangor, Malaysia. *Atmospheric Research* 58(1), 41–54.

Oberndorfer, Erica, Lundholm, Jeremy, Bass, Brad, Coffman, Reid R., Doshi, Hitesh, Dunnett, Nigel, Gaffin, Stuart, Köhler, Manfred, Liu, Karen K.Y. and Rowe, Bradley. 2007. Green roofs as urban ecosystems: ecological structures, functions, and services. *BioScience* 57(10), 823–833.

Olgyay, Victor. 2015. *Design with Climate: Bioclimatic Approach to Architectural Regionalism.* Princeton, NJ: Princeton University Press.

Olgyay, Victor, and Seruto, Cherlyn. 2010. Whole-building retrofits: a gateway to climate stabilization. *ASHRAE Transactions* 116(2), 1–8.

Philippines Atmospheric Geophysical and Astronomical Services Administration. 2016. *Climate of Philippines.* Retrieved 20 July 2016 from www.pagasa.dost.gov.ph/index.php/climate-of-the-philippines.

Polo, J., Martínez, S., Fernandez-Peruchena, C. M., Navarro, A. A., Vindel, J. M., Gastón, M., ... Olano, M. 2015. *Maps of Solar Resource and Potential in Vietnam.* Ministry of Industry and Trade of the Socialist Republic of Vietnam. Retrieved on 29 July 2016 from http://renewables.gov.vn/Uploads/documents/tailieu/Maps of Solar Resource and Potential in Vietnam REPORT FOR PUBLISHING .pdf.

Ramírez, Reynaldo Pascual and Pech, L. A. C. 2013. *Reporte del Clima en México.* Mexico: Conagua.

Ribeiro, M., Kortchmar, L. and Slama, J. G. 2001. Case study of building diagnostics: acoustic post-occupancy evaluation of buildings in tropical climates. *Building Acoustics* 8(3), 213–222.

Roaf, S. 2003. *Ecohouse 2: A Design Guide*. Oxford; Burlington, MA: Architectural Press.
Sahabuddin, Mohd Firrdhaus Mohd. 2012. *Traditional Values and Their Adaptation in Social Housing Design: Towards a New Typology and Establishment of 'Air House' Standard in Malaysia*. MSc dissertation. University of Edinburgh, UK.
Santamouris, Matheos. 2006. Ventilation for comfort and cooling: the state of the art. *Building Ventilation: The State of the Art* 27(217).
Skias, I. and Kolokotsa, D. 2007. Contribution of shading in improving the energy performance of buildings. Paper read at 2nd PALENC Conference and 28th AIVC Conference on Building Low Energy Cooling and Advanced Ventilation Technologies in the 21st Century.
Slama, J. and Silva, A. M. P. Da. 2000. Noise control and natural ventilation in dwellings in humid hot countries. In Societe Francaise d'acoustique (ed.), *The 29th International Congress and Exhibition on Noise Control Engineering* (pp. 1–4). Nice, France: SFA.
Szokolay, Steven V. 2014. *Introduction to Architectural Science: The Basis of Sustainable Design*. London: Routledge.
Ta, Tiun Ling. 2009. Managing high-rise residential building in Malaysia: where are we. Paper read at 2nd NAPREC Conference, INSPEN.
Vatal, S. 2015. Indoor air quality – making buildings better. Sustainable Construction Group. BSRIA. Retrieved on 11 April 2017 from www.slideshare.net/BSRIA/indoor-air-quality-45522186.
WBCSD (World Business Council for Sustainable Development). 2008. *Energy Efficiency in Buildings: Business Realities and Opportunities*. Geneva, Switzerland: WBCSD.
Wines, James, and Jodidio, Philip. 2000. *Green Architecture*. London: Taschen.
Zubair, L. 2002. Diurnal and seasonal variation in surface wind at Sita Eliya, Sri Lanka. *Theoretical and Applied Climatology* 71, 119–127.

4 Historical evolution of buildings in tropical regions

Nor Haniza Ishak, Nur Farhana Azmi and Noor Suzaini Mohamed Zaid

After studying this chapter you should be able to:

- understand the general characteristics of buildings in response to tropical climates;
- describe the typology and morphology of tropical buildings; and
- appreciate various building examples from different tropical regions.

Introduction

The historical evolution of buildings in tropical regions has not been documented fully since its emergence. The international conference on tropical architecture held at University College, London in 1953 was critical in the development of architectural approaches for hot climates. Seen as something other than colonial architecture, the contemporary approaches to tropical building show a remarkably cohesive attitude towards design that responds to climatic conditions. The approaches to the concept, however, have been replaced by a multitude of attitudes to design. It was reflected in various building designs and characteristics throughout the tropical regions. This is further developed through the evolution of buildings from vernacular architecture which look very traditional, to ultra-modern or even high-tech designs. The evolution in building form in tropical areas has seen an initial shift away from vernacular forms to the more generic international styles and technologies, although there is now a trend for reversal towards bioclimatic design adopting passive environmental control features.

Overview

The purpose of this chapter is to generate an understanding of the general characteristics of buildings in response to tropical climates according to their anatomy in particular roofs, main and supporting structures. This will be followed by description of the typology and morphology of buildings in tropical regions. The former classifies different categories of buildings ranging from vernacular to contemporary architecture. The evolution of each type of the building is described by its morphology. Appreciation of various typology and morphology will be illustrated by giving several building examples from different tropical regions.

General characteristics of buildings in response to tropical climate

Tropical, hot and humid climates are those located within the region between latitude of 23 degrees north and south of the Equator that have high humidity. Surrounded by thick rainforests of stunning varieties of flora and fauna, vernacular buildings in this geographical zone are primarily built from readily available natural resources such as local hardwood and softwood timber, bamboo, rattan, clay and rocks. According to Watson and Bentley (2007), at least until the nineteenth century, transport limitations ensured that most buildings throughout the world had to be constructed from locally sourced materials, whilst contributing to the uniqueness of the building, or more generally a place. As argued by English Heritage (2010) and Parks Canada (2011), the materials used present one of the key features of a building that significantly contribute to its unique character and, therefore, make them worthy of being preserved and maintained. Materials, in addition to people's needs and climate, represent one of the prime considerations that distinguish buildings as belonging particularly to the zone.

In addition to physical environment, climate also has significant influence on the architectural design of a building in a particular area. For example, buildings in tropical climates which enjoy almost constant daily amounts of sunlight and a high rainfall are obviously different from those constructed in cold climates. The buildings may differ not only in the planning and exterior massing but also in general appearance, where the latter are likely to be severe, bleak and formal, while the former may be exotic and informal. In fact, buildings in the same region may also adapt to climate conditions very differently. The tropics are a region of great richness and diversity in vernacular architecture and therefore posits no single recognizable style of tropical architecture.

Despite the differences, buildings in the tropics also share many common characteristics as the work of tropical architecture is known for its remarkable cohesive approach towards design that responds to climatic conditions. The purpose of constructing a building is to provide protection from extremes of climate and other sources of danger and discomfort. Responding to climatic exigencies, tropical buildings can thus be considered as naturally sustainable. According to Jamaludin *et al.* (2014), a sustainable building is one which is designed to harmonize with the local climate, traditions, culture and the immediate environment. For Cairns Regional Council (2011), consideration of environmental concerns in designing and constructing buildings would effectively reduce the environmental impacts and, also, ensure the health and comfort of the occupants. Rapid technological expansion since World War II (1939–1945), however, had significant implications on traditional architecture of countries in tropical regions. Whilst subject to the threat of disappearance, few studies relating to traditional tropical buildings that were suited to the region's environment have been performed to date. This is really unfortunate as the truly traditional buildings were actually intelligent buildings and part of a wider environment-friendly way of life.

Therefore, it is important to study climatic conditions and their influence on the characteristics of buildings in the tropics. This chapter attempts to discuss these characteristics for three main parts of the buildings, specifically the roof, main and supporting structures. Like individuals, buildings are highly organized structures made up of different components; like individuals, these components are interdependent on each other and work together to provide structurally sound and environmentally controlled spaces to house and protect occupants and contents. An attempt has also been made throughout the chapter to cite the examples of traditional architectural and constructional techniques from different countries in the region. It is hoped that this information can help architects, planners, designers, builders and other related stakeholders in tropical countries to continue a sustainable practice.

Roof structure

The roof has been defined by the Cambridge dictionary as the covering that forms the top of a building. This covering serves to enclose the space and shield the building from weather exposure. As one of the most important elements, the roof should always be carefully designed with regard to climatic conditions. Required to withstand heavy rainfall, roofs for buildings in tropical countries are traditionally pitched and sloped to quickly drain off rainwater. The definition of pitched roof, however, varies in different parts of the region. For example, in accordance with Section 2 of the Malaysia Uniform Building By-Laws 1984, pitched roof means a roof having an inclination of more than seven and a half degrees with the horizontal. In India, a roof with a pitch greater than forty degrees is considered as a sloping roof (Bureau of Indian Standards 2005). For some countries such as Singapore, the Philippines and Indonesia, there is no single designated angle that a roof needs to follow in order to meet the definition of pitched roof. The roof can be considered as pitched as long as it displays some slope.

Throughout the tropical countries, people have traditionally combined roofs of different angles or slopes in a single building. This practice results in at least two or more tiers of roof. This type of roof can be seen through several traditional houses in Java Island, Indonesia. One of the examples is the roof of the *joglo* house which is typically used for the residences of the high-ranking nobility and also public buildings such as palaces, mosques and government offices. According to the Javanese philosophy, the shape of roofs represents the hierarchy structure of Javanese community from the lowest to the highest: *kampung*, *limasan* and *joglo* roof. The ordinary *kampung* roof and *limasan* roof consist of two and four aslant roofs respectively. As illustrated in Figure 4.1, the *joglo* roof is the trapezium-shaped roof that is made up of two or three *limasan* roofs. It is steepest in the middle part of the roof and therefore looks like it soars upwards shaped like a mountain (Subiyantoro 2011). This shape indirectly creates a high attic volume in the building. Apart from storing goods and incorporating a 'maid room', the space also enables the circulation

Figure 4.1 View of buildings with *joglo* roof in Java Island, Indonesia.

of air and, thus, reduces the effect of heat gain on the roofs during sunny days. The roof plays a significant role in the prevention of heat build-up as well as rainwater dispersion.

Some of the traditional houses in other countries such as Malaysia and Cambodia also hold attic rooms in their roof spaces. These spaces can be differentiated from the *joglo* roof by having a triangular or slightly rounded gable wall on their sides (Figure 4.2 illustrates the former example). The ventilation panels (a set of sloping planks fixed horizontally at regular intervals) which are often found on the gable walls not only enhance the air movement but also allow the input of natural light into the spaces. There have also been ventilation grilles used at its gable ends called *tebar layar* (Joo-Hwa and Boon 2006). For ventilation and lighting purposes, some buildings in tropical countries like Malaysia are also equipped with a jack roof which is an elevated small-size roof placed over the main roof as illustrated in Figure 4.3. According to Wan Ismail (2005), the use of this roof style can be traced back to China in the 1800s.

Another climatic responsive design feature that can be seen in the roofs of tropical buildings is a wide overhang. Similar to the roof slope, the eaves projection is also available at varying lengths and heights across the region. Roof overhang is critically important in tropical regions in order to protect the majority of the walls including the openings such as windows and doors from direct sunlight. This is especially important for east and west facing openings as the sun rises and sets respectively in these directions. Furthermore, keeping the eaves away from the wall also helps to protect the house from heavy downpours which minimizes the splashback of water flowing from the roof on the facade. Although the problem could be controlled by

Figure 4.2 An ordinary house in the state of Negeri Sembilan, Malaysia, with traditional Minangkabau roof style; note the fixed ventilation panels at the triangular gable ends of the roof.

Figure 4.3 One of the surviving traditional Malay houses in Kuala Lumpur, Malaysia, called the *Rumah Penghulu Abu Seman* with a jack roof above its main roof.

the use of gutters as part of the roof system, the installation of rain gutters has been traditionally argued as not suitable for buildings in tropical zones. Buildings in these zones often do not have any gutters as the buildings are surrounded by heavily shaded trees and covered with vegetation. In addition there is great potential for heavy rainfall to surcharge the gutters and downpipes in storm conditions.

The eaves, which often end in a fascia board, are often decorated with ornamentation and carvings. There are various types of design patterns used such as floral and geometric designs, thus making the buildings interesting and aesthetically valuable. Figure 4.4 demonstrates an example of roof eaves with flowery motif designs in one of the Muslim mosques in Thailand. In addition to its functions in keeping the rain water off the walls, providing shade and beautifying the building, the decorative carved eaves also allow for good lighting of the building. The ornamentation or decoration elements can also be found in other parts of the roof such as the roof ridge and its ending (Figure 4.5).

Figure 4.4 Decorative carved roof eaves on the multi-tiered roof of a Muslim mosque in Pai, Thailand.

Figure 4.5 The National Museum building in Phnom Penh, Cambodia, shows various decorative elements on its multi-tiered roof; note the top roof tier near the central spire towers is designed in an ornamental V shape.

The fact that traditional buildings in tropical regions rely on nature for their resources is evidenced by looking at the materials used to build the roof. Taking Malaysia as an example, the most common roofing material used for the traditional Malay house is the *attap* which is a kind of thatch roofing made from the fronds of the nipah, rumbia, bertam and other palm trees found in the area where the local people live. Mohd Sahabuddin and Gonzalez-Longo (2015) argued that the thatch roofs not only protect the buildings from rain but also give a cooling effect as thatch does not retain heat due to its low thermal capacity. For Joo-Hwa and Boon (2006), roof coverings of palm leaf thatch allow natural ventilation through the palm fronds. While made up of safe, environmentally friendly and sustainable roofing materials, the thatched roof is also very durable since it can last up to five or more years. A similar effect can also be achieved with the use of thin *senggora* tiles. In particular, these tiles are made up of clay, another indigenous material that is commonly used in the Malay Peninsular East Coast. The *senggora* tile is purposely thin for four reasons: it is light to carry around; it would not burden the weight of houses on slender stilts (another dominant feature of vernacular buildings in the tropical region); its thinness reduces its heat absorption and re-radiation capacity; and if damaged it could be more easily removed and replaced. Above all, it is all very practical and has environmentally friendly qualities.

Due to unprecedented problems relating to the depletion of natural resources, the thatch has been increasingly replaced by corrugated sheets of asbestos cement or galvanized steel (exposed zinc). During the twentieth century, the latter has become popular due to its lower cost, ease of installation and relative longevity. In contrast to the thatch, the danger of being home for rats and the risk of fire are not issues that owners of a building with exposed zinc should be worried about. The use of zinc for roofing buildings in the tropics also demonstrates serious disadvantages particularly in increasing heat and producing irritating noise during heavy rain. The substitution of this kind of roof, however, does not change the overall traditional roof form. In fact, the use of traditional roof materials is still noticeable nowadays across the region. Interestingly, this old ancient practice is still used in a diverse range of new buildings. Traditional architectural forms, elements or materials can still be relevant for modern buildings. Figures 4.6 and 4.7 illustrate the use of thatch roof in one of the homestays in a small town of Pai, Thailand, and the use of plain clay tiles of *senggora* in a resting hut near Jalan Persisiran Pantai, Bachok, Kelantan. For some buildings, the traditional and modern materials are also used in combination (Figure 4.8).

Above all, roofs for buildings in tropical regions should have sufficient pitch, wide overhanging eaves and also be designed using materials that minimize heat retention as they are greatly affected by the great amount of solar radiation and high rainfall. While responding to climatic conditions, these characteristics also play an important role in making the building distinguishable visually.

Main structure

The main structure of a building consists of the space between the roof and the ground. Similar to the roof, in tropical regions, it is always convenient to wall this space using readily available materials from the surroundings. Traditionally, materials such as bamboo, rattan and sometimes a special type of grass are often used for walls weaving.

Figure 4.6 One of the homestays with thatch roof in Bamboo Village, Pai, Thailand.

Figure 4.7 The use of traditional plain clay tiles (commonly known as *senggora*) in a small resting hut near Bachok, Kelantan.

Figure 4.8 The original thatch roofing has been replaced by corrugated zinc sheets in some parts of the houses found in aboriginal settlements in Cameron Highland, Pahang.

Figure 4.9 A traditional house in Pai, Thailand, has walls made from bamboo.

Figure 4.10 Overlapping arrangement of wooden planks at the perimeter wall of a building found in Kampung Kepayang, Perak.

Figure 4.11 An aboriginal house in Cameron Highland, Pahang, with woven *bertam* wall in zigzag pattern.

Clay and various wooden panels arranged in horizontal, vertical or diagonal directions are also used extensively in building the walls in traditional buildings. On the other hand, materials that are exceptionally stiff and strong, such as hardwood timber, are typically used for the frame of the main structure. All these natural and renewable resources feature low thermal mass, thus they perform better as they minimize heat transfer into the building during the day. The extent of heat gain in a building can be reasonably controlled by the appropriate selection of materials.

Due to their high porosities the predominant use of these materials further improves air movement in the building. In tropical regions, this is very much needed to reduce indoor temperature and provide comfort for occupants. Similarly, the gaps between the ubiquitous wall matting also allows air penetrations through the wall which eventually help to cool the building. Thatched or woven walls, however, are better known because of their aesthetic values. Repeat or random arrangements of various traditional and abstract motif patterns can be designed on the walls using different types of local materials that can be woven into thatch wall. The decorative effect of walls in making buildings recognizable and having a unique character of their own is therefore undeniable.

Unfortunately, more dense and heavyweight materials such as brick and concrete are now used for building the walls following industrial logging and urbanization. Few buildings with wall thatching or those which were built from jungle products are still standing, although it is still possible to see remains of these buildings especially in rural areas. The unprecedented urbanization and depletion of naturally existing resources changes the urban climate from natural to man-made and alters the atmospheric composition that differs from the suburban and rural natural environment. In addition to difficulties in obtaining constant supplies of natural resources, the practice of using these materials is also discontinued because of its high maintenance costs.

Unlike traditional materials, concrete or brick walls are known to take a long time to release most of the heat received during the day due to their high thermal capacity. This has resulted in high dependency on modern services such as fans and air-conditioning.This disadvantage can be lessened by placing large quantities of openings in perimeter walls of the building. This quality of openness eventually demonstrates another common characteristic that is geared towards providing effective ventilation for buildings in tropical regions (Lim 1998; Mohd Sahabuddin and Gonzalez-Longo 2015). This is a key principle of bioclimatic design. The openings which generally comprise the doors, windows, slatted panels and also porous carvings or ornamentation encourage air movement through the building while maximizing the effect of natural lighting. Careful consideration of glare, driving rain and tropical pests should be given in the provision of these openings at body level. The operable openings such as casement and louvre windows should be used to overcome this problem so that they can be adjusted to be opened or closed depending on weather conditions. This is important as different individuals have different levels of desire for privacy and personal, local ventilation. In achieving effective natural ventilation in the tropics, these openings should also be oriented towards the directional flow of the wind, located on both the windward and leeward sides of the building (to induce a sucking effect), or near the ceiling level to vent out warm air which is less dense than surrounding air (Cairns Regional Council 2011).

Figure 4.12 Large numbers of windows running horizontally at the Sungai Lembing Museum building in Pahang, Malaysia, allowing good ventilation qualities.

Figure 4.13 Bungalow of Syed Kamarul Ariffin in Perak, Malaysia with various wall decorative elements; note the coloured glass and louvre panels below the gable screens reflect the importance given to ventilation and lighting.

Another characteristic of tropical buildings is reflected through open and fluid interior space which further encourages natural air flow through the building. Traditional houses in Malaysia can be a good example of this. The house is usually divided into several areas rather than rooms, thus using few numbers of permanent partitions or barriers. The house also accommodates very little furniture such as chairs, beds, stools or tables. This quality maximizes ventilation over the building.

All these traditional climate responsive approaches are not really an option in modern commercial buildings. However, designers are slowly making a comeback to the passive approach in modern building design. For example, tinted or laminated glass is often used running horizontally or vertically across the building to enable daylight penetration into the building while protecting the occupants from glare, and also reducing the amount of heat transmitted through the building. Besides incorporating openings, modern buildings also accommodate open verandahs or balconies to offer the occupants immediate contact with the external environment. Notwithstanding the fact, there are certain limitations in terms of the availability of natural material for today's building.

Supporting structure

Another notable feature of buildings in tropical regions is further demonstrated by their supporting structure, in particular for those buildings which are raised on stilts or columns. Many early settlements were situated along rivers or near the coasts as the waterways served as major transportation networks, especially for trading activities. Therefore, the practice of constructing the building higher than the ground became an ideal solution for coping with the ground dampness. The system also

Figure 4.14 Traditional Malay house on timber stilts that is believed to have been the first house constructed in the small town of Kuala Kubu Bharu, Selangor, Malaysia.

became a dominant feature for these buildings as flash floods frequently happen in tropical regions during heavy rains. The amount of elevation usually varies depending on the expected flood levels of an area. Meanwhile, there are unique ways of determining the spacing between stilts throughout the region.

The raised floor construction plays a huge role in ventilating the building and thus exhibits another quality of openness in tropical buildings (Cairns Regional Council 2011). The effect becomes even greater if there are gaps or cracks on the floor. Furthermore, the fact that the building is elevated lessens the contact of inhabitants with wild animals and other potentially harmful vermin or pests. Being raised, the buildings also enable these animals to roam freely without hindering their movement and leave the local flora untouched and growing underneath the buildings. This is therefore very much aligned with an environmentally conscious approach. The practice shows great respect to animals, plants and generally the earth itself as only a minimum part of the buildings, particularly the stilts, are touching the soil. In addition, buildings that are raised on stilts undoubtedly create areas of shade which can eventually serve many different activities or purposes. The underneath space can provide shelter for the livestock, vehicles, equipment, working space, and also a clothes-drying area during rainy days. Since tropical climates allow outdoor activities to occur throughout the whole year, it also becomes a place for children to play and for communities to meet, chat and mingle with each other.

The use of stilts to raise the level of a building requires a fixed ladder as a means of access from ground to floor level. The ladder is normally made of hardwood timber and placed under the roof eaves or within its projection to protect the ladder from rotting in the rain. Similarly, hardwood is also commonly used for constructing the posts as they are more durable and impervious to termites. Notwithstanding this fact, it is still necessary to take reasonable precautionary steps to make the posts last longer. For instance, the posts or stilts usually rest on stones to avoid having direct contact with the soil which generally contains more moisture and dirt. Figure 4.15

Figure 4.15 The hardwood timber post rests on granite to avoid having the post in direct contact with a lime mortar base and the ground, to prevent it from rotting.

illustrates that granite is used as a liner between a timber post and a lime mortar base as the latter is composed of few ingredients, one of which is water. Alternatively, the timber base can also be painted or treated using preservative.

Although it is still common in some regions, most newer buildings no longer practise the system and simply build directly on the ground. In fact, buildings which still retain the original raised floor construction often have their stilts replaced by concrete or other non-organic materials. In addition to this, the underneath space is also sometimes permanently closed and covered with bricks or plywood. Besides affecting the eco-friendly nature of the system, the practice also wastes the utilization of the space under the buildings.

Building in context

In an architectural context, the external elements will influence the characteristics of the building in terms of its design (orientation, internal space layout, form and shape, etc.) and the materials used. These elements are physical and non-physical. Roads, buildings and land contours are examples of physical elements while non-physical elements are weather conditions, local culture, as well as political and economic constraints. Thus they are contexts that are influencing contemporary architecture design. Design for a house in four-season countries is different than the design in tropical countries. Understanding the local context is important for designers to produce a good building design. The context determines the architectural style, building material selection and site layout, which are very important in creating an effective design. All these promote continuity between the building and local circumstances (Mohd Sahabuddin 2011).

Orientation

In the context of tropical climates, orientation is the positioning of a building in relation to seasonal variations in the sun's path as well as prevailing wind patterns. Well considered orientation can increase the energy efficiency of a house, making it more comfortable to live in. Orientation of building on a site will influence the amount of sun it receives and will affect the building comfort. It therefore will maximize wind exposure and natural lighting when required but also deflect unwelcome monsoonal winds and rains. A study in tropical climates by La Roche *et al.* (2001) found that buildings must avoid large apertures on the east and west where they receive approximately twice the amount of radiation compared to north and south elevation (Figure 4.16).

Other considerations with regard to the orientation of buildings in tropical climates include access to views and cooling breezes. Orientation and layout will also be influenced by topography, wind speed and direction, the site's relationship with the street, the location of shade elements such as trees and neighbouring buildings, and vehicle access and parking. Ideally, the consideration of orientation, location and layout in the design stage is not only from the beginning of the design process but immediately from the time the site is being selected. Once a building has been completed, it is impractical and expensive to reorientate later.

Good orientation of a building keeps out unwanted sun and hot winds while ensuring access to cooling breezes. Normally, the orientation of a tropical house has

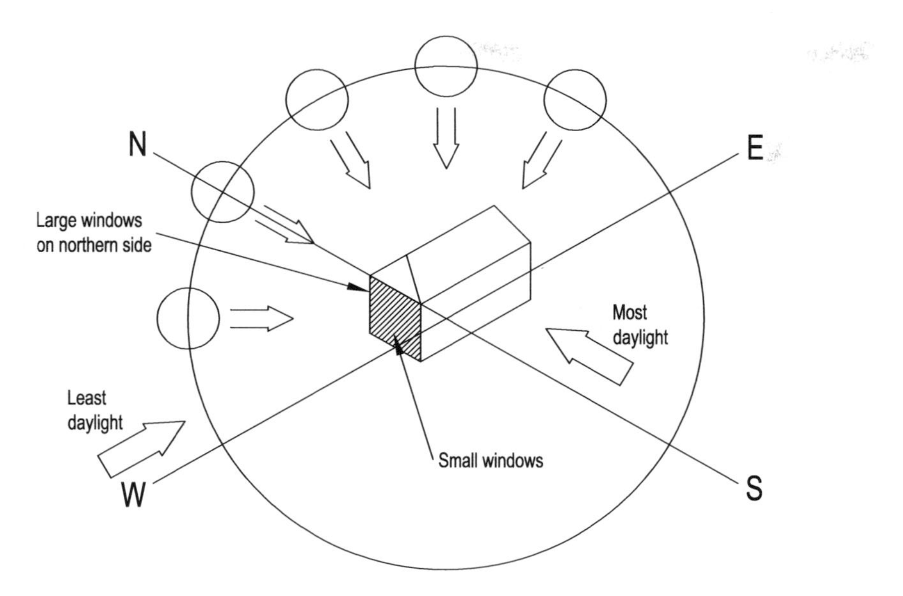

Figure 4.16 Building orientation towards sun's path in tropical climate (after La Roche *et al.* 2001).

Figure 4.17 Orientation of building on site will influence the amount of sun it receives and will affect the building comfortability.

Figure 4.18 Orientation and layout influenced by topography.

a maximum exposure to cooling breezes and limited but less exposure to direct sun, depending on climate. Narrow, elongated or articulated buildings facilitate passive cooling. North and south wings are designed for habitable rooms, but if buildings are in shade, variations are possible to provide maximum airflow. Large sized eaves are used to provide permanent shade to north and south windows and walls. Mostly, east and west wall areas are minimized with less provision of openings to prevent low morning and afternoon sun heating up the house.

Figure 4.19 shows the design of louvres and fixed openings in a traditional timber house. There are maximum openings to allow maximum breeze for cross ventilation in opposite walls and internal partitions. This can be achieved by having a maximum area of windows (e.g. louvres) that can be opened. Higher pitched roofs or double roofs with louvres will draw heat out from the roof space. Traditional buildings in tropical climates were designed for natural ventilation, as are many newer residential and small scale commercial buildings.

Layouts and form

The layout of space involves the interior and exterior of the building. Interior design is all about experiencing space and its function. It affects how people live, work, play and other related activities. Tropical interior design can involve the use of natural materials, such as wicker, rattan, bamboo and teak. Buildings are usually separated and scattered with free spaces between them to utilize airflow. Individual structures should be freely elongated; rooms preferably single banked with access

Figure 4.19 Maximum openings design for maximum natural daylighting and ventilation.

from open verandahs or galleries. It may be advantageous to raise buildings on stilts. Open spaces link to a long narrow floor plan which catches the breeze best. Open spaces can also be located between spaces in the form of courtyards, terraces and shaded balconies.

Internal layouts and the arrangement of the spaces of a building can be planned effectively according to the orientation of the building. Normally the main living spaces such as living, family and dining rooms should be facing north if possible. North-facing rooms will have good daylight most of the day. East-facing rooms are most suited for kitchen areas as they can benefit from early morning solar gain and will be cooler in the late afternoon. As west-facing rooms get low-angle, late afternoon sun, they usually require some shading to prevent overheating and excessive glare, particularly during the summer. South-facing rooms are not suitable for habitable spaces as they have lower levels of daylight during parts of the year, with little or no heat gain. They are most suited for the location of the garage, laundry, bathroom, toilet, workroom and stairs, where people spend little time and/or use infrequently. In general, outdoor living areas should be north-facing so they receive the sun when they are in use.

Current research shows that the objective of architecture in hot and humid regions has been to moderate the most significant climatic components of the region, i.e. the high humidity and temperature. Pursuit of this objective has yielded solutions for creating shadows and minimizing the effect of sunshine on the interior of the buildings as well as using the natural air current, prevailing winds and local breezes. Characteristics resulting from such solutions define the configuration of architecture in this region.

Figure 4.20 Maximum opening concept at traditional timber house with raised floor and louvres for cross ventilation.

Figure 4.21 Higher pitched and louvres at gable area to draw heat out from the roof space.

Figure 4.22 Double roof design with fixed glass louvres act as air well.

Figure 4.23 Open space and water element in the courtyard.

Figure 4.24 Buildings separated and scattered with open spaces to utilize airflow.

Figure 4.25 Open space between building and road.

Surrounding/outdoor areas

The surrounding context is very important and needs to be taken seriously through the process of design. Thorough site analysis will lead to a good building design, by understanding all aspects including weather, current context (i.e. existing buildings, parking, roads), accessibility, vegetation, circulation, orientations, services and facilities. In the climate-oriented house pattern, any particular space obtains its physical features under the influence of the climate, as a result of which, the given ensemble establishes the most desirable connection with its surroundings, and would tend to run on renewable energy supplies (Shemirani and Nikghadam 2013).

Most of the buildings in tropical climates are shaded by vegetation and green. Types of vegetation planted should not block free passage of air. Adequate storm water drainage must be provided to prevent flash floods. In any developments, no substantial trees on the site should be destroyed and should be in harmony with nature. An outdoor area will be more lively and meaningful if surrounded by a garden and non-reflective landscaped surface and the outdoor area shaded with grass cover rather than paving. Trees and shrubs act to cool the air passing through the house. Buildings are spaced to provide least breeze disruption and tall trees planted on the east and west sides of the house to shade walls.

In tropical climates, response to the surrounding contexts is very important because it determines the success of a building design. In response to these contexts it will create communication channels between buildings and people around (Mohd Sahabuddin 2011).

Figure 4.26 Connection of buildings with the surroundings.

Evolution of tropical buildings typology and morphology

Tropical buildings can be categorized either by type, function or time. Considering the latter, Fry and Drew (1964), in *Tropical Architecture in the Dry and Humid Zones*, divided them into three main areas, particularly early tropical in construction of pre-1955, mid-tropical from 1955 to the early 1960s, and late tropical from the late 1960s to the early 1970s. This way of categorizing tropical buildings suggests that there is no single recognizable style of 'tropical' architecture. However, buildings in the tropics typically share some common characteristics as the work of tropical architecture is known for its cohesive attitude towards design that responds to climatic conditions (Le Roux 2003).

The term morphology in architecture generally means the evolution of form and describes changes in syntax or composition of buildings and cities. Morphology of architectural form can be influenced by climate or cultural aspects, where building form in tropical climates has evolved in response to exposure of sun radiation and heavy rain all year around. This section identifies the morphological evolution and development of buildings in tropical regions, from the simplest built sheds to the technologically sophisticated skyscrapers of the late twentieth century.

Vernacular and traditional

Vernacular architecture can be defined as everyday dwellings of the local people. The word vernacular is derived from the Latin *vernaculus*, meaning domestic or indigenous. Vernacular architecture reflects the technology and culture of the indigenous society and environment, and acknowledges no architect or formal training. To simplify, vernacular architecture is architecture without architects, and examples of vernacular forms include the North American Tipi (Figure 4.27), the Central Asian and Mongolian Yurts or Yurta (Figure 4.28) and the Tropical Region huts (Figure 4.29) and Malay House (Figure 4.30).

Subsequently, traditional architecture can be described as building that represents symbolic forms of a particular culture or society in a specific location and can also be characterized by a period. Traditional architecture can belong to a grand tradition such as fortresses, mosques, churches or public assembly buildings, where formal architecture and highly skilled builders are required.

For the purpose of this book, we will follow the morphology of the Malay House that evolved from its vernacular form into traditional, classical and contemporary built form. The vernacular Malay House with pitched roof and elevated floors morphed into a bigger scale such as public assembly halls or schools. Figure 4.31 is a photo of a traditional school built in 1909, with references to the Malay House built form, with the same pitched roof and highly elevated floor from the ground.

Classical architecture, on the other hand, is the expression of traditional architecture of a society or place or period that has achieved the highest, most articulate and refined appearance. This classical form should not be confused with the classical architecture style of ancient Greece and Rome, where the style is defined by big columns. The classical form concerned in this book is related to the morphology of the architectural form which evolved from vernacular to traditional and into classical form, which can be used to represent a defining architectural form of a region or culture. This morphology can be represented in the architectural form of classical

Figure 4.27 Oglala Lakota American Tipi.

Photo credit: John C. H. Grabill (1891) from United States Library of Congress.

Malay Palaces, grand mosques and museums, which have derived inspiration from the vernacular Malay House.

Consequently, in the decade after World War I in the early twentieth century, architectural form began to evolve based on principles of functionalism where the "form of a building should be determined by practical considerations such as use, material, and structure" (Encyclopedia Britannica 2016) and the rejection of historical patterns and ornament. The modern architectural form began to use modern technologies and materials such as iron, steel, glass and concrete. The modern architectural form of the pre-Industrial Revolution era represented the ideas of 'the machine age' with minimal exterior expressions and open floor plans.

During the nineteenth century, Malaysia was colonized by the British Empire and therefore the architectural form was defined by the colonial era, which mixed modern architecture with the traditional form. Such examples can be seen in the Selangor Club House and the iconic Sultan Abdul Samad building. The architectural form evolved from using traditional timber and wood into materials such as brick, concrete and mortar.

Figure 4.28 Turkmenistan Yurks.

Photo credit: Prokudin-Gorskiĭ and Sergeĭ Mikhaĭlovich (between 1905 and 1915) from United States Library of Congress.

Figure 4.29 Tropical Hut in Ecuador.

Figure 4.30 The Malay House.

Figure 4.31 Traditional Islamic School in Ipoh, built in 1909.

Figure 4.32 Malacca Sultanate Palace (replica of the palace built in 1984).

Figure 4.33 Malacca Sultanate Palace (replica of the palace built in 1984).

Figure 4.34 Kampung Laut Mosque (rebuilt in 1966).

Photo credit: Islamic Tourism Centre of Malaysia.

Figure 4.35 Teratak Za'ba (Memorial of late national writer laureate Zainal Abidin Ahmad) built in 1996, completed in 2005.

Subsequently, contemporary architecture reflects the general or standard architectural form of the moment, and is not limited to a single stylistic form. Contemporary architecture is of the present time, and therefore is innovative, dynamic, and constantly changing. Contemporary forms borrow bits and pieces of the different architectural forms preceding the current, and can be quite eclectic and diverse for this reason. The contemporary Malay vernacular house exhibits a subconscious response to the rich ethnic and cultural diversity of the region. Examples of vernacular Malay House morphology into contemporary form can be seen in the 'palace of culture' building, more commonly known as Istana Budaya (Figure 4.36), a performance arts centre, and in the National Museum, known as 'Muzium Negara' (Figure 4.37), with the traditional pitch roof form.

In the statement below, Paul Ricoeur suggests the necessity for societies to sometimes abandon their cultural past in order to participate in a 'modern' civilization, but stresses the difficulty of reviving old dormant civilization.

> In order to take part in modern civilization, it is necessary at the same time to take part in scientific, technical and political rationality, which often requires the pure and simple abandonment of whole cultural past. There is the paradox; how to become modern and to return to your sources; how to revive an old dormant civilization and take part in universal civilization.
>
> (Ricoeur 1965, p. 277)

> From now on, the places visited by the traveler become ever more similar to the commodities that are part of the same circulatory system. For twentieth century tourism, the world has become one big department store of landscapes and cities.
>
> (Schivelbush and Beng 1995, p. 26)

Figure 4.36 Istana Budaya, built in 1995.

Photo credit: Istana Budaya (2016).

Figure 4.37 Muzium Negara, built in 1963.

Photo credit: Department of Museums Malaysia (2016).

Vernacular, traditional and classical architectural forms are usually interlinked because they can be lined in a single spectrum where the most humble and basic architectural form begins with vernacular and evolves into traditional architecture and finally into classical form, using the most precious materials and grandly scaled (Figure 4.38). However, not all vernacular architectural forms are morphed and evolved to traditional or classical form.

Tropical typologies

Typology is generally defined as a study or classification of types and categories such as categorization of residential, commercial, institutional or public buildings. The following sub-section briefly uncovers the typical tropical typologies in Malaysia, for example.

Residential typology

i The Malay Kampong houses

'Kampong' is a Malay word meaning 'village'. Traditionally, the Malays lived in separate dwellings grouped into Kampongs which were separated social entities. The

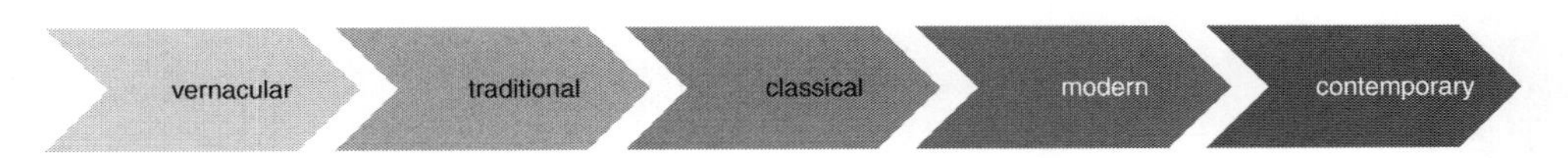

Figure 4.38 General morphology of architectural form.

coastal and riverine Malay Kampong house is very simply constructed with a rectangular pitched roof shelter raised about two metres above the ground. This traditional form was described in the section on morphology.

ii Indigenous longhouses

The longhouse is the oldest architecture form in Sarawak, going as far back as the history of the ethnic groups that inhabit the area. Among the Iban, Bidayuh and Orang Ulu, the basic concept of the long house is similar. It is a linear arrangement under one common roof of separate apartments, the doors of which open out to a common hall or gallery. From this corridor there is often a single exit. The main door, through which everything entering or existing must pass, is very important. The evil spirits must be prevented from passing through. Threatening faces are carved into the doors, often accompanied by stylized motifs.

Sited usually near a river or stream because of the available water supply, fishing and ease of transportation, and on high ground for strategic defence purposes, each longhouse varies in size from about twenty to eighty apartments (Figure 4.39). Sometimes, because of site constraints, two parallel longhouses were constructed as the community grew. The average house could accommodate between 200 and 300 people. The size of the building is impressive, and the hive of activity in the common gallery or verandah can be overwhelming: adults chattering, children running about along the timber flooring, dogs barking. As the longhouse is raised on wooden stilts or piles at least a full storey above the ground, the area below is where the pigs and chickens are kept, foraging around for whatever rubbish or food that is thrown down from the family kitchens above.

Figure 4.39 Maranjak Longhouse Lodge, Kudat Sabah.

Commercial

i Shophouses

Shophouses and their residential counter-part, rowhouses, were not indigenous forms but evolved from cultural circumstances (Chinese immigrants) and climatic considerations. Malaysia has not had a long history of urban settlements, being a predominately rural and Kampong (village)-based society. The large urban settlements such as Malacca and Penang were established as strategic locations for trade routes, or – as Kuala Lumpur, Taiping and Ipoh – as centres of tin-mining activity. The original residents and their architects/builders derived inspiration and ideas from Europe, blending them with architecture seen in neighbouring countries.

The shophouses evolved to allow merchants to live and work in the same building. Basically, the design follows the same floor plan to the present day. A covered colonnade forms the transition from the street, the shop is in front, with storage and the kitchen at the rear. Upstairs are living/dining/sleeping areas. A central air well provides light and ventilation, and facilitates the collection and disposal of rain water. These long narrow buildings are repeated to create comprehensible streets and squares of human scale.

Blending with the European concepts of urban life were the perceptions of the merchants, most of whom were Chinese. What emerged was a building that minimized the effects of heat, rain and glare in a tropical climate by using thick, brick walls with high ceilings, a roof with ventilation, an interior with an air well and a shop

Figure 4.40 A short row of shophouses.

Figure 4.41 Federal Territory Shariah Court.

front with a verandah. The early shophouses were purely utilitarian adaptations to the tropical climate. However, by the early 1900s, European, Chinese and Malay motifs were intricately executed on the facades, creating the illusion of a 'false front' to the simple structure.

Public and institutional buildings

According to the Malaysian Federal Department of Town and Country Planning (2006), public and institutional buildings consist of buildings for the purpose of education, health, religious activities, cemetery, security and for government administration. This typology includes buildings such as schools, public libraries, post offices, public halls, museums, government agencies and so forth.

References

Bureau of Indian Standards. (2005). *National Building Code of India*. New Delhi, India: Bureau of Indian Standards.

Cairns Regional Council. (2011). *Sustainable Tropical Building Design: Guidelines for Commercial Buildings*. Queensland, Australia: Cairns Regional Council.

Department of Museums Malaysia. (2016). Muzium Negara. www.jmm.gov.my/en/museum/muzium-negara.

Encyclopedia Britannica. (2016). *Functionalism Architecture*. Encyclopedia Britannica Inc. https://global.britannica.com/art/Functionalism-architecture.

English Heritage. (2010). *Good Practice Guide for Local Heritage Listing*.

Federal Department of Town and Country Planning. (2006). *Kod Warna Klasifikasi Guna Tanah Rancangan Tempatan* or Local Plan Land Use Classification Colour Code. http://103.8.161.36:28150/c/document_library/get_file?uuid=d3e2e426-0060-40a0-a4b1-e2cd4a1c9895&groupId=103624.

Fry, M. and Drew, J. (1964). *Tropical Architecture in the Dry and Humid Zones*. London. BT Batsford.

Jamaludin, N., Khamidi, M. F., Abdul Wahab, S. N. and Klufallah, M. M. A. (2014). *Indoor Thermal Environment in Tropical Climate Residential Building*. Paper presented at the Emerging Technology for Sustainable Development Congress.

Joo-Hwa, Bay and Boon, Lay Ong (2006). *Tropical Sustainable Architecture: Social and Environmental Dimensions*. Oxford: Architectural Press.

La Roche, P., Quiros, C., Bravo, G., Gonzalez, E. and Machado, M. (2001). Keeping cool: principles to avoid overheating in buildings. In Szokolay, S. V. (ed.), *PLEA Notes, Passive and Low Energy Architecture International: Design Tools and Techniques*. New South Wales: Research, Consulting and Communications (RC&C).

Le Roux, H. (2003). The networks of tropical architecture. *The Journal of Architecture*, 8(3), 337–354.

Lim, J. C. S. (1998). *Tropical Architectural Studies Towards a Sustainable Future*. Australia: Northern Territory University.

Mohd Sahabuddin, M. F. (2011). *How Important is Context in Contemporary Architectural Design*. The University of Edinburgh, Edinburgh School of Architecture and Landscape Architecture (ESALA).

Mohd Sahabuddin, M. F. and Gonzalez-Longo, C. (2015). Traditional values and their adaptation in social housing design: towards a new typology and establishment of 'air house' standard in Malaysia. *International Journal of Architectural Research*, 9(2), 31–44.

Parks Canada (2011). *Standards and Guidelines for the Conservation of Historic Places in Canada*.

Ricoeur, Paul (1965). *Universal Civilization and National Cultures: History and Truth*. Trans. Charles A. Kelbley. Evanston, IL: Northwestern University Press.

Schivelbush, Wolfgang and Beng, Tan Hock (1995). (Re)presenting the vernacular/re(inventing) authenticity: resort architecture in Southeast Asia. *Traditional Dwellings and Settlements Review (TDSR)*, VI(11), 26.

Shemirani, S. M. M. & Nikghadam, N., (2013) Architectural Objectives in Tropical Climates (Comparing Climatic Patterns in Vernacular Houses of Bandar-e-Lenge and Dezful). *International Journal of Architecture and Urban Development*, 3(3), 57–64.

Subiyantoro, S. (2011). The interpretation of Joglo building house art in the Javanese cultural tradition. *MUDRA Journal of Art and Culture*, 26(3), 221–231.

Wan Ismail, W. H. (2005). *Houses in Malaysia: Fusion of the East and the West*. Johor, Malaysia: Universiti Teknologi Malaysia.

Watson, G. B. and Bentley, I. (2007). *Identity by Design*. Jordan Hill, Oxford: Elsevier Ltd.

5 Construction technology for tropical regions

Michael Farragher

After studying this chapter you should be able to:

- appreciate the selection of materials and construction techniques used in tropical buildings;
- recognise the design factors that must be considered in building structure and fabric;
- understand how the design and construction of buildings can integrate passive design features to cater for tropical conditions;
- identify the various technical solutions to the formation of structures, building elements and internal environments in tropical climates;
- describe the connection between these factors and integrated building design; and
- appreciate the influence of the choice of materials and construction techniques on whole life performance of tropical buildings.

Introduction

As discussed in Chapter 2 there are several differing climatic conditions in tropical regions: Central South America has a hot and humid climate with seasonal differences; Central Africa a hot and wet climate; Asia and Australasia is hot, humid and seasonally wet. The microclimates in these broad ranging areas can differ in terms of nocturnal and diurnal temperatures and humidity. In general terms, the hot and humid tropical regions have little difference in temperature during the day in comparison to night, whereas hot and wet climates can cause a considerable drop in temperature during the night. These varieties of weather conditions and temperatures cause a variety of building typologies and form which differ from region to region.

Hot dry tropical climates lying between the two annual isotherms of 20°C consist of areas where vapour pressures are below 25 millibars and the temperature in the hot season may reach 43°C or more (Fry 1964; Breheny 1992). This can vary in highland areas where the temperature drops by approximately 2° per 1,000ft and humid areas where vapour pressure increases because the microclimate is closer to sources of water, wind variance and foliage. The description of the 'hot, dry tropics' is in itself misleading in terms of rainfall. These areas, although throughout most of the year they have deep blue skies and the sun-glare is blindingly overhead, do have a rainy season where for around 30 days the climate is susceptible to flash rainfalls where as much as 50 mm of rain can fall in one hour. This variance in climatic conditions has a direct influence on both building form and materials where shrinkage and contraction can produce pathological problems to the structure as well as the more direct effect to deal with in the shedding of large quantities of rain over a short period.

The humid tropical climate is again different where regions can achieve a relative humidity of up to 90 per cent, heavy rainfall and a mean temperature of 18°C all year round. There are usually two wet seasons, dark green vegetation all year round and at coastal areas conditions where rain can fall at a 90° declination at varying times (Fry 1964).

The accompanying map (Figure 5.1) shows the variance in climatic conditions in the tropics around the world based on five zones. It is intrinsically likely therefore that these variances will result in an ever different or changing building form both domestically and commercially.

Overview

Historically shelter has provided humankind with protection from the elements such as cold, heat, rain and the sun. With the development of architecture and building physics, we have been more concerned with not only shelter but also human comfort. It is with this in mind that there needs to be an understanding of heat loss from the human skin. The human can lose heat by radiation, convection and evaporation. The first two conditions are dependent on the temperature of air around the skin, whereas evaporation happens when the surrounding air is dry enough to absorb further moisture. It is important therefore that in tropical regions where relative humidity is high, there should be an increase in airflow around the skin, which is essentially achieved by good ventilation. Passively this can be aided by the correct orientation of the building.

Orientation

The following diagrams (Figures 5.2 and 5.3) show the difference of the path of the sun during a day in December. Figure 5.2 is an example from Kuala Lumpur in Malaysia and Figure 5.3 is Stansted airport near London. We see that the sun in Kuala Lumpur is overhead in December whereas the winter sun in the temperate region of the UK has a shallower altitude. In fact, the sun in tropical regions tends to be overhead during most of the year and follows an east west route at almost a 90° angle from the earth. It is therefore necessary that the orientation of the roof axis of tropical buildings tracks the path of the sun as the majority of the sun penetrates the building from the east and west. It is better therefore that the building is situated on an elongated axis from east to west so that ventilation and light into the building can be achieved along the north and south elevations.

The north-south orientation of the long facades in hot and humid climates allows windows to be fully opened and a larger surface area for ventilation in conditions where wind speeds can be low unless there are tropical storms. There are multiple forms that can help the shading and passive cooling of buildings. Passive cooling is not only a sustainable solution as opposed to air-conditioning but also is a result of poorer tropical regions. The majority of tropical regions are situated in developing

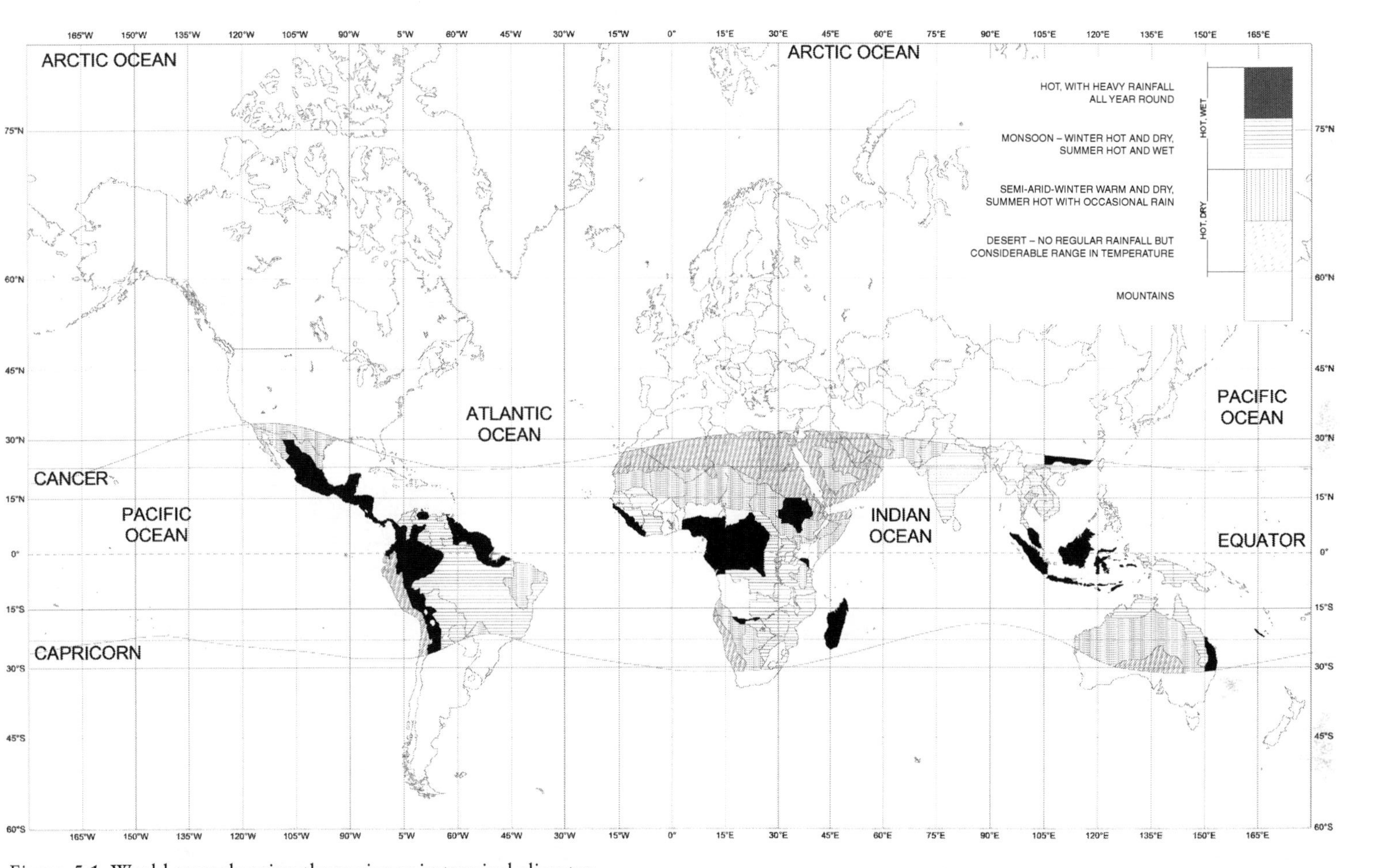

Figure 5.1 World map showing the variance in tropical climates.

countries and there are usually vernacular building types that have been developed over time to indicate ways that buildings can be constructed to passively cool.

The traditional Malay house is elevated from the ground on timber columns. This was originally to prevent the ingress of animals into the house and to ward against flooding which is a trait of humid tropical climates. In combination with a lightweight building skin, this elevated form improves ventilation through the walls, floors and roof. The elevated floor also helps control excessive winds in the event of

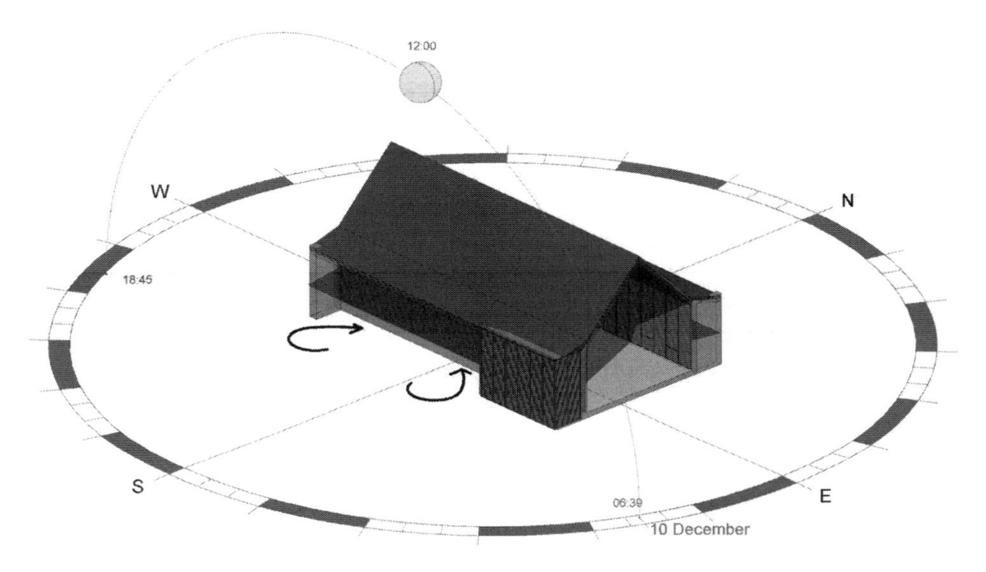

Figure 5.2 Sun path Kuala Lumpur, Malaysia, December.

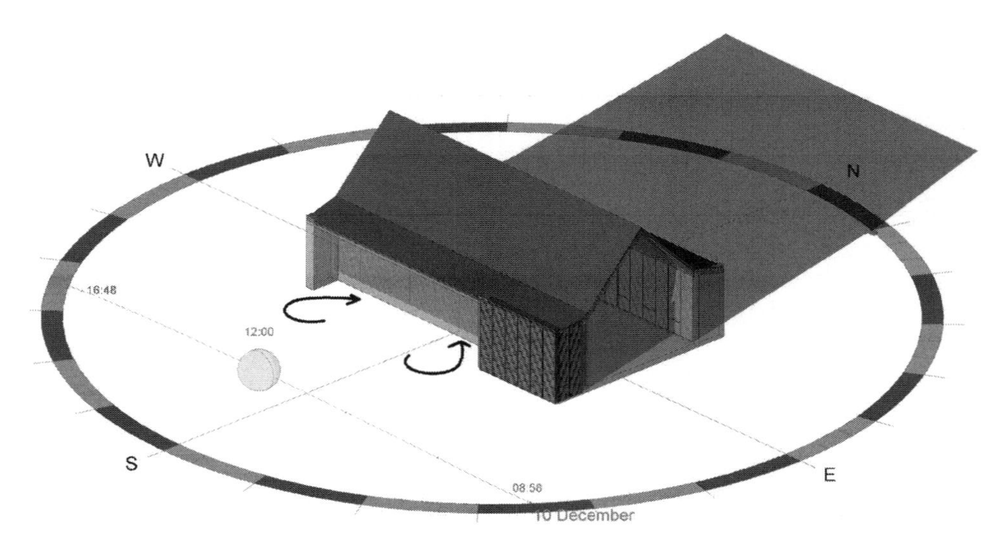

Figure 5.3 Sun path Stansted Airport, London, UK, December.

tropical cyclones. The steeply pitched main roof throws off tropical rainstorms and the shallower overhanging roofs protect from the sun and deflect the rain from the facades of the building. Figure 5.4 illustrates this.

Building form: the humid tropics

The Malay house also has aspects of form that separate different areas; for example, living and cooking. The smaller areas in the house help to separate the hot task of cooking from the main house and to shade entrances from the sun.

There are other examples of lightweight construction in the tropics, which are built for coolness but also due to the nature of the ecological surrounding of the people who build them.

The Bakossi tribe are a hill people of Cameroon and build conical shaped buildings of timber and leaves (Figure 5.5). They are a farming subsidence tribe and use materials that are part of the farming process as the traditional wattle and mud daub cannot bond due to the earth being volcanic. The houses are conical to allow heat to escape and to dry firewood out in the higher levels of the coned roof for heating the house at night.

These are formed from timber saplings with intermediate giant ferns for the walls, tied together with bamboo thongs or vines. The positioning of the Bakossi houses is on the eastern lea of the hill to afford some protection from the

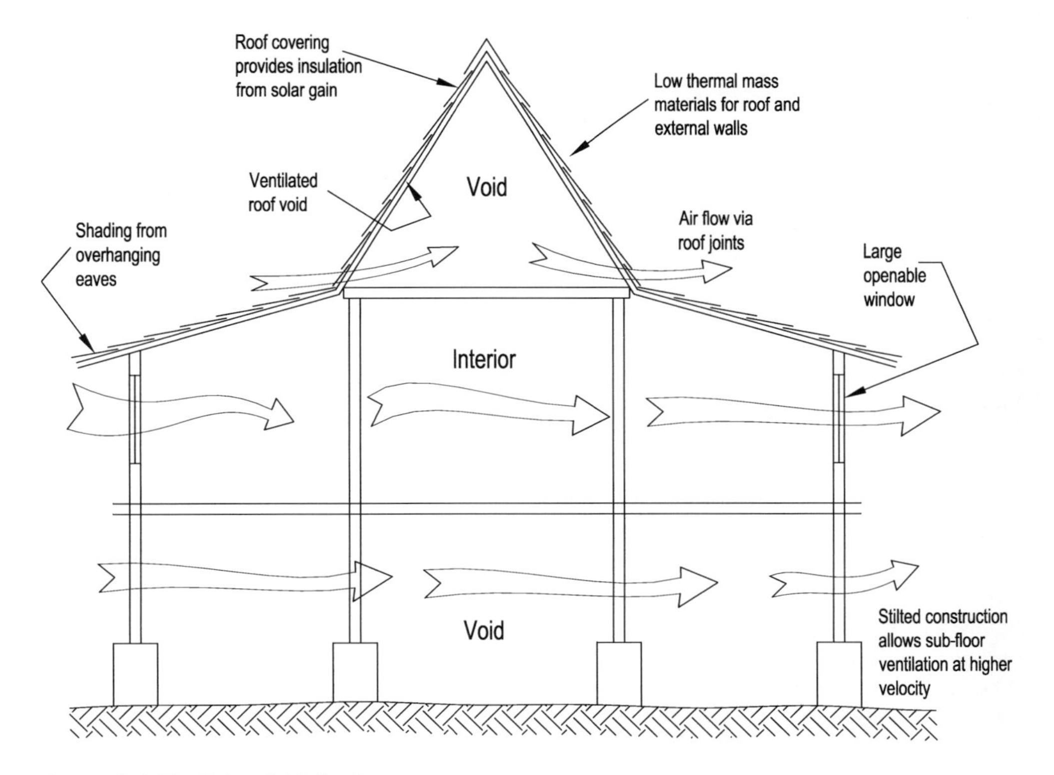

Figure 5.4 Traditional Malay house.

Figure 5.5 Bakossi tribe houses, Cameroon, Africa.

north-easterly winds (Oliver 1971). It is important, as can be seen in the following chapter, that wall types for hot and humid climates should be breathable, without any insulation and achieve little or no heat resistance. High- to medium-rise building form in tropical environments would differ little in building form between hot and humid and hot and dry areas. The main difference being in the choice of materials for walls and infill panels.

The main ingredient in the design of tall buildings between these two classifications would be to prevent the sun from falling directly on walls. The additional application in hot and humid areas would be the shedding of heavy downpours of rain. Aia *et al.* studied the effect of balconies on tall buildings in the humid tropics and proved that they not only provide shade from the sun but also help to stabilise air flow into and within the structures in question (Aia *et al.* 2011). It can be determined therefore that the balcony has a bigger role to play in facades in the humid tropics than merely providing external high-level space. Le Corbusier, in his projects for Algiers, proposed a high-level semi-enclosed street that was a shading and dust prevention device to stabilise the temperature and airflow within the building envelope.

The balcony can be a useful mechanism for keeping the sun off the façade of the building and reducing glare in the spaces within. In Algiers we see Le Corbusier propose a protected internal/external space, which would be a space for community and shaded from the sun. Windows and doors can open onto these semi-internal spaces. We can see the use of atria also to shade and cool buildings internally. This works well where the building has a deep footprint and the building needs to be cooled internally. The Hong Kong Shanghai Bank by Norman Foster and Partners (Figure 5.8) includes a ten-storey atrium, which not only cools the building through stack ventilation but also creates shade both internally and at ground floor level where the public can interact in the blistering humidity of Hong Kong summers.

Figure 5.6 Tribal houses in flood plain, Africa.

In hot and dry tropical climates, there is still the requirement to place the building in an east west axis so that the roof tracks the path of the sun. The important issue with hot and dry climates is keeping the sun away from the walls as solid walls prevent some of the heat radiating into the building but a solar shade of the roof over the elevations would result in a cooler fabric (Sing Saini 1980). The solar shade should reach over the building by a minimum of 900 mm to prevent the radiation of the sun falling on the wall fabric. This example is applied to a single storey building and would increase for each storey. Another solution would be to provide intermediate shades on each storey along the wall height.

In the example below (Figure 5.7) taken from a model of a hypothetical building in Kampala, Uganda, in November, the benefit of the oversailing balcony can be addressed. Most of the floorplates are in shade with some areas of the balconies being clipped by the sun. This opens up a variety of solutions to the solar problem. Balconies can be combined with fin walls to form vertical and horizontal shades; they can also be combined with a series of floors and horizontal solar shades.

Figure 5.9 shows how wind can be funnelled into the central façade by vertical fins that also act as shading devices together with overhanging floors to govern the horizontal radiative reaction with the glass.

The overall form of the tall building needs to cope with the azimuth and altitude angles of the sun in both humid and hot and dry climates. A series of step backs is not adequate, as this will produce larger surface areas where the sun's rays can land. More importantly, tall buildings should overhang either in form or with solar shading elements. The model in Figure 5.11 shows the façades of the building bowing outward to

Figure 5.7 Model showing overhanging balconies.

form a larger plan footprint on the top floor as opposed to the ground floor. This mechanism could be utilised on the south and north elevations only so that the elevations track the sun in the southerly azimuth in August and the respective northerly position in December; in the tropics the winter solstice sees the sun in a northerly altitude and azimuth. The depth of such overhangs depends on the shallowest altitude angle of the sun. A solar study of the building should be carried out to determine this.

Declination is the angular distance of the sun north or south of the earth's equator. The earth's equator is tilted 23.45 degrees with respect to the plane of the earth's orbit around the sun, so at various times during the year, as the earth orbits the sun, declination varies from 23.45 degrees north to 23.45 degrees south. This gives rise to the seasons. Around 21 December, the northern hemisphere of the earth is tilted 23.45 degrees away from the sun, which is the winter solstice for the northern hemisphere and the summer solstice for the southern hemisphere. Around 22 June, the southern hemisphere is tilted 23.45 degrees away from the sun, which is the summer solstice for the northern hemisphere and winter solstice for the southern hemisphere. On 22 March and 23 September are the fall and spring equinoxes when the sun is passing directly over the equator. Note that the tropics of Cancer and Capricorn mark the maximum declination of the sun in each hemisphere.

Declination is calculated with the following formula:

$$d = 23.45 * \sin [360/365 * (284 + N)]$$

Figure 5.8 Hong Kong Shanghai Bank. H.K. PRC. Norman Foster Associates.

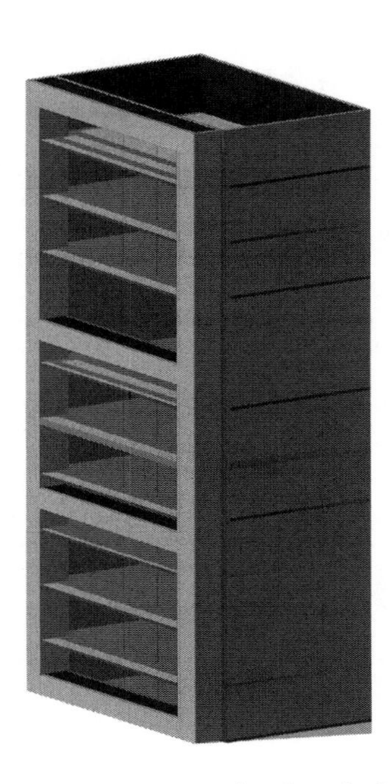

Figure 5.9 Example of vertical wind funnelling.

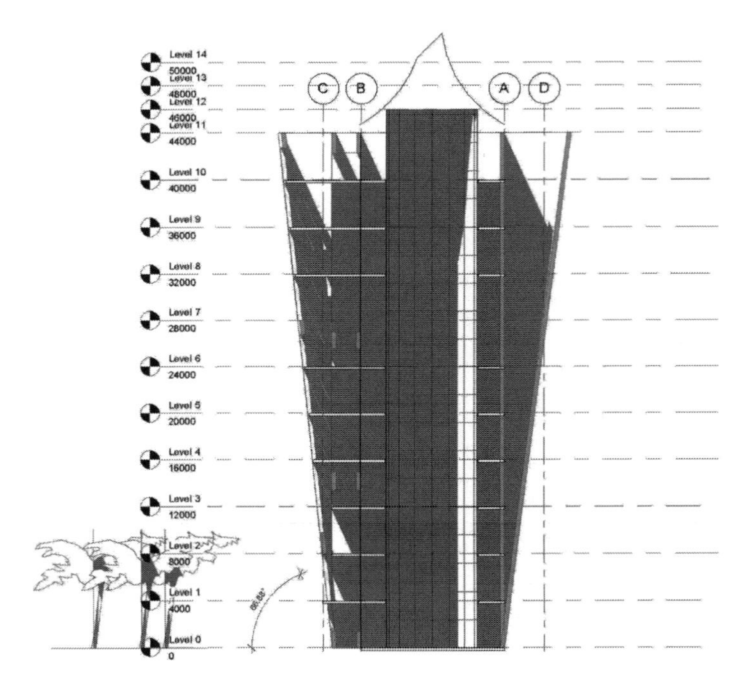

Figure 5.10 Hypothetical project.

Where:
d = declination
N = day number, 1 January = day 1

The hour angle

To describe the earth's rotation about its polar axis, we use the concept of the *hour angle* (ω) (o (*omega*)). As shown in Figure 5.11, the hour angle is the angular distance between the meridian of the observer and the meridian whose plane contains the sun. The hour angle is zero at *solar noon* (when the sun reaches its highest point in the sky). At this time the sun is said to be 'due south' (or 'due north', in the Southern Hemisphere) since the meridian plane of the observer contains the sun. The hour angle increases by 15 degrees every hour.

The azimuth angle is the horizontal angle at which the sun's rays are falling on a wall. It is always measured with respect to due south. Thus, it is always given as east of south or west of south. The solar azimuth angle is the angular distance between due South and the projection of the line of sight to the sun on the ground.

A positive solar azimuth angle indicates a position east of south, and a negative azimuth angle indicates west of south. Note that in this calculation, Southern Hemisphere observers will compute azimuth angles around +/– 180 degrees near noon.

The azimuth angle is calculated as follows:

$$\cos(Az) = (\sin(Al) * \sin(L) - \sin(D))/(\cos(Al) * \cos(L))$$

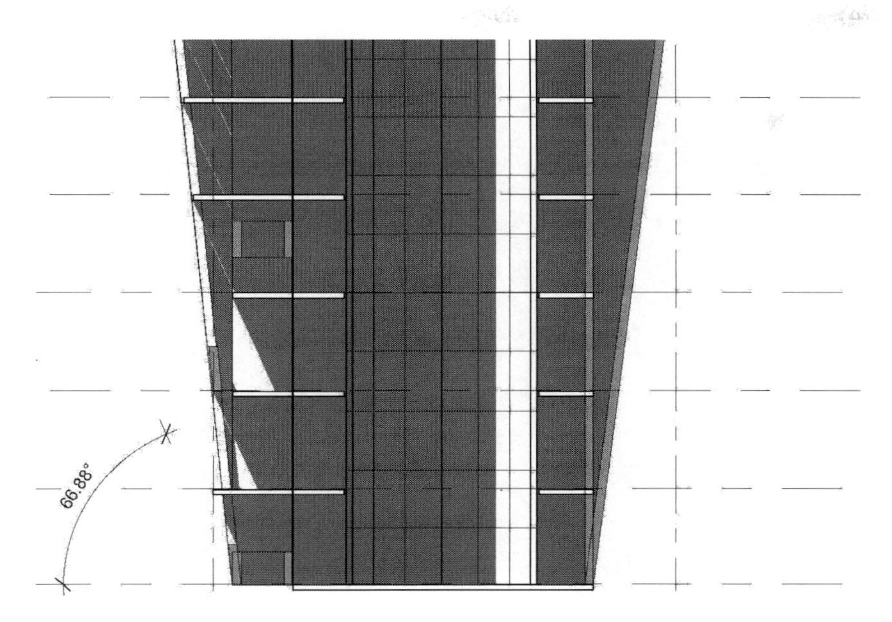

Figure 5.11 Example diagram of altitude angle calculation.

Where:
L = Latitude (negative for Southern Hemisphere)
Az = Solar azimuth angle
D = Declination (negative for Southern Hemisphere)
Al = Solar altitude angle

The sign of the azimuth angle also needs to be made equal to the sign of the hour angle when using the above equation.

The altitude angle is the vertical angle between the sun and the wall section. It is always measured with respect to an imaginary line, running perpendicular to the wall. The altitude angle (sometimes referred to as the 'solar elevation angle') describes how high the sun appears in the sky. The angle is measured between an imaginary line between the observer and the sun and the horizontal plane the observer is standing on. The altitude angle is negative when the sun drops below the horizon. In this graphic, replace 'N' with 'S' for observers in the Southern Hemisphere.

The altitude angle is calculated as follows:

$$\sin(Al) = [\cos(L) * \cos(D) * \cos(H)] + [\sin(L) * \sin(D)]$$

where:
Al = Solar altitude angle
L = Latitude (negative for Southern Hemisphere)
D = Declination (negative for Southern Hemisphere)
H = Hour angle

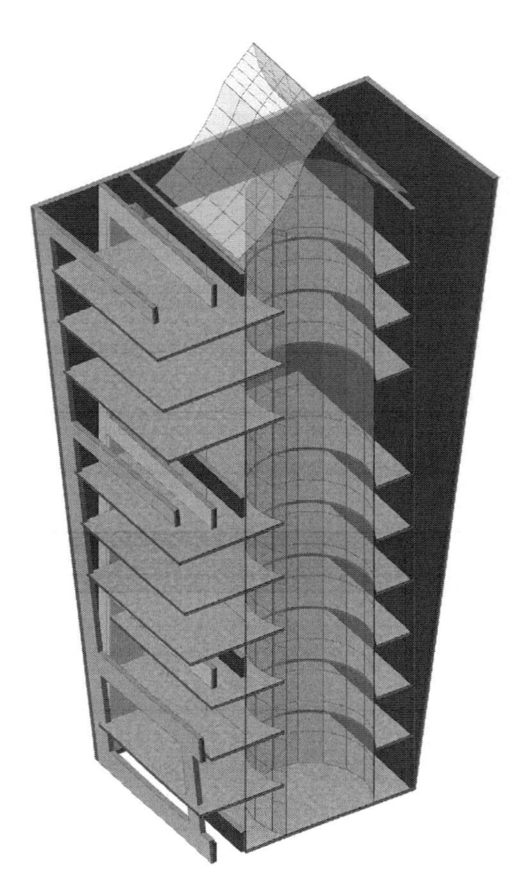

Figure 5.12 Example of multiple grouped floors as solar shades.

Example calculation to determine depth of balconies/solar shades based on a building in Kampala

It is possible to calculate the altitude angle of the sun using Building Information Modelling software such as Autodesk Revit. However, it is important to understand the mathematical method in order to accurately dimension the solar shading device or balcony, otherwise we have to rely on a shading analysis which in Revit is based on physically measuring shadows in section view, which can be somewhat ad hoc. The methodology for the following calculation is to find the altitude angle of the sun at 12 noon on the summer solstice in Kampala, which is 21 December. This is when the sun is at its hottest in this location, although the sun would be as hot on the Winter Solstice on 22 June although it would be due north. As Kampala is very near the equator the winter solstice sun is at a similar angle due north as the summer solstice due south on 21 December. Therefore, the north facades at this latitude need to be similar in shading form as the south. For southerly tropical regions (e.g. Rio de Janeiro, Brazil), the sun will be due north at noon to differing extents.

The altitude angle is calculated as follows:

$$\sin(Al) = [\cos(L) * \cos(D) * \cos(H)] + [\sin(L) * \sin(D)]$$

where:
Al = Solar altitude angle
L = **Latitude_**0.3476°N
D = **Declination_**23.45° (Northern Hemisphere).

The declination angle is calculated by the following formula:

$$d = 23.45 * \sin[360/365 * (284 + N)]$$

Where:
d = declination
N = day number, 21 December = day 355
$d = 23.45 * \sin[360/365 * (284 + 355)]$
$d = 23.45 * \sin[0.986 * 639)]$
$d = 23.45 * \sin[630.054)]$
$d = 23.45 * -0.9999$
$d = 23.44°$
H = **Hour angle_** at noon the hour angle will be 0°

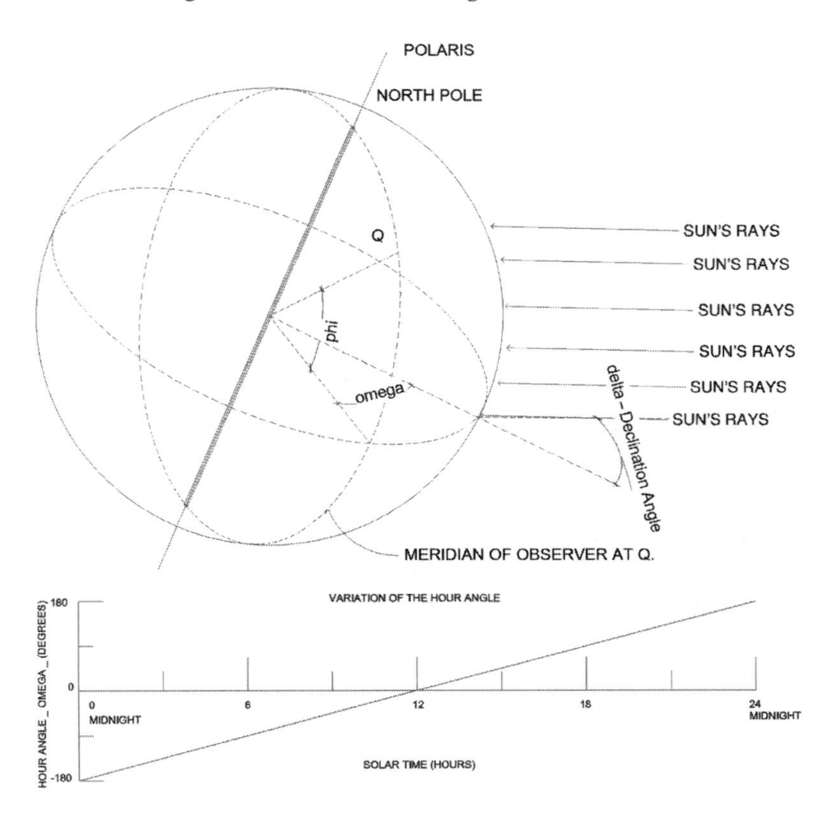

Figure 5.13 Example of the solar angles around the Earth.

Therefore:

$$\sin(Al) = [\cos(L) * \cos(D) * \cos(H)] + [\sin(L) * \sin(D)]$$
$$\sin(Al) = [\cos(0.3476°) * \cos(23.44) * \cos(0°)] + [\sin(0.3476) * \sin(23.44)]$$
$$\sin(Al) = [0.9999 * 0.9174 * 1] + [0.00606 * 0.3979]$$
$$\sin(Al) = [0.9173] + [0.00241]$$
$$\sin(Al) = 0.91971$$
$$Al = \sin^{-1} 0.91971$$
$$Al = 66.88°$$

Furthermore, using Pythagoras' theory:

The balcony width can be determined by calculating this as the opposite side to the Pythagorean triangle (Figure 5.14). Based on a finished floor to ceiling height of 3,800 mm, the following format can be used:

Tan x° = balcony width ÷ 3800 mm, where x equals the alternate to the altitude angle = 90° − 66.88° = 23.18°.

∴ **Balcony width** = tan 23.18° * 3800 mm = 1622 mm

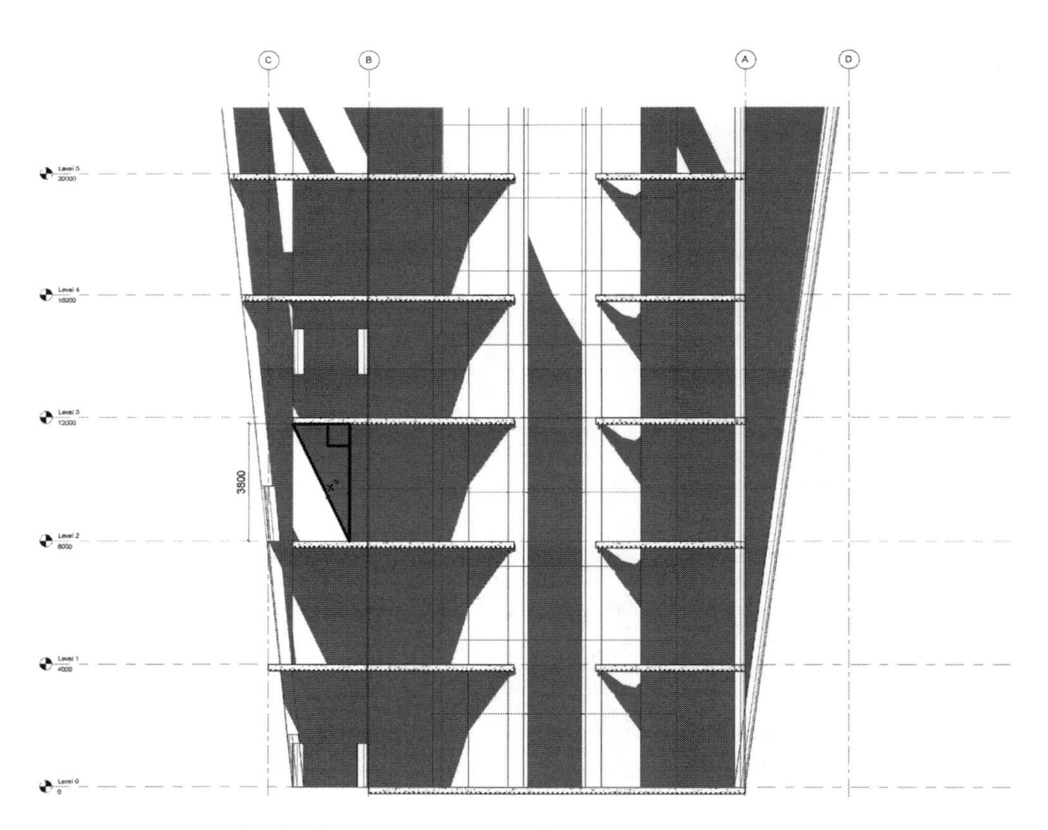

Figure 5.14 Example of balcony overhang calculation.

In the example building model, the form of the sloping overhang of the building allows the glass or lightweight building envelope to be set back on each floor so there is no need for each balcony to cantilever the full 1,622 mm distance. If floors have alternating arrangements with balcony-flush sloping skin-balcony and the inclining envelope is rotated to the altitude angle of the sun, then a more efficiently structural configuration will be achieved. The balcony cantilever can also be reduced if there is a ventilated skin at low level on each storey so that the finished floor to ceiling height can be reduced in the calculation.

Building form: the hot and dry tropics

Many of the characteristics that occur in buildings in the humid tropics can be used in the hot dry tropics. The key building design element to keep in mind is the fact that the sun will be due north in the tropics between the equator and Capricorn. A masterpiece in designing for hot dry climates was completed in La Plata, Argentina by Le Corbusier (Figure 5.15). Although Argentina overall is not recognised as tropical, characteristics of the design can be applied to hot dry tropical regions. The first mechanism that Le Corbusier used in this house was to bring the roof off the walls of the house to achieve ventilation at high level. Other nuances of his design indicate the architect's recurrent theme of providing shaded secondary spaces at all levels to achieve deeper building plans that form an open intermediate area to cool the building and upon which habitable rooms overlook. These spaces can be seen in his work in Algiers in 1931 where the deep plan was

Figure 5.15 House at La Plata, Argentina (Le Corbusier).

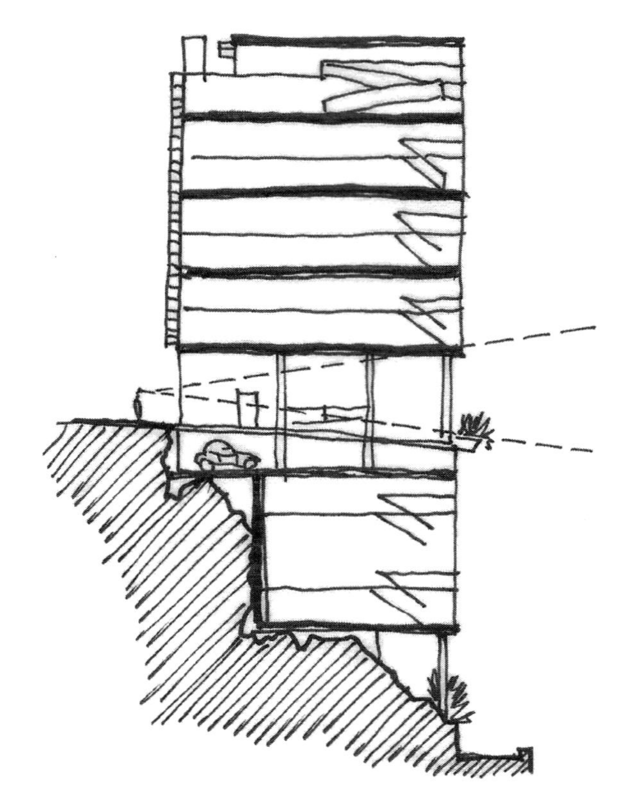

Figure 5.16 Housing at Algiers, Africa (Le Corbusier, 1930s).

afforded the comfort of a shaded external space that could be used for multiple communal activities, views, planting and cooling.

Tropical areas that are hot and dry typically lie near and beyond Tropics 20–25° N and S and 15–30° N and S. The reduction of heat will take precedence over air movement as it has been proven that the human feels more comfortable in hot dry climates than humid ones where air can be stifling and the cause of perspiration. Shade from the sun is a similar solution as humid regions but there are other factors such as reduction of glare and removal of dust from the atmosphere. Building form should take into account the benefits of deep plan types, which is what we try to avoid in temperate climates in our quest to ensure adequate solar gain. Courtyards and buildings that are placed close together can work in unison to remove dust and glare. Courtyards will provide internal shaded areas from which to draw air and will prevent swirling dust storms from entering the building.

As discussed earlier there are several ways to achieve solar shading to walls, but it is a little more difficult to shade roofs from the sun, especially where a concrete structure lends itself to reducing heat gain by storing heat in its fabric. The house in La Plata by Le Corbusier uses a pergola type roof to allow some social interaction on the roof and prevent overheating of the roof deck. The path of the sun in dry regions near the equator leads to building forms that track these but also allow some protection. A similar example to La Plata is Le Corbusier's house in Ahmedabad, India (Figure 5.17) where the facades and roof are protected by concrete solar shades that allow ventilation from the rooms below in the shaded and partially shaded rooftop courtyards. The utilisation of rooftops in India for activities such as relaxing and sleeping at night time has been documented for the benefits of higher relative temperatures at night on the higher floors as the heat of the day dissipates through the fabric of the building (Bose 2015).

There have been multiple experiments with form for these 'over-roofs' in hot and dry tropical climates and if the budget of the building project warrants, there is evidence to show that these are the best options for keeping the heat of the sun from the main building fabric. Le Corbusier's work at Ahmedabad was a master-stroke in that there was variation in the amount of ventilated over-roof. Deep wall recesses help produce places of interaction and solar shading devices. Due to the heat capacities of stone, mud and concrete, the buildings in these climates tend to look monolithic with small window openings. The heavy mass of the building helps to absorb heat from the air that can then be slowly released during the cooler evenings and nights. In Figure 5.29 below a design option can be seen for a courtyard building with a veranda over-roof. The courtyard would cool the centre of the building whilst the over-roof with intermittent openings would protect from the rays of the sun.

The roof had an area for open-air use and an area for indoor use. This had a threefold effect in that the higher level ventilated roof, shown in Figure 5.18, cooled the rooms at third floor level and the lower level shaded the outdoor roof

Figure 5.17 House at Ahmedabad, India (Le Corbusier).

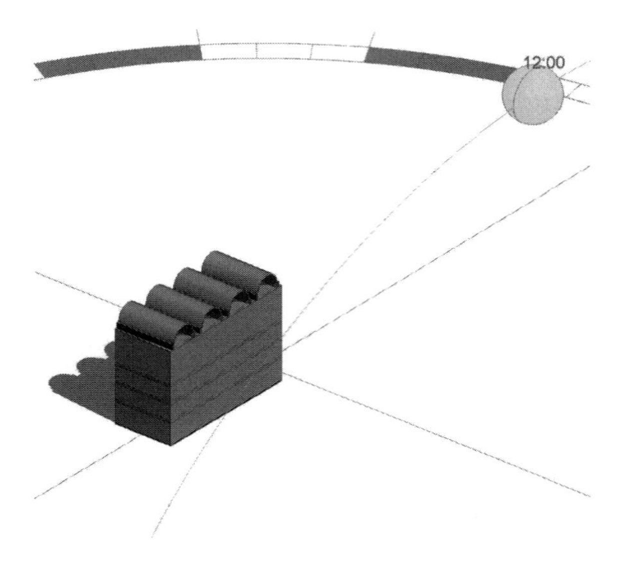

Figure 5.18 Hypothetical residential block.

space which still reduced the intensity of the sun on the rooms at second floor level; the third being purely architectural in that the variation produced an interesting aesthetic solution, showing that there is an intrinsic link between function and form.

The functional characteristics displayed in Le Corbusier's buildings at La Plata and Ahmedabad need further explanation; the example shown in Figures 5.18 and 5.19 (which is a hypothetical scheme for Chad, Africa, at a latitude of 15.4542° N) explores a barrel vaulted over-roof. The shading analysis obviously shows the fact that the sun will be kept off the primary flat roof at level four with the exception of a small area at the southerly edge, which could easily be removed by high-level solar shades. This example has no usable space at roof level and is solely to be used as a ventilated over-roof. The barrel vaults would keep the majority of sun away from the level 4 primary roof by cooling the air in the semi-circular voids so that the flat roof would not be susceptible to the convective and radiative forces of the sun's rays. One problem with this solution is the structural 'gaps' in the ventilation spaces at gridlines 2, 3 and 4. These are areas where direct convection/radiation is possible. The barrel vault is an achievable and economical option in the tropics and therefore there is some study required in comparing the physical build end energy saving costs in order to determine the roof's viability.

Figure 5.19 shows the direction of convective forces of the sun's rays. Some will penetrate the barrel vault and this heat will be diverted by the ventilated space shown by the blue arrows. The other issue with the repetitive barrel vault and similar roofs such as saw-tooth and wave shapes is that they provide a greater surface area than a flat over-roof that would catch the rays of the sun. There is little research in the physics of what is the most energy efficient shape for such roofs. Sing Saini (1980) has some answers in that he states that barrel vaulted roofs (which

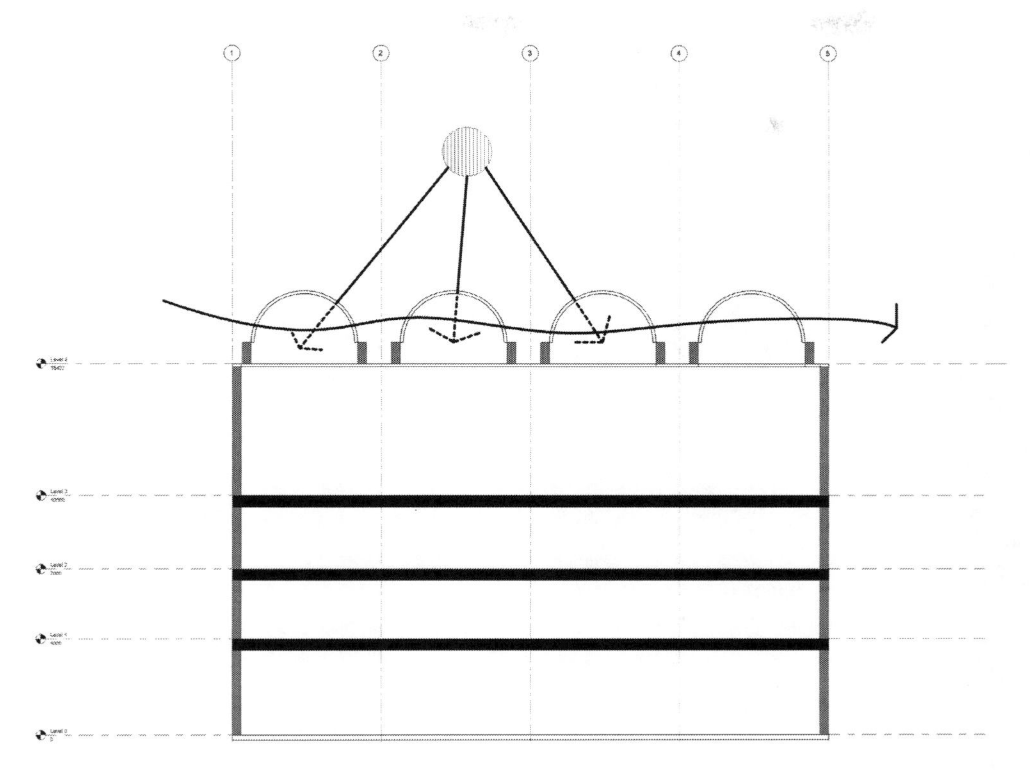

Figure 5.19 Hypothetical residential block.

relate to other repeating type roofs) are storage vessels for heat. However, if designed correctly to capture the prevailing winds, there is evidence to suggest that barrel vaults made from clay bricks are an economical solution compared to flat in-situ concrete flat roofs which have a smaller surface area ration to building footprint. A further solution can be seen in Figure 5.20 which shows a raised barrel vaulted roof, which is supported only by piloti that provide a small surface area in contact with the primary flat roof. This configuration would allow ventilation from all directions and human activity at level 4. Roux (1948), Roux *et al.* (1951) and van Straaten (1961) studied the effect of ventilated ceiling spaces, which are discussed in more detail later in the chapter. Although they stated that ventilated ceiling spaces had no scientific heat reduction of the storey below, there is little academic evidence that raised 'parasol' roofs increase convection to the storey below. Fry (1964) and Kéré (2013) give multiple examples where repeating double roofs are effective; the US Embassy, Baghdad, Iraq (architect José Luis Sert), the Palace of Justice, Chandigarh, India (architect Le Corbusier) and Gando Village School in Burkina Faso (architect Diébédo Francis Kéré).

In order to explore the issues with this type of roof further, the cross section needs to be studied. There will be some heat that is transferred into the roof terrace by convection. Fry (1964) cites that there is an interval or 'lag' in time between the upper and lower temperatures in the slab (in this case a barrel vaulted clay brick slab). For a 50 mm slab this is 85 minutes and for a 200 mm slab, it is five hours.

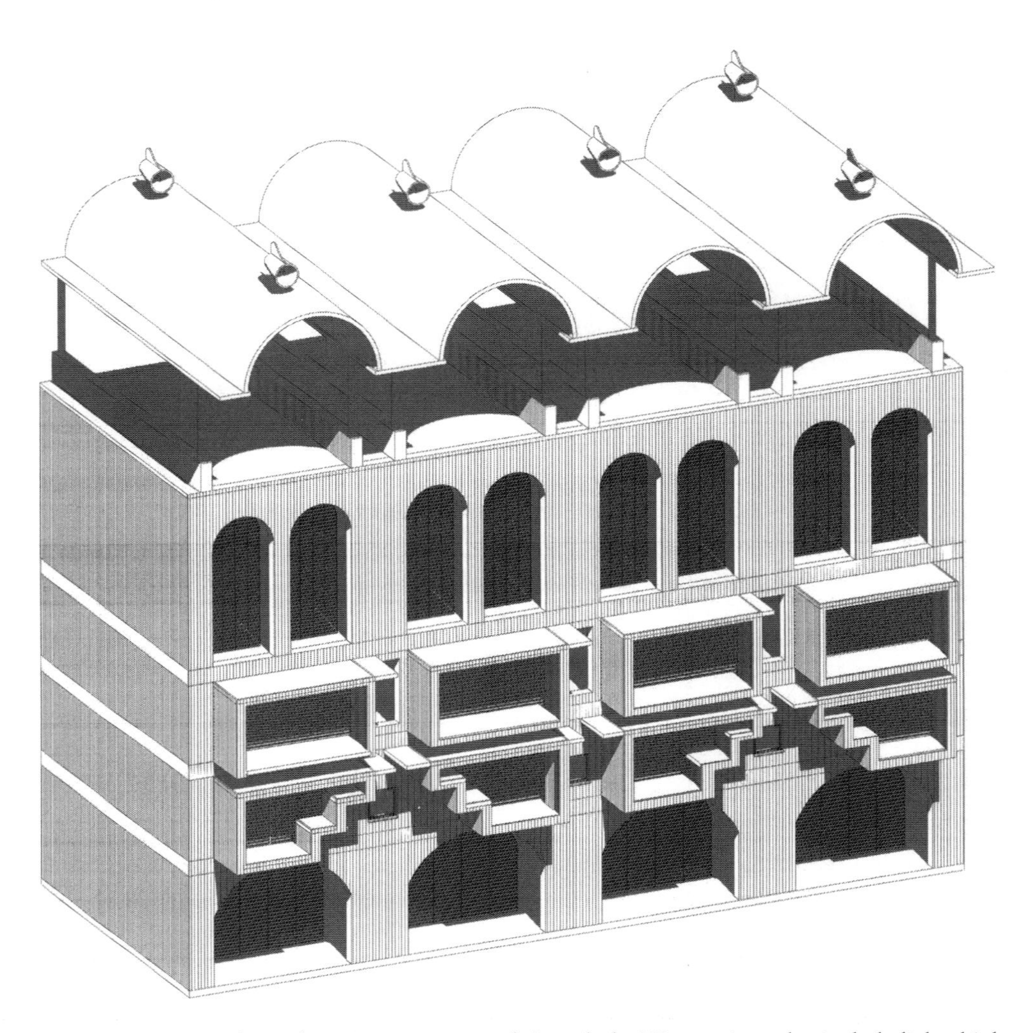

Figure 5.20 Hypothetical project courtesy of Autodesk. Warm air to be 'exhaled' by high-level wind-drawn roof vents.

Effectively this would cause heat to dissipate from the mass over these times after the sun goes down, which is a problem for a warm preliminary roof slab but not for the barrel vaulted secondary roof as night breezes would remove this 'exhaling' heat. In fact, in the day, the barrel vault or repeating secondary roof would produce a natural heat sink for warm air under its canopy; vents at high level would reduce the detrimental effects of this on the slab. These vents would be of the wind-catcher type as used in Bed Zed by Bill Dunster (Snell 2005).

There are issues of overheating in buildings that require large floor to ceiling heights in the hot and dry tropics due to the larger surface area of external walls and larger external openings that are required. An example would be a ground floor entrance area to a hotel, hospital or government building or the upper floors of a conference centre or sports facility. If no or small external openings are to

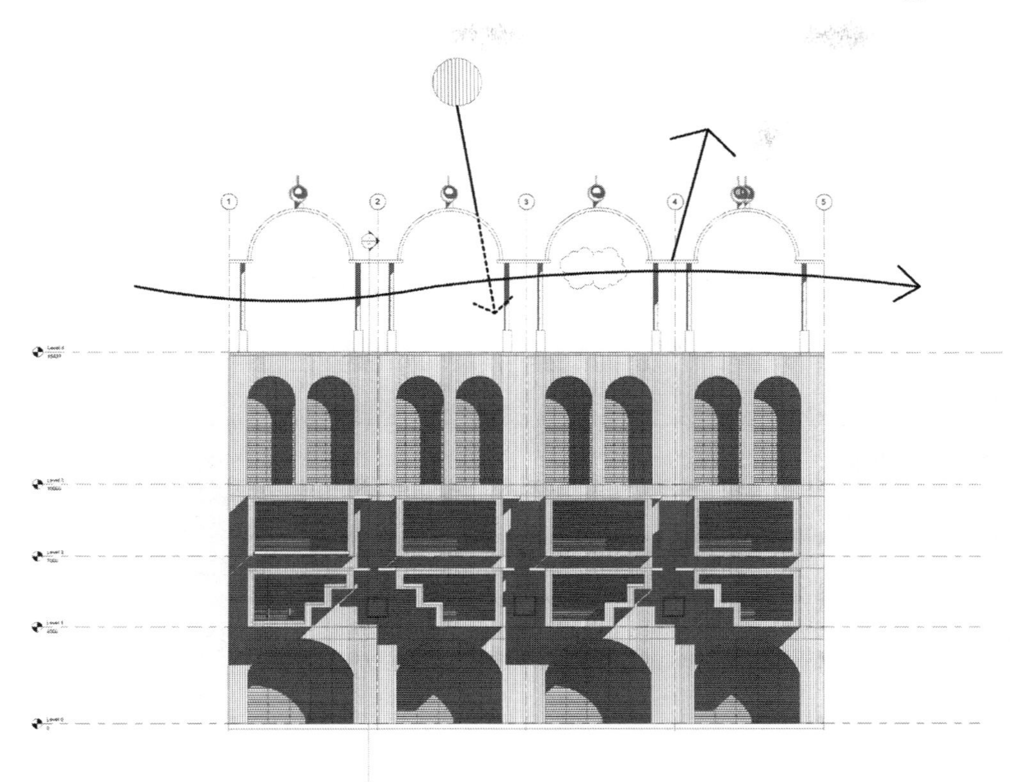

Figure 5.21 Hypothetical project.

be avoided, there will need to be some solar shading or ground floor or clerestory colonnades.

An example of this treatment is the Sultan Abdul Samad Building in Kuala Lumpur, Malaysia (although this building is in the hot and humid tropics, it is a good example) by the Liverpool architect, Arthur Benison Hubback. Hubback, who served in the British Army in Malaya before becoming an architect, recognised the problem of solar irradiance and convection into loadbearing buildings with distinction. In the Abdul Samad building, he produced a form that had variance and texture. He recognised that a government building such as this would need recessed fenestration and therefore provided a large portico at this position to shade totally the external envelope, as can be seen in the background in Figure 5.24.

The stair towers are also building elements that are used as a mechanism to shade and cool the interiors, with warm air being removed by rising thermal convection currents. High-level clerestory colonnades at a smaller scale produce articulation to the facades and cool the upper floors at the same time as creating spaces for human interaction. Balustrades and the brick and stone reveals of the openings provide heavy mass for interval heat lag and architectural variation.

The example in Figures 5.22 and 5.23 shows a hypothetical project for a building in Chad, Africa, at a latitude of 15.4542° N which shows the south elevation on

Figure 5.22 Hypothetical project for Chad, Africa.

Figure 5.23 Hypothetical project for Chad, Africa.

Figure 5.24 Abdul Samad building, Kuala Lumpur.

which the sun would fall in this location. Large colonnades at low level are split to form narrower openings at the upper floor. Intermediate floor openings can be small to prevent sun ingress; horizontal solar shades at window head level would encourage cooler interiors and add some tectonic interest.

Figure 5.26 is an interpretation of a building element seen at the Scottish Parliament building built by Spanish architect Enric Miralles. A stepped window was developed that projected out from the building onto the western elevation of the parliamentary complex, inspired by a combination of the repeated leaf motif and the traditional Scottish stepped gable. In each office, these bay windows have a seat and shelving and are intended as 'contemplation spaces'.

Constructed from stainless steel and framed in oak, with oak lattices covering the glass, the windows are designed to provide MSPs with privacy and shade from

Figure 5.25 Abdul Samad building, Kuala Lumpur.

the sun. This tectonic mechanism would be well suited to hot and dry tropical climates as the project shades the facade and other surrounding windows and a deep overhang to protect the interior from solar rays.

The windows themselves could have semi-continuous horizontal solar foils and this subject is discussed later in the chapter. In the Miralles building, solar shades are used in a random pattern and different variations of the unit are used. It is not the purpose of this chapter to produce every solution that can be adopted but many alternatives could be developed with further study.

Figure 5.28 shows an example for Kolkata, India, with a latitude of 22.5726° N. A total over-roof is used which was parametrically developed from a BIM software which demonstrates that a multitude of building forms can be arranged using such software as Revit Dynamo by Autodesk and Archicad by Graphisoft.

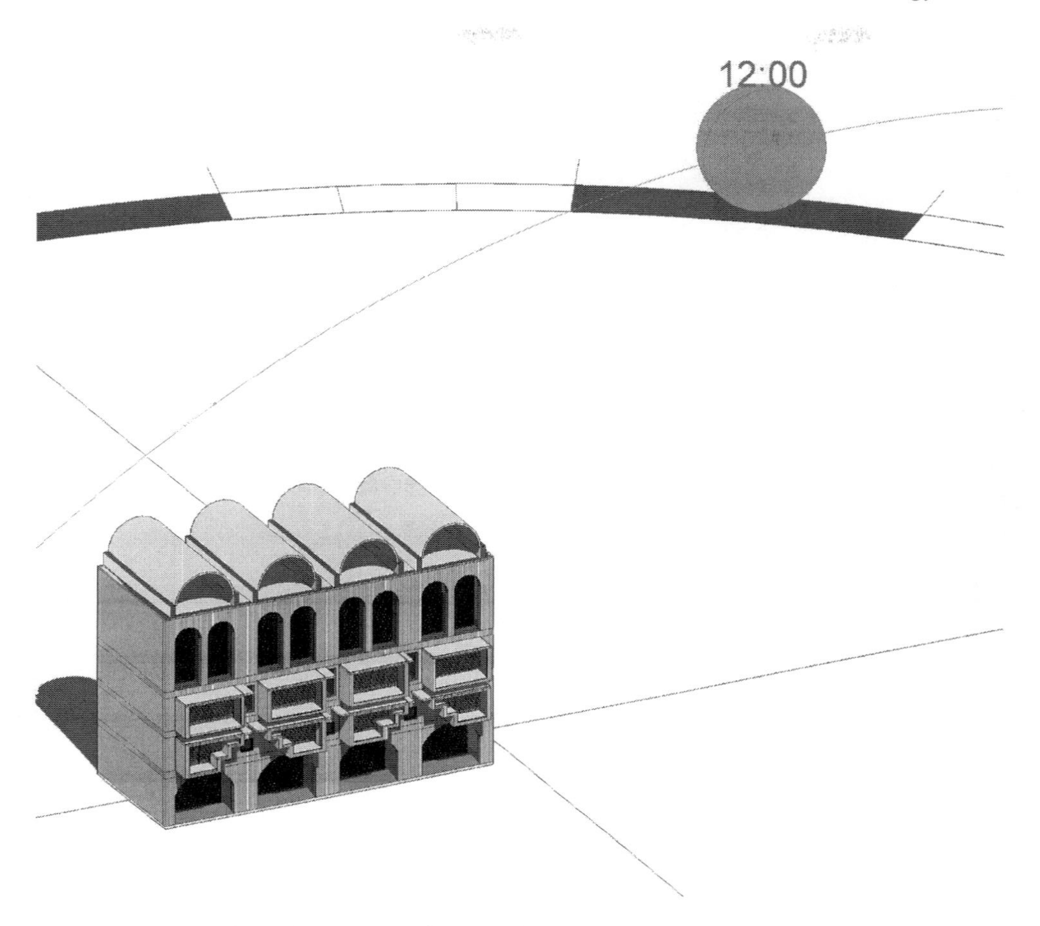

Figure 5.26 Hypothetical project, Kolkata, India.

With a total over-roof, cooling exterior spaces and a freedom of building type below can be provided. In this example an internal courtyard is shown which also cools the deep plan building below; missing over-roof panels also allow some sunlight to enter the courtyard; these can be solar shaded with pergolas and, if the project budget allows, these can electronically track the position of the sun to produce diffuse light. In the Kolkata example, the sun at this latitude has a shallower altitude angle so this must be kept apparent in design calculations.

The incidence of courtyard architecture in hot and dry climates is vast as a protected internal space in buildings provides solar and dust storm protection. Sing Saini (1980) introduces scientific study on the matter with contributions by McCabe (1961) on air pollution. More recent studies into outdoor ambient air pollution include WHO (2016). Design mitigation for the effects of dust and smoke particles has been documented by Bagnold (1941). Saini and Borrack (1967) produced documentation from a Town Planning point of view, together with the effect of sand and dust on the individual building envelope, and it is the latter that is explained in this section. The size of particles, regional incidence,

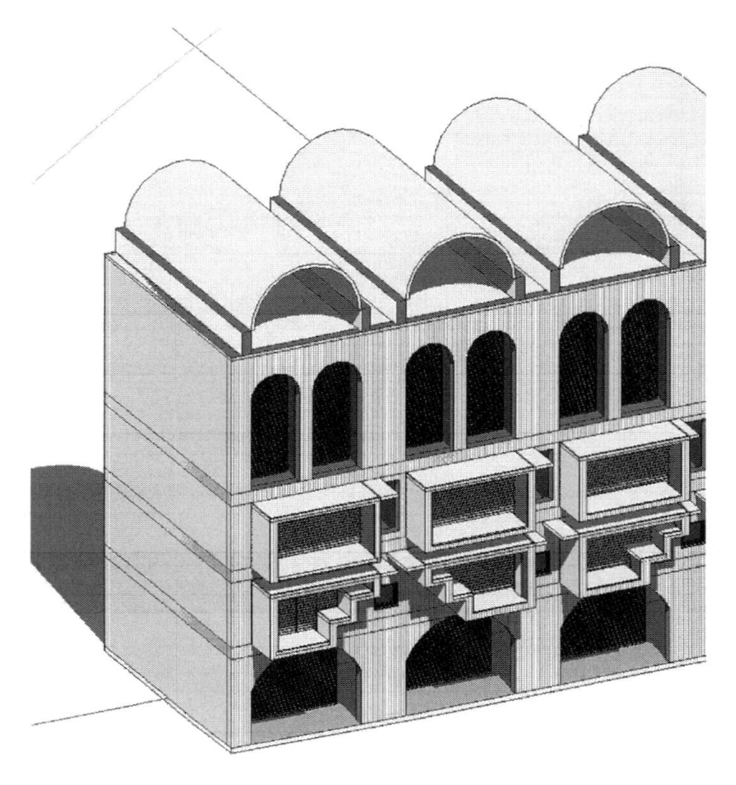

Figure 5.27 Hypothetical project, Kolkata, India.

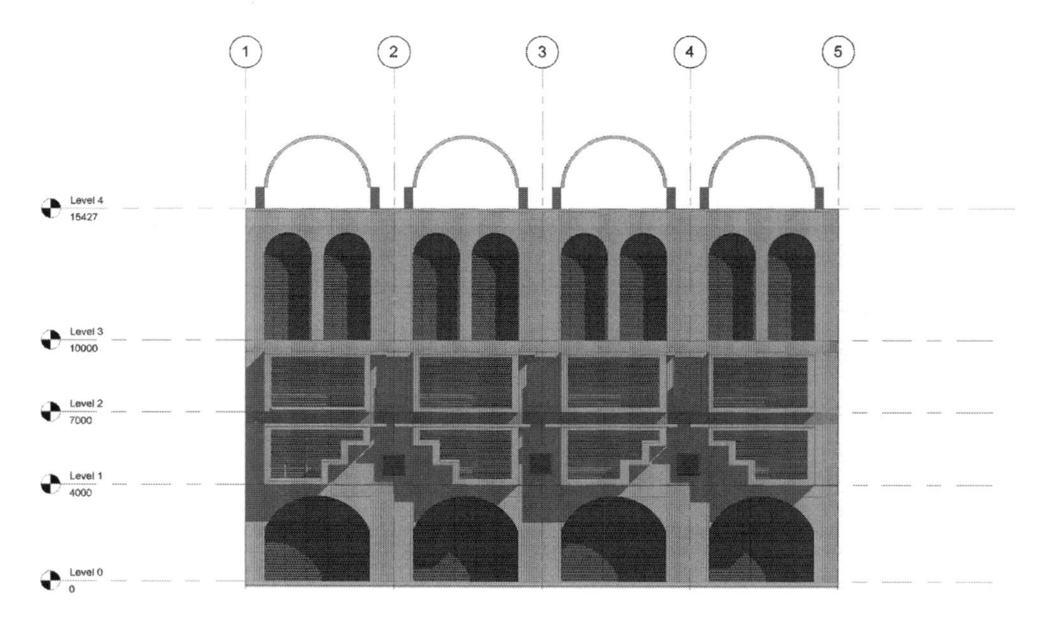

Figure 5.28 Hypothetical project for Kolkata.

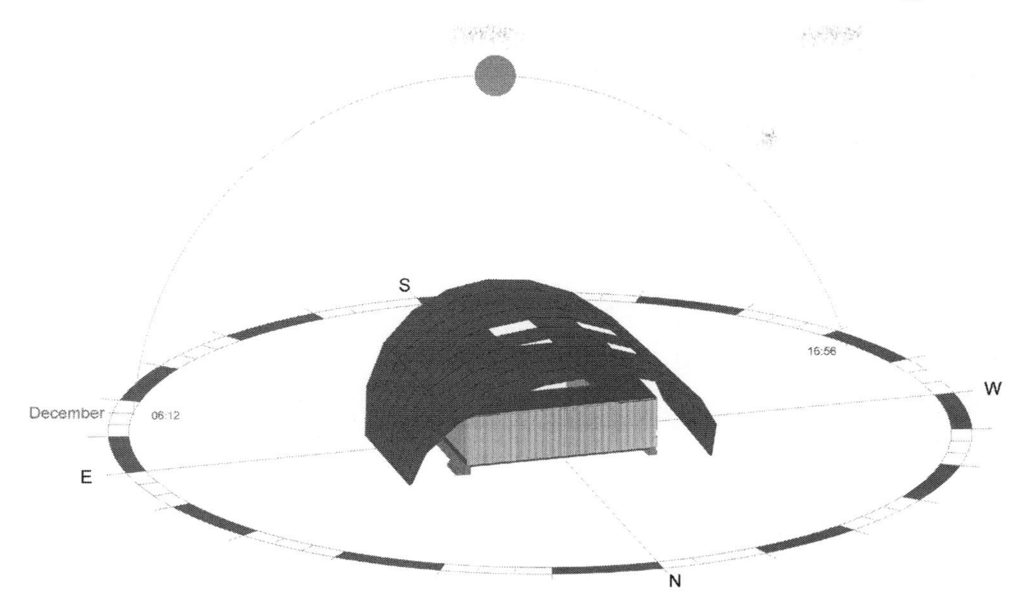

Figure 5.29 Example of courtyard building with parasol over-roof.

wind direction and speed are all aspects to consider in the design of protective measures. Marshak (2008) and Sobel (2012) offer data on wind direction and speed for global tropical environments in determining plan profiles and aspect ratios of buildings in affected regions. The texture of the ground is also a contributing factor to the movement of particles (Figure 5.30) in that dust storms are reduced in abrasive surfaces by friction.

The wind pressure over a building exponentially reduces the length of dust clouds at the lea side, the larger the footprint. A smaller dust cloud appears if wind movement is allowed to pass around the building; this exponentially increases, the wider the protective facade. Air pollution in courtyards is shown in Figure 5.31. At (a) a central square court gives protection from sand and most of the wind-borne dust: orientation is unimportant; whereas at (b) a central rectangular court gives good protection for depths up to 3A. For depths exceeding 3A, the building should have the long axis perpendicular to the wind direction. At (c) the perimeter square court gives protection as in (a) against overhead dust but eddy whirls caused by sideslip will allow dust to enter from the sides.

The depth of courts in this instant is limited to 3A with both receiving the same protection. At (d) perimeter rectangular courts provide some protection to a maximum depth of 3A but would be more suited to positioning on the lea side of the building. Barriers are required to fully exposed building forms to provide protection from overhead dust and side swirls; protection is determined functionally by the length and height of the barrier and the distance from the face of the building in question.

The historical use of courtyards for cooling and sand storm protection are well documented.

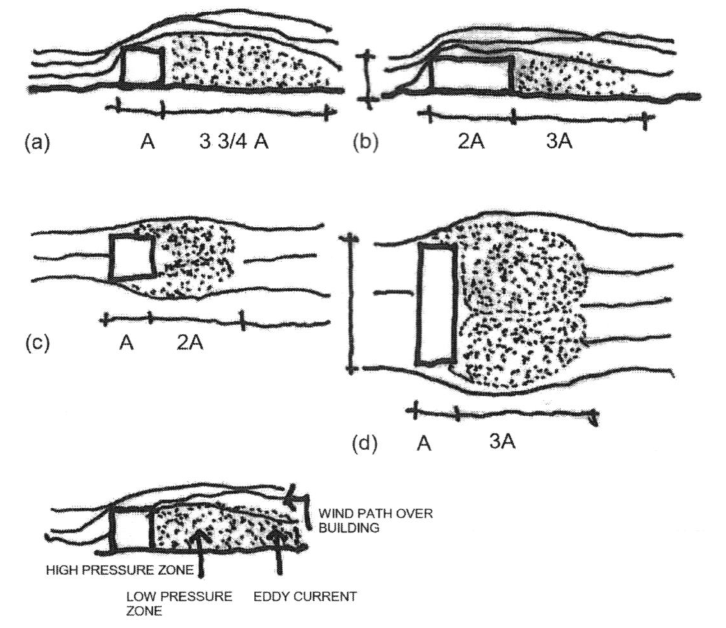

Distribution of low presure zone in the lee of a building for a laminer surface wind of constant speed and direction: (a) Low pressure zone due to air passage over building (elevation). (b) Low pressure zone due to sideslip of air round building (plan): depth of zone is less than in (a). (c) Depth of low pressure zone decreases as depth of building increases (elevation: cf. (a)). (d) Depth of low pressure zone increases with increase in length of the building (plan, cf. (b)).

Figure 5.30 Movement diagrams for dust and sand.

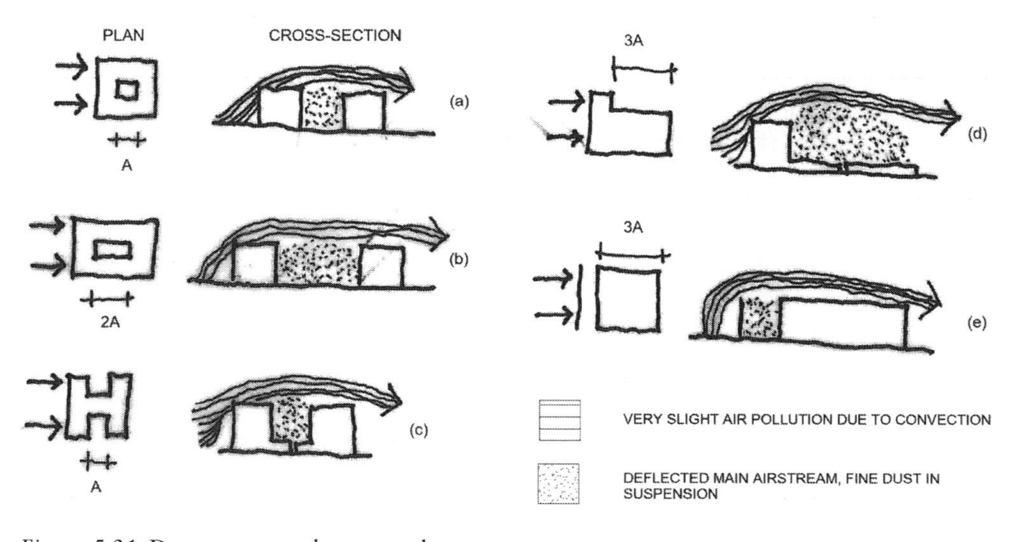

Figure 5.31 Dust storms and courtyards.

Urban morphology

Although African villages initially were planned by cultural principles, ecology and climate does form some of the responsibility for their layout. As in the Malaysian kampong (Figure 5.32), houses were laid out to form a more organic arrangement in order to improve ventilation in hot humid climates and allow wind flow.

In the tropics, there is a broad zone of low pressure, which stretches either side of the equator. The winds on the north side of this zone blow from the north-east (the north-east trades) and on the southern side blow from the south-east (south-east trades) (Sobel 2012).

In humid conditions, houses can be arranged to direct wind towards the north-south facades. This can be done by deflecting screens or the external footprint of the house (Figures 5.33 and 5.34).

South of the equator, certain rooms can be placed so that south-easterly prevailing winds can be funnelled into the centre of the house and vice versa north of the equator.

Many scientific studies have already discussed and pointed out urban morphology as being a key issue in determining overall energy consumption in cities (Owens 1986; Breheny 1992; Williams *et al.* 2000; Ratti *et al.* 2003; Steemers 2003; Droege 2007; Batty 2008), as well as its potential for producing energy locally (Grosso 1998; Martins 2014a, 2014b; Serralde *et al.* 2015). In the previous sections of this chapter, the subject of urban morphology has been described with respect to the

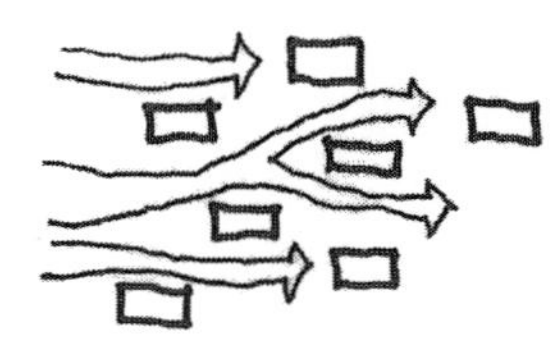

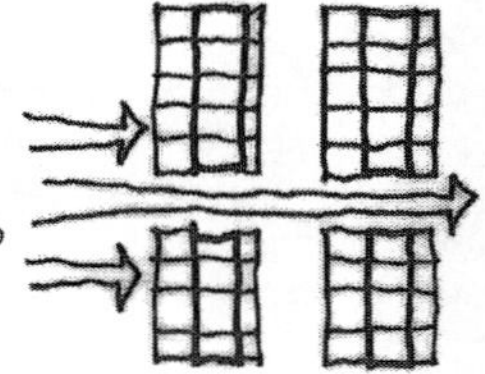

Figure 5.32 Kampong planning, Malaysia.

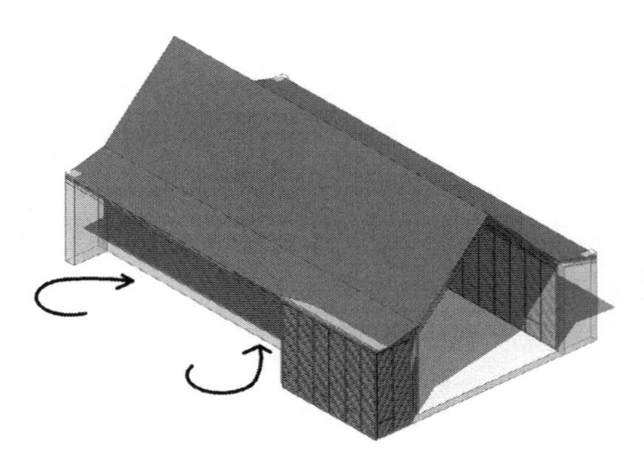

Figure 5.33 Wall fins to direct wind to the centre of the facade.

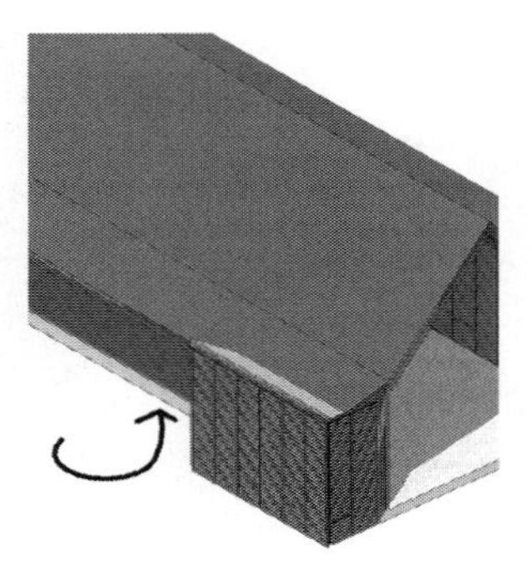

Figure 5.34 Outrigger rooms can form wind foils.

individual building. The overshadowing and wind barrier effects of adjacent buildings, to name a couple, have been mentioned. Urban morphology is the area of analysis of the size, shape and grouping of buildings in an urban setting. This subject is increasingly important with the growing urbanisation of cities in tropical regions which corresponds directly to the development of informal and poor settlements. In a study of the effect of solar radiation in a city of northern Brazil, Maceió, a country where 47 per cent of the population lives in cities, a system of analysis has be carried out based on the well-known density-related (e.g. Plot Ratio, Floor Area Ratio, Building Height) and climate-related (e.g. Aspect Ratio, Shape Factor) parameters as documented by Sanaieian *et al.* (2014) and Agra de Lemos Martins *et al.* (2016). Table 5.1 shows the parameters of built form that need to be addressed when analysing the effects of solar radiation in the tropics and also temperate climates.

It is fairly simple to analyse a singular building in that sensors and testing measures can be attached to one edifice and measurements taken. The difficulty is when we look at the microclimate caused by a group of buildings or even an urban district to see what effect surrounding structures have on an individual structure in terms of solar gain compared to energy usage. Table 5.1 shows these different factors and describes what they are. However, there are other elements that need to be considered, such as the effect of solar radiation on the surrounding land, and reflectance and overshadowing by adjacent buildings.

Agra de Lemos Martins *et al.* (2016) studied a grouping of 25 hypothetical buildings for the site in Maceió which has a latitude of 9.6499° South. The study is important in that the parameters mentioned above can be determined by using a

Table 5.1 Agra de Lemos Martin's study of morphology parameters

Main urban form factors considered	
Morphological parameters	*Definition*
Shape factor [/] SF	Ratio of the non-contiguous building envelope to its built volume, over an urban fabric
Floor area ratio [/] FAR	Ratio of the total built floor area to the plot area where the building is located. This is a common metric used by planners, regulators, and developers to discern the intensity of a development.
Plot ratio [/] PR	Ratio of the total footprint area of a building to the plot area on which the building is located
Aspect ratio [/] AR	Ratio of building height to the width of the distance between buildings (street width + building setbacks)
Verticality [m^{-1}]	Vert Mean height of buildings of an urban scene weighted by their built footprint area
Distance between buildings [m] BS	The minimum spacing between adjacent buildings
Building depth [m] BD	The average depth of buildings in an urban area
Building width [m] BW	The average width of buildings in an urban area
Building height [m] BH	The average height of buildings in an urban area

software (Citysim) to alter the shape of buildings and streets so that the most beneficial design aspects can be adopted. To model this in the laboratory without using a computer algorithm software is impossible for many reasons; the main one being our inability to model the power of the sun without a computer. The findings were extraordinary in that they showed scientifically the importance of solar irradiance and thermal radiation on surfaces and roofs of buildings. The simple format of the 25 buildings is shown in Figure 5.35 below.

The statistical results obtained for the overall student t-test indicate that the aspect ratio (AR), the distance between buildings (SW) along both axes (north-south and east-west) and the albedo (solar reflectance on a surface) are the urban factors producing the most significant effects on the modification of the three response-variables considered: the solar irradiation, the emitted thermal radiation and the illuminance levels on building surfaces. In Table 5.1 there is a correlation between the Aspect ratio (the ratio of building height to the width between buildings) and the location of Maceió which is close to the equator. Effect size indicates the relationship between the factor and the response variable: negative values indicate that the relationship is inversed. The sun is slightly to the north in the day but generally overhead, so there will be a higher relationship of reducing solar irradiation on the east, west and north facades whereas the south facade is some 40 per cent lower in its reducing capabilities (168.8, 143.8 and 159.7 compared to 98.7 in the south). These figures are the incidence algorithm values for raising the Aspect ratio to its highest amount of 3.9 in the study. A negative response is better as this indicates that the solar irradiation or thermal radiation is reduced as the algorithm factor is increased. The result presented below shows that increasing the AR (e.g. by introducing very narrow streets) for an east-west street axis (buildings' north and south orientations) may thus be less relevant than for a north-south street axis (buildings' east and west orientations).

Figure 5.35 Typical urban layout – Maceió, Brazil.

Moreover, the importance of the results for the west facade is shown to be greater in the study. This is especially true for the thermal radiation factors as this correlates to the fact that the sun rises in the east and the building heats up in the day. In the afternoon the sun in the west has a greater influence on the building facade as it has the combined effect of a pre-heated building fabric (from the morning) and the irradiance of the sun from the post-noon direction (west).

There is an interesting relationship between Aspect ratio and Plot ratio with regard to the differences between solar irradiation and thermal radiation factors of roofs. Increasing the Aspect ratio of a roof has less of an effect on thermal radiation as it does on solar irradiation. This is obviously the case as the higher the building, the more sun the roof will fall on its surface (–130.3 SI compared with –39.3 TI). Thus thermal radiation factors will be similar for differing building heights. A similar correlation is seen with Plot ratio although with its increase solar irradiation has an increased direct effect on the roof rather than inverse (25.8 SI compared to –3.4 TI). It is the latter figure that leads into the realms of urban heat sinks.

The lower the AR and the greater the distances between the buildings, the smaller will be the sky obstructions produced and the shadows cast over these surfaces, which will increase the direct and diffuse contribution of the incoming solar energy. In contrast, the lower the AR, the greater the amount of thermal radiation emitted (Figure 5.36) by the heated roof surfaces, which can help mitigate the urban heat island effect by reducing radiation trapping in the urban structure. So the taller the buildings and narrower the street canyons, the more likely the occurrence of heat sinks. An example of this is shown by the results highlighted in bold in Tables 5.2 and 5.3. The

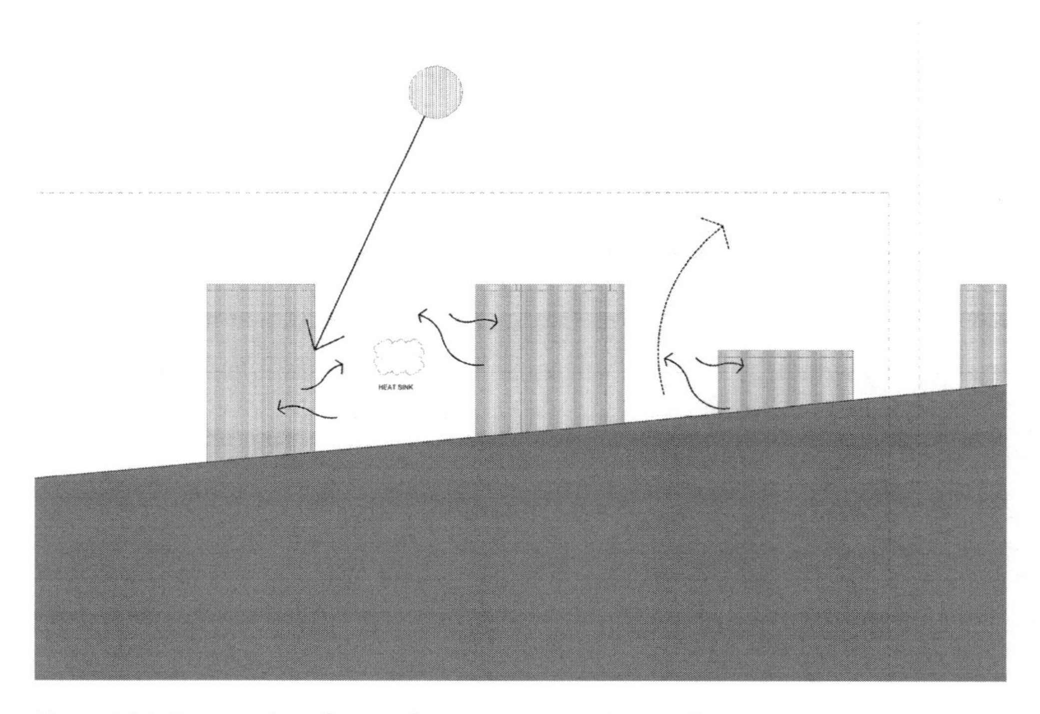

Figure 5.36 Cross section of streets for project in Maceió, Brazil.

Table 5.2 Effect size of urban morphology factors on the solar irradiation of building surfaces (roof and facades)

Factors	*Effect size (kWh/m²/year)*	*Roof*	*East facade*	*West facade*	*North facade*	*South facade*
1	AR	−130.3	−168.8	**−143.9**	−159.7	−98.7
2	Albedo	67.6	259.5	413.3	228.5	247.4
3	SW–north-south	84.6	144.8	129.0	45.5	23.4
4	SW–east-west	89.0	62.2	54.6	144.9	88.9
5	Shape factor	33.1	84.5	77.6	73.4	40.0
6	Building depth	−26.7	−79.1	−92.3	2.6	2.0
7	Plot ratio	25.8	−30.3	**−58.1**	−21.3	−10.2
8	Building width	19.8	−7.2	−11.2	−74.4	−38.4
9	Floor area ratio	8.2	−89.2	−74.1	−55.4	−37.7
10	Vert.	−7.0	−61.3	−55.3	−38.8	−27.1

north-south street width factor is an important one as this is the direction that will be most effective in the tropics. Facades are likely to be west or east facing with streets on the north-south axis. By increasing this street width we see that solar irradiation increases, which is expected, but thermal radiation reduces (129.0 SI and –48.5 on TR on the west facades). This is due to the fact that heat is allowed to escape from the urban structures the wider the air-gap between buildings. Although the wider the street the better it is to remove unwanted thermal radiation through surfaces, there needs to be a weighting to determine whether this compares to the reductions in solar irradiance by overshadowing. Exponentially there is a greater difference in reduced energy by increasing the Aspect ratio for solar irradiance compared to the savings that would be made by increasing the Plot ratio for better thermal radiation figures (shown in bold).

Table 5.3 Effect size of urban morphology factors on the thermal radiation of building surfaces (roof and facades)

Factors	*Effect size (kWh/m²/year)*	*Roof*	*East facade*	*West facade*	*North facade*	*South facade*
1	AR	−39.3	56.7	**58.2**	58.8	58.8
2	Albedo	16.3	168.8	167.1	168.5	169.8
3	SW–north-south	−11.6	–47.9	–48.5	–27.0	−26.7
4	SW–east-west	−11.5	−25.2	−25.1	–48.4	–46.9
5	Shape factor	−5.6	–43.3	–45.0	–44.5	–43.2
6	Building depth	4.2	24.0	24.8	6.3	6.7
7	Plot ratio	−3.4	24.3	**25.2**	24.9	26.1
8	Vert.	−2.4	20.0	16.4	17.6	16.4
9	Building width	−1.6	24.5	24.3	43.4	42.4
10	Floor area ratio	−0.6	31.9	28.0	27.6	29.7

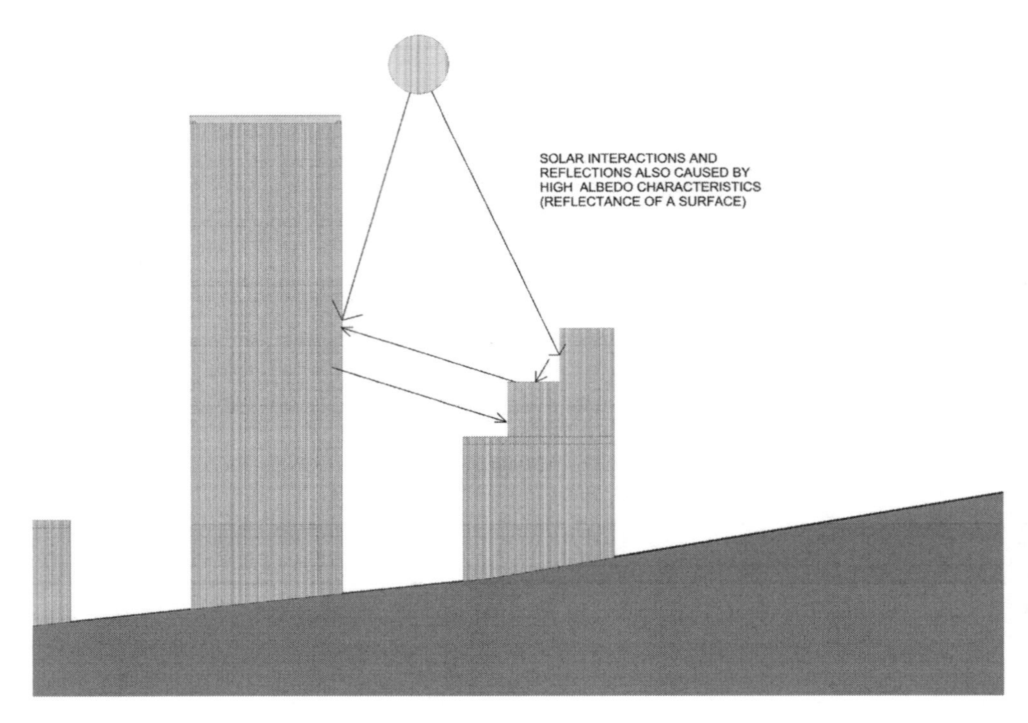

Figure 5.37 Set-back effects on urban street in the tropics.

Another aspect to consider would be the building set-back laws by planners which would affect the Shape Factor. The Shape Factor is the ratio of the non-contiguous building envelope to its built volume, over an urban fabric. Therefore the more facets that a facade has the more surface area per volume there is; giving the sun more angles and surfaces to fall upon. With a rise in the Shape Factor ratio, there is a direct (positive) effect on solar irradiance, whereas the thermal radiation results drop to an inverse (minus). These would indicate that the more surface area, the more there is a likelihood that heat can escape and allow energy savings to occur. This study therefore shows that planners and urban designers in tropical regions can be better informed by analysing a simulation of the urban form.

The below diagram (Figure 5.38) shows different characteristics between hot and dry and humid and hot climates. In hot and dry climates the large buildings would provide useful shade from the sun and dust storms in order to protect smaller buildings. In hot and humid climates this building configuration would prevent the flow of air to the smaller buildings below which is not desirable.

Climate analyses

Wladimir Koppen was a climatologist of Russian-German descent, who introduced his climate classification system in approximately 1900 (Marshak 2008).

The following list (Table 5.4) shows various countries with their Koppen climate classification.

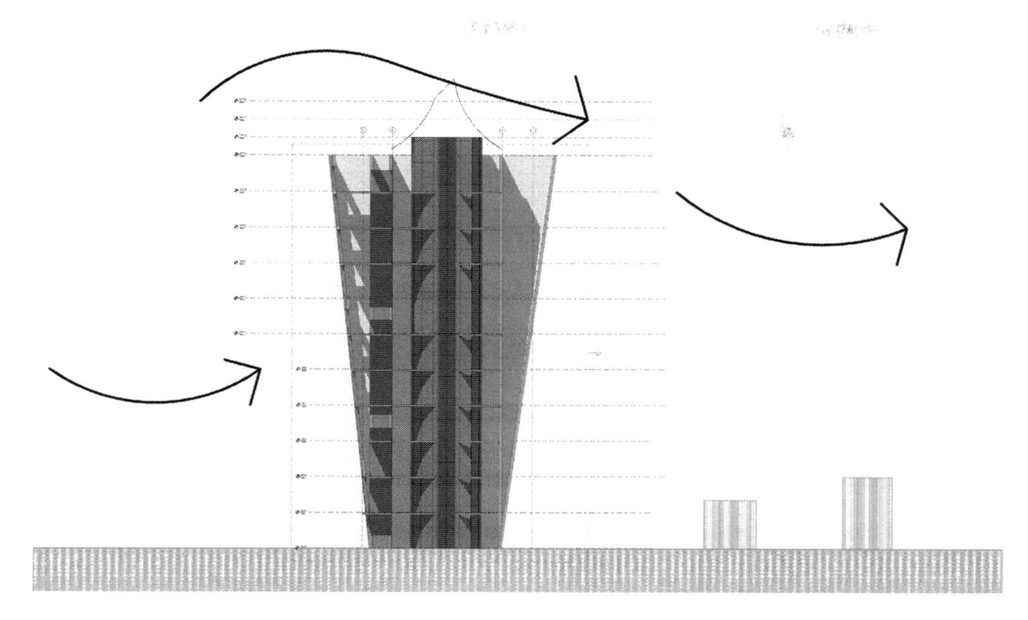

Figure 5.38 Dust storms and air flow around buildings in the tropics.

Domestic building elements in the tropical regions

The previous discussions have generally been about form but there is a need to look at tropical buildings with respect to individual construction elements. The use of certain types of construction is affected by climate, availability of materials and resources, labour and economy.

Foundation

Concrete products for foundations are widely used in the hot and humid tropics and most cases suffer little from climatic damage. The main issue with concrete is the supply of the ingredients in areas that may not have access to the raw materials of cement sand and aggregate. This might be caused by logistical or economic reasons where building budgets cannot cope with the exorbitant cost of transportation. The most important ingredient is the supply of clean water as saline, salty water and water with phosphate content and organic matter is not suitable for use in concrete. Water is used not only in the mixing of concrete but also to wash the aggregate and in the constant wetting of concrete surfaces in hot climates to prevent cracking or too rapid curing of the mixture. This process can continue for up to two weeks until adequately timed curing has occurred.

There are two types of cracking that can occur: crazing cracks and plastic shrinkage cracks. Crazing cracks are those that cause too rapid drying, and shrinkage on the surface and the lower part of the foundation can occur. Plastic shrinkage cracks are deeper and more widely spaced. They develop during finishing and are caused by rapid loss of mix water to the subgrade, by dry aggregate or by dry, hot air that is caused by the sun or extremely dry or windy conditions (Sing Saini 1980).

Table 5.4 Climate analysis of some arid countries

Country	*Climate*
Afghanistan	• arid to semiarid; • cold winters and hot summers.
Algeria	• arid to semiarid; • mild, wet winters with hot summers along coast; • drier with cold; • winters and hot summers on high plateau and experience sirocco; • a hot, dust/sand-laden wind especially common in summer.
Angola	• semiarid in south and along coast to Luanda; • north has cool, dry season (May to October) and hot, rainy season (November to April).
Australia	• generally arid and semiarid, temperate in south and east; tropical in north.
Azerbaijan	• semiarid steppe.
Bahrain	• arid; mild, pleasant winters; • very hot, humid summers.
Bolivia	• varies with altitude; • humid and tropical to cold and semiarid.
Botswana	• semiarid; • warm winters and hot summers.
Cameroon	• varies with terrain; • from tropical along coast to semiarid and hot in north.
Djibouti	• desert; torrid, dry.
Egypt	• desert; hot, dry summers with moderate winters.
Eritrea	• dry desert strip along Red Sea coast; • cooler and wetter in the central highlands (up to 61 cm of rainfall annually, heaviest June to September); • semiarid in western hills and lowlands.
Haiti	• tropical; semiarid where mountains in east cut off trade winds.
Iran	• mostly arid or semiarid, subtropical along Caspian coast.
Jordan	• mostly arid desert; • rainy season in west (November to April).
Kazakhstan	• arid and semiarid; • cold winters and hot summers.
Kuwait	• dry desert; intensely hot summers; short, cool winters.
Mali	• subtropical to arid; hot and dry (February to June); • rainy, humid, and mild (June to November); cool and dry (November to February).
Mauritania	• desert; constantly hot, dry, dusty.
Mexico	• varies from tropical to desert.
Mongolia	• desert; continental (large daily and seasonal temperature ranges).
Namibia	• desert; hot, dry; rainfall sparse and erratic.
Niger	• desert; mostly hot, dry, dusty; tropical in extreme south.
Oman	• dry desert; • hot, humid along coast; • hot, dry interior; • strong south-west summer monsoon (May to September) in far south.
Pakistan	• mostly hot, dry desert; • temperate in north-west; arctic in north.
Peru	• varies from tropical in the east to dry desert in west; • temperate in Andes.
Qatar	• arid; • mild, pleasant winters; • very hot, humid summers.

Saint Barthelme	• practically no variation in temperature; • has two seasons – dry and humid.
Saudi Arabia	• dry desert with great temperature extremes.
Somalia	• principally desert; • north-west monsoon (December to February); • moderate temperatures in north and hot in south; • south-west monsoon (May to October); • torrid in the north and hot in the south, irregular rainfall, hot and humid periods (Tangambili) between monsoons.
South Africa	• mostly semiarid; • subtropical along east coast; sunny days, cool nights.
Sudan	• tropical in south; • arid desert in north; • rainy season varies by region (April to November).
Syria	• mostly desert; • hot, dry, sunny summers (June to August) and mild rainy winters (December to February) along coast; • cold weather with snow or sleet periodically in December.
Tunisia	• temperate in north with mild, rainy winters and hot, dry summers, desert in south.
United Arab Emirates (UAE)	• arid subtropical climate due to location within the northern desert belt; • cooler in eastern mountains. • cold offshore air currents produce fog and heavy dew.
Western Sahara	• hot, dry desert; • rain is rare.
Yemen	• mostly desert; • hot and humid along west coast; • temperate in western mountains affected by seasonal monsoon; • extraordinarily hot, dry, harsh desert in east.

The optimum temperature of concrete at the time of placement is lower than can be generally attained in hot weather. An upper limit of 32.2°C should be considered reasonable although temperatures approaching this are likely to be deemed inappropriate. The best way to keep the temperature of concrete down is to control the temperature of its ingredients. As a guide, CSIRO (1968) has calculated that concrete containing six bags of cement per 0.76 cubic metres with a water/cement ratio of 0.5 by weight will be changed by 0.5°C by any one of the following changes in the temperature of its ingredients: cement 5°C, water 2°C, aggregate 0.9°C. From this it can be seen that the temperatures of aggregates and mixing water exert the most pronounced effect on the temperature of concrete. Water reducing retarders can be used as an admixture subject to seeking the advice of local trade suppliers, or concrete ingredients can be kept in the shade for cooling. Other methods of pouring concrete include keeping the mixers in the shade and out of hot areas, laying concrete soon after mixing, and spraying the subgrade with water, after laying constructing shading devices to allow the concrete to cure slowly. The curing process is critical and after laying the concrete should be wet cured with water for 24 hours. Extreme temperature differences between curing water and concrete should be avoided as this can cause cracking.

Generally if concrete is available and within budget then the standard foundation types can be used dependent on soil conditions such as strip, pad, raft and piled (Riley and Cotgrave 2008). It is always important to obtain a soil investigation prior to designing the foundation. Broadly speaking, common soil is made up of sand and clay in variable proportions. In any sample of soil a conglomerate of grains of different size and texture is present, representing gravel, sand, clay, silt and colloids. All these are classified according to their size. Often in more rural and developing regions a simple soil sieve analysis can be carried out to determine larger sizes of soil grains. To determine what is present, the soil is passed through a series of different sieve gauges with the help of water. In many Middle East, North African, Indian and Pakistani regions unstable soils are strengthened by adding straw to the earth; preventing cracking in moist and weathered conditions. In other regions in Africa, blood, urine from cattle and horse manure have been known to have stabilised the soil.

Concrete foundations, due to their high cost, are not always used for domestic buildings and quite often stone slab or dense concrete blocks are used as a footing bearing plate for the foundation in cohesive soils. If these are used it is important that the bearing slabs of stone or concrete block are wider than the width of the wall and take into account the standard 45° angle of forces that are associated with the traditional strip footing. See Figure 5.39.

In the traditional Malay and Indonesian House, the foundations are generally similar to pads; traditionally in igneous stone although more recently in concrete. Figure 5.40 shows the typical arrangement. The house is constructed in a certain order. The first column, the *tiang seri*, is placed in the centre of the house in a traditional cultural ceremony and the remaining columns are erected in the positions that

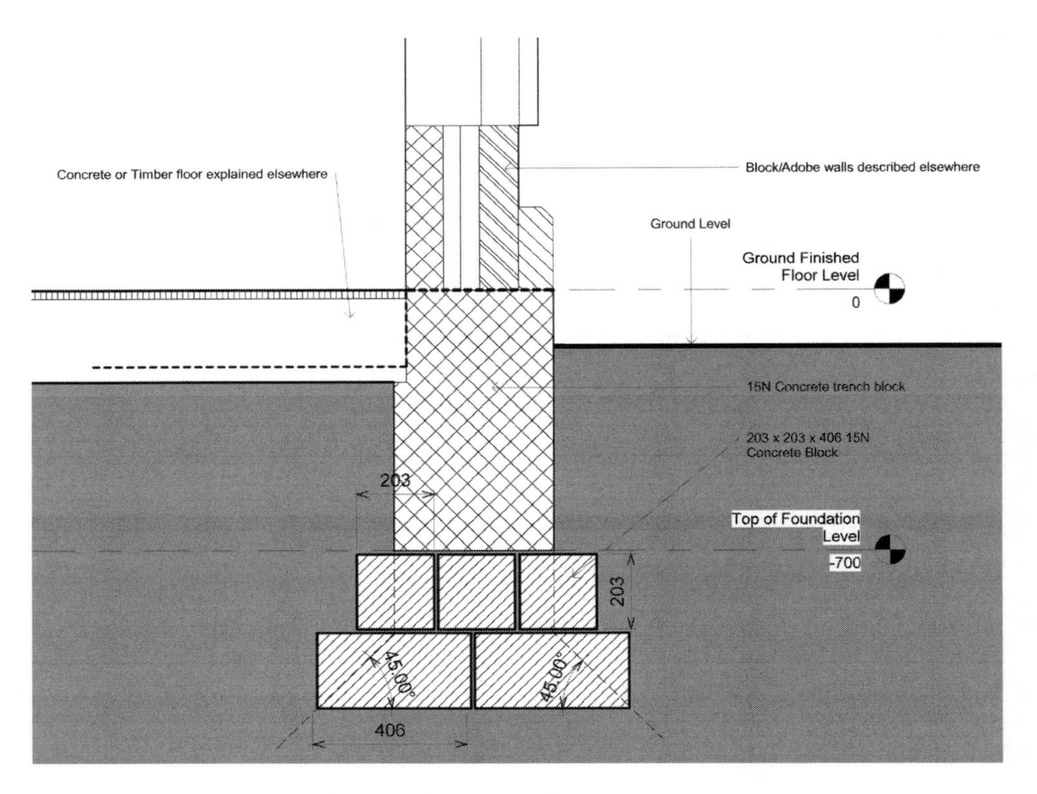

Figure 5.39 15N concrete block foundations in Africa.

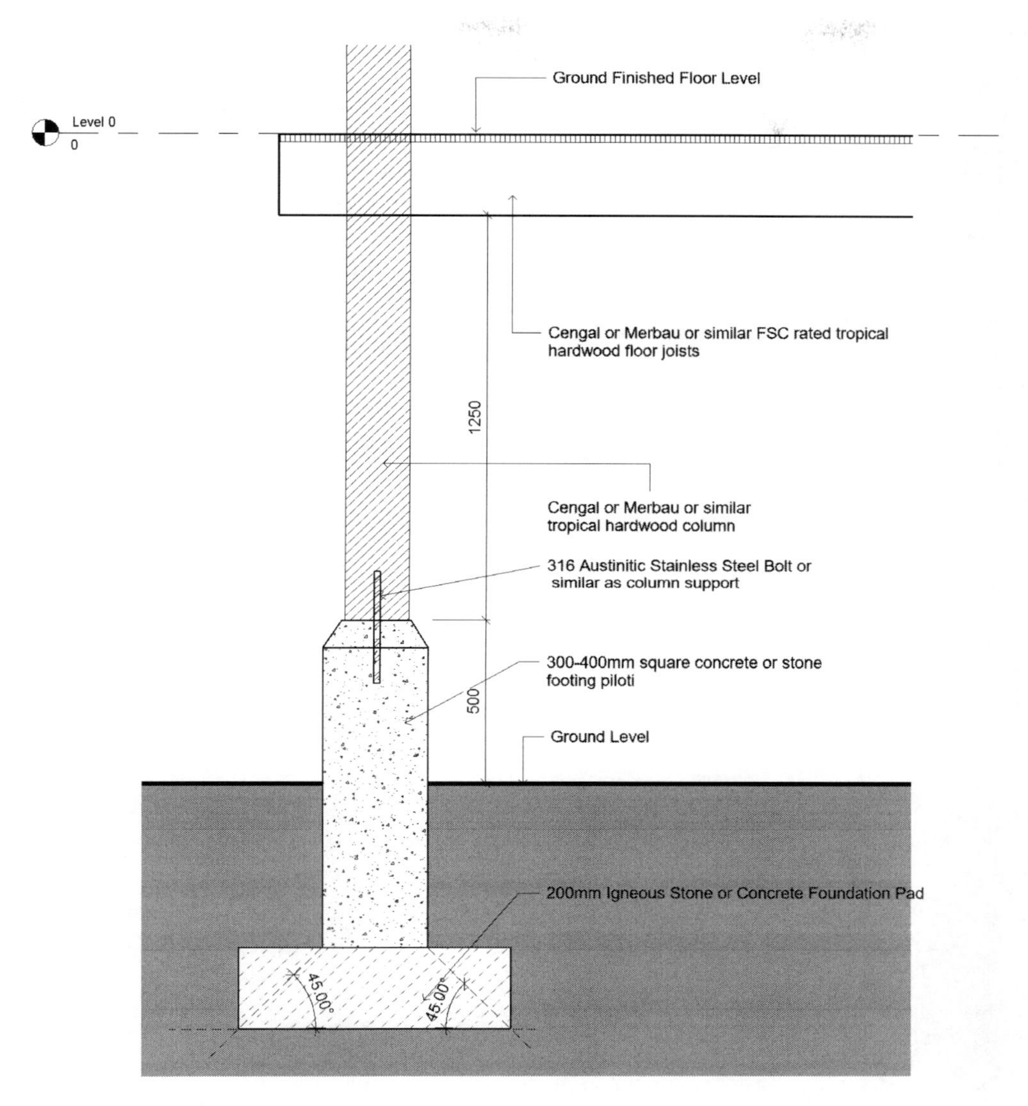

Figure 5.40 Malay House foundation.

they are removed from the trunk of the tree from which they were hewn. The base footing rests on a stone or igneous rock pad and is made traditionally from igneous rock itself. The purpose of this plinth is to raise Cengal or Merbau away from a position that could be attacked by termites. The overall reason for raising the ground floor is to mitigate flooding and avoid the dangers of animals and snakes; moreover to allow welcome cool breezes to ventilate the underside of the house. It is still important to keep the 45° load splays to the pad. The 20 mm steel rod locks the timber stilt to the stone/concrete plinth. Currently stainless steel would be used to prevent corrosion in what is a humid, damp environment but traditionally this fixing would be in cast iron or hot rolled steel.

In sub-Saharan Africa traditionally mud bricks would be made to construct the house. Sun dried mud bricks have a problem in that they attract the infestation of termites. Rurally a system of firing the mud bricks on a pyre can still be seen on the roadside today (Figure 5.41). This consolidates the bricks from termite attack. Bricks are simply stacked formally in a pile and a gap is left at the base to light the fire left burning for four to six days. The mud is dug from a hole at least a metre deep to avoid the vegetation matter which is common in rural Africa. Previously fired bricks surround the fresh mud units to protect them and keep in the heat and the fire is lit with wood and straw. In rural Africa there is also a profitable business for recycling the remaining charcoal into briquettes for fuel. Mud bricks are generally sold for 3–7 Kenyan shillings each in the East African Rift which amounts to less than 1US$.

For single storey buildings in Africa generally mud bricks are used as foundation blocks. Ideally stone should be used below these foundation blocks as a footing. Sun fired bricks are also used; not only in Africa but throughout the tropical world including historically South America where it is said that the Incas were the first to use this technology. Generally called Adobe after the Egyptian word 'dobey'; the word was introduced to the Hispanics via Egypt only to find that they were already working in this method. The word was then adopted by the Spanish language for sun dried earth bricks. Pisé walls are those that in the United Kingdom are known as rammed earth. The earth mixed with sand is laid damp and rammed in timber formwork in 4 inch layers that are reduced to 2.5 inch. The system is formed in layers so

Figure 5.41 Stack mud brick pyre.

that the layer below can dry and become stable in order to receive the next layer. Pisé has better weathering qualities than Adobe and less cracking characteristics. Pisé walls are only really used in Africa for temporary or informal settlements due to termite burrowing (Sing Saini 1980). Nevertheless, there are foundation details that can help to prevent this infestation. This usually means adopting an in situ concrete foundation reinforced for soil stabilisation. Figure 5.42 shows a typical T-shaped reinforced in situ concrete foundation for this condition. There are other types of details available for strip foundations for similar situations.

Earth in hot and wet climates is quite durable but very durable in hot and dry climates. In humid climates earth is not a good option as it can degrade quickly under the moist atmosphere.

Tropical regions are susceptible to land and mudslides in numerous areas. This is mainly due to tropical cyclones, monsoons and the fact that the underlying igneous rock weathers over time and goes through an in situ decomposed process; i.e. saprolitic soil (Lacerda 2007). Mudslides are more common in tropical regions due to the weathering of the underlying bedrock which, with heavy tropical rainfall, can form into a slurry. Main examples of this type of slide are well documented in terms of the loss of life and built property, not least that in Rio de Janeiro in 1988 (Marshak 2008). This can also occur more so in poorer tropical countries where the provision of adequate storm drains is scarce. Volcanic and earthquake areas should be mentioned in that these are different conditions where landslides can occur. It is deemed that these are not as common as mudslides in tropical regions and therefore are omitted from the scope of this chapter. In general principles, mitigation exercises to stop the flow of mud or to protect the built environment consist of some kind of

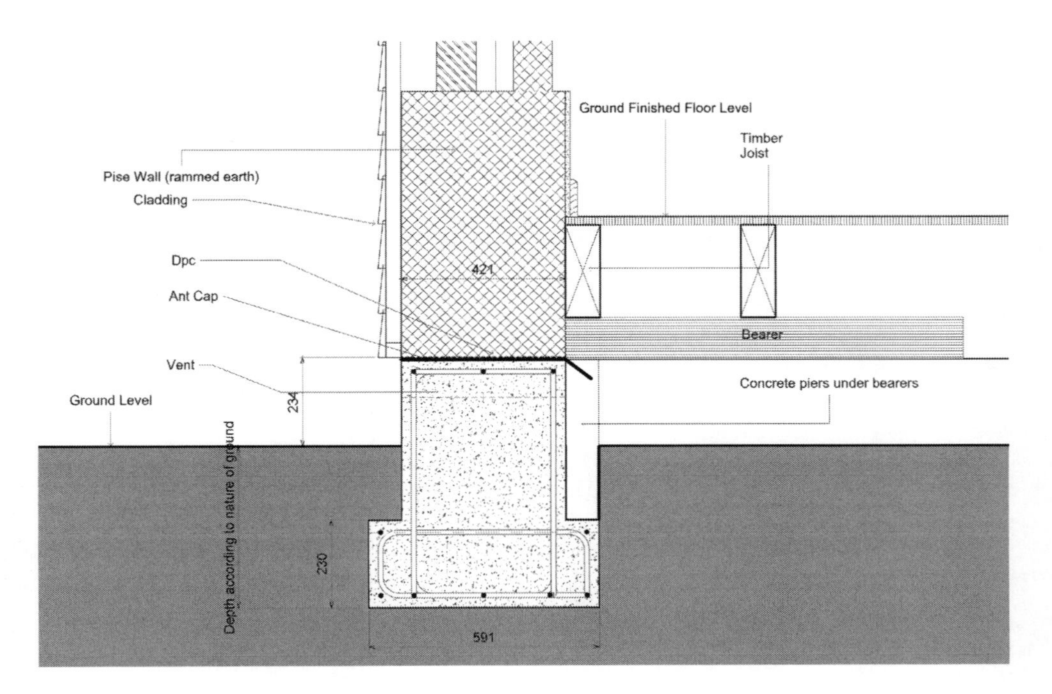

Figure 5.42 Typical soil stabilised foundation.

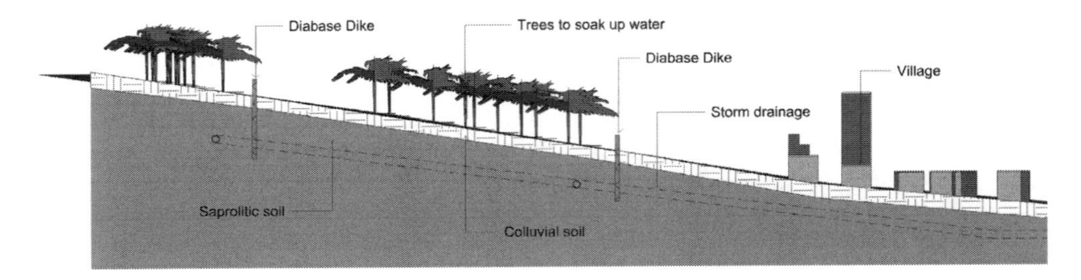

Figure 5.43 Landslide mitigation project.

containment, if indeed it is humanly possible so to do. At the very least, if building in such known disaster areas, there needs to be input from government or developer led projects; protecting a single property will render little effect as the amount of earth being moved needs to be dealt with at village, town or even city level. Other than in volcanic areas, there needs to be a gravitational force to initiate the slide. Therefore this alludes to the fact that care needs to be taken when building on slopes.

Other than these governmental strategies, the individual houses could be built on piled foundations to the solid bedrock if the depth can be absorbed by the construction budget. The depth of the saprolitic soil in tropical regions can extend to 100 m deep in some situations, making a piled foundation unfeasible.

Floors

The traditional Malay and Indonesian House has been described previously in that they are formed from Cengal or Merbau structurally and that floor decks and wall claddings are made from softer hardwoods such as meranti. This is due to the strength of the materials but also the ability for meranti to move with temperature differentials. These houses tend to creak after years of thermal movement. However, modern houses in Asia and indeed the African and South American continents are tending to be built with in situ concrete floors where construction budgets and availability of materials make it feasible. Timber ground floors in most tropical regions, especially Africa, are not advisable because of termite attack. In rural locations in Africa there are no ground floors; simply compacted earth bases internally. To prevent ingress of water there is a bund that is formed in fired mud bricks as a moisture barrier.

Early West African bungalows were built on stilts to provide ventilation from the radiative heat forces from the sun baked earth. However, if concrete is used there is no need for insulation to a ground floor slab as this would prevent heat entering from the earth but would also stop heat from escaping through the slab. More interestingly, the material treatment to the surrounding landscape to the ground floor slab provides an opportunity to cool the floor by protecting the perimeter from the convective solar heat. Treatments such as permeable paving, permeable grass surfaces and trees all allow moisture into the surround ground, with the latter two actively absorbing heat and returning it back to the atmosphere. Therefore having a healthy planting scheme around the house will not only provide shade but remove radiative forces from the ground. Such a planting scheme warrants the collection and reuse of rainwater and in this chapter harvesting systems will be described. Also using materials with a high Albedo effect such as polished concrete will reflect heat although to a lesser extent as there will also be some reflection interacting with the building itself.

There are temperature variables in all climates but the differentials are greater in the tropics. In the dry tropics, sun temperatures can reach 71° or 77°C with corresponding night temperatures of 27–32°C; a drop of 44°C in a few hours. In the humid tropics a sun temperature of 63°C can be reduced to 24°C in about an hour by heavy rain. These phenomena can last for weeks and the stresses on structure that they cause can result in cracks. Small unit construction such as bricks and blocks can easily be guarded against these conditions. It is monolithic concrete that is the main exception and as the majority of new constructions have concrete floors if they are at ground level a study of the behaviour of this material is paramount. The expansion of 30 m of reinforced concrete over a temperature difference of 28°C is about 10 mm. If the stresses are allowed to remain the floor total force on the floor slab will be about 20 tons per 300 mm width. This is likely to crack the concrete, but more importantly if there is a ground beam or an upper floor, the structure causing the restraint is likely to crack; thus having serious implications for the whole building. If a concrete slab is supported on blockwork, there must be a sliding joint between these elements. It is also important to keep the slab in question away from the sun so that only shade temperatures affect the concrete. As mentioned earlier the concrete must also be cured slowly and kept cool with water for the first two weeks after pouring. Expansion joints can be incorporated in the slabs. Overall with a 44°C temperature differential, approximately two no. 15 mm joints will be needed over a 30 m length and these should correspond to movement joints in the masonry walls. Figure 5.44 shows a typical internal slab/partition detail for a hot and dry tropical region. A similar detail is used for hot and humid regions except that the partition wall will be a lightweight partition such as bamboo or meranti boarding so that good ventilation is maintained throughout the building.

Figure 5.45 shows a typical 15 mm slab to slab expansion joint that should be placed a minimum of 15 m apart. This will allow movement as the slab shrinks. Another situation where expansion is needed is at the junction between the ground floor slab and an external wall. This again may have varying wall types depending on region. The distinguishing features of this detail are the fact that position of the damp proof course may be higher than the standard 150 mm as the building could be likely to be in a flood zone. Also there may be an addition of an adequate storm drain adjacent to the wall. The reinforced T beam can be substituted for concrete block subject to soil stabilisation issues.

Roofs

An earlier section dealt with the form and shape of roofs in tropical climates, whereas within this section the make-up of the roofing materials and to some extent shape in the domestic situation will be explored. Again there are two main macroclimates that we need to appreciate in determination of a correct specification for

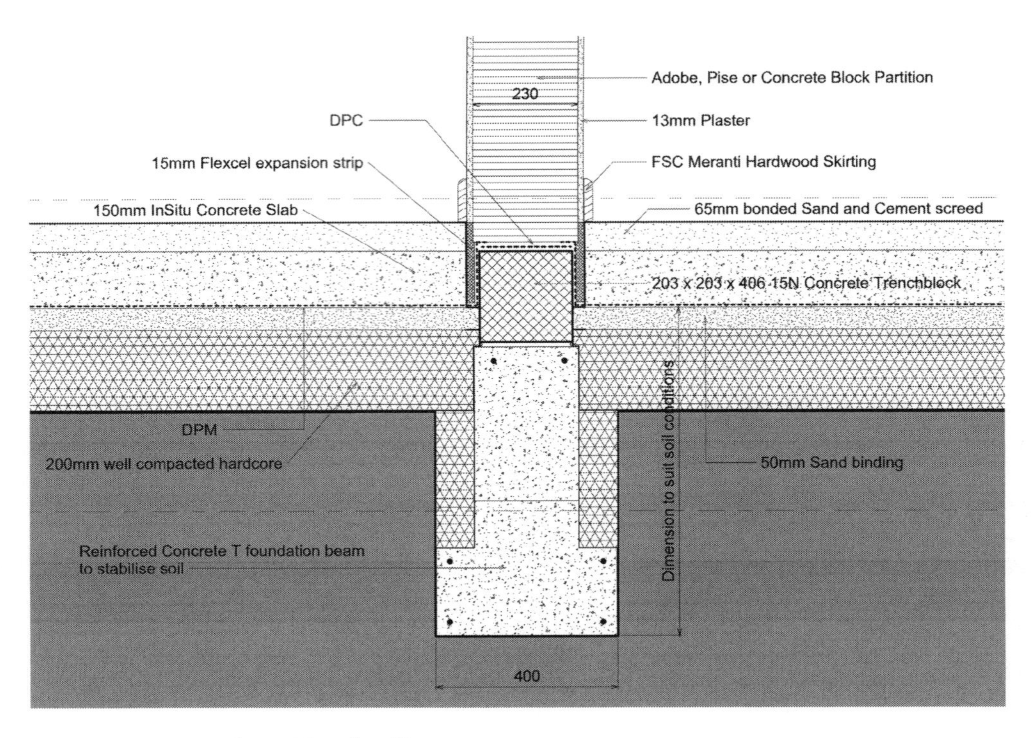

Figure 5.44 Internal partition detail.

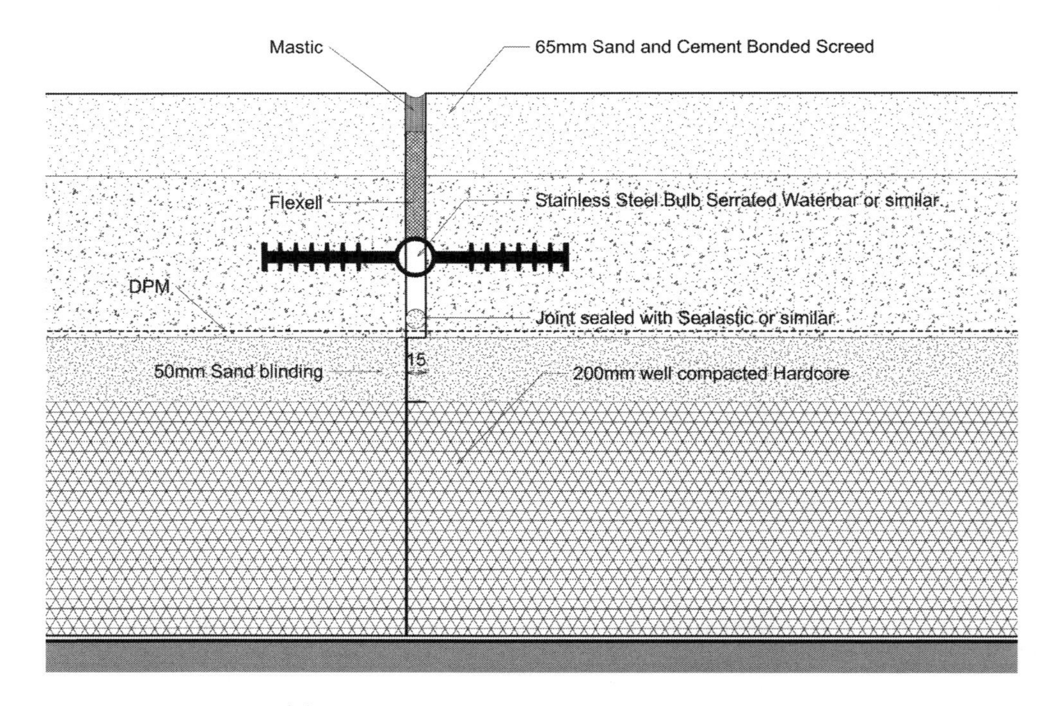

Figure 5.45 Concrete slab movement joint.

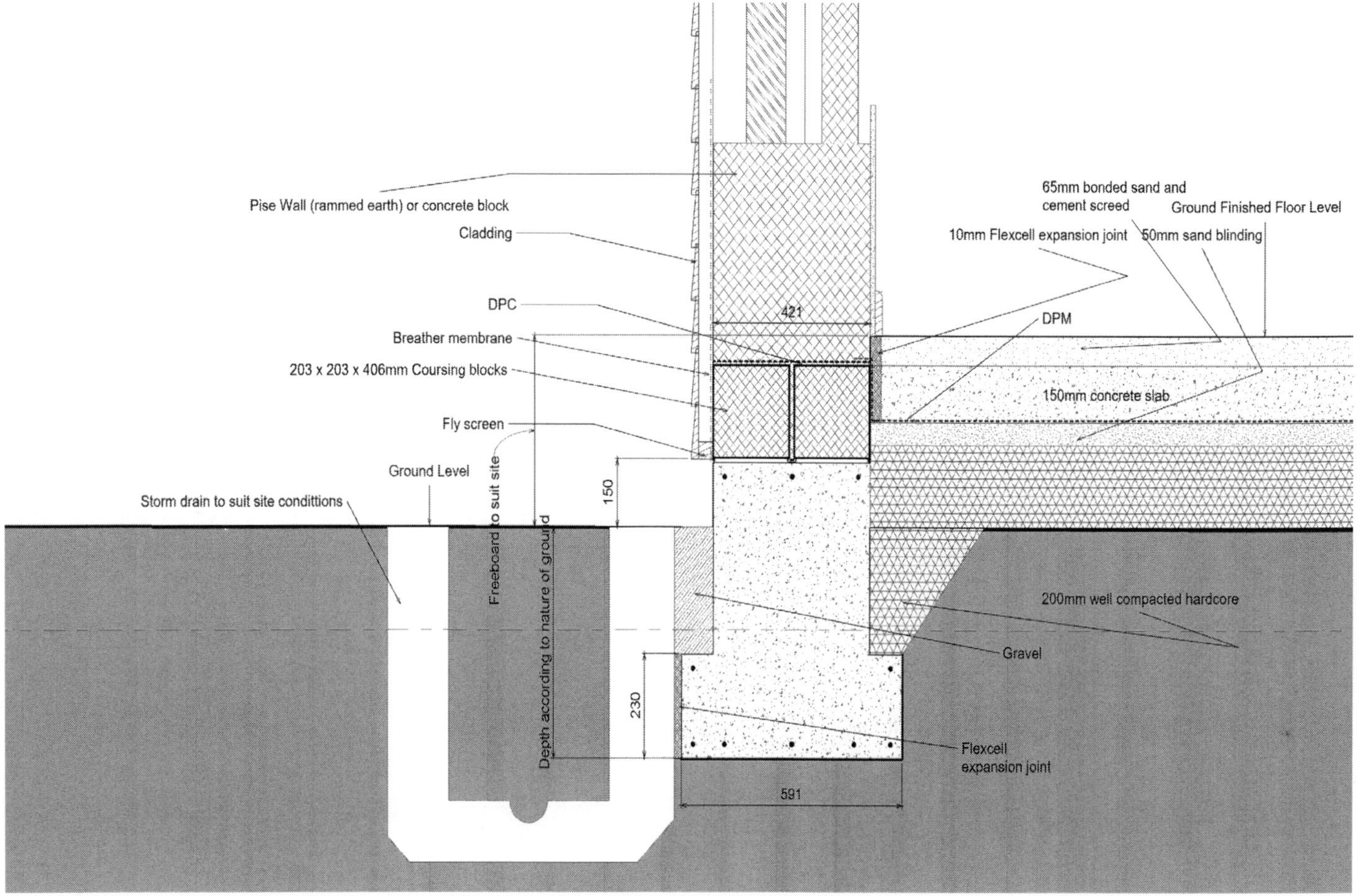

Figure 5.46 Standard slab edge detail in flood region.

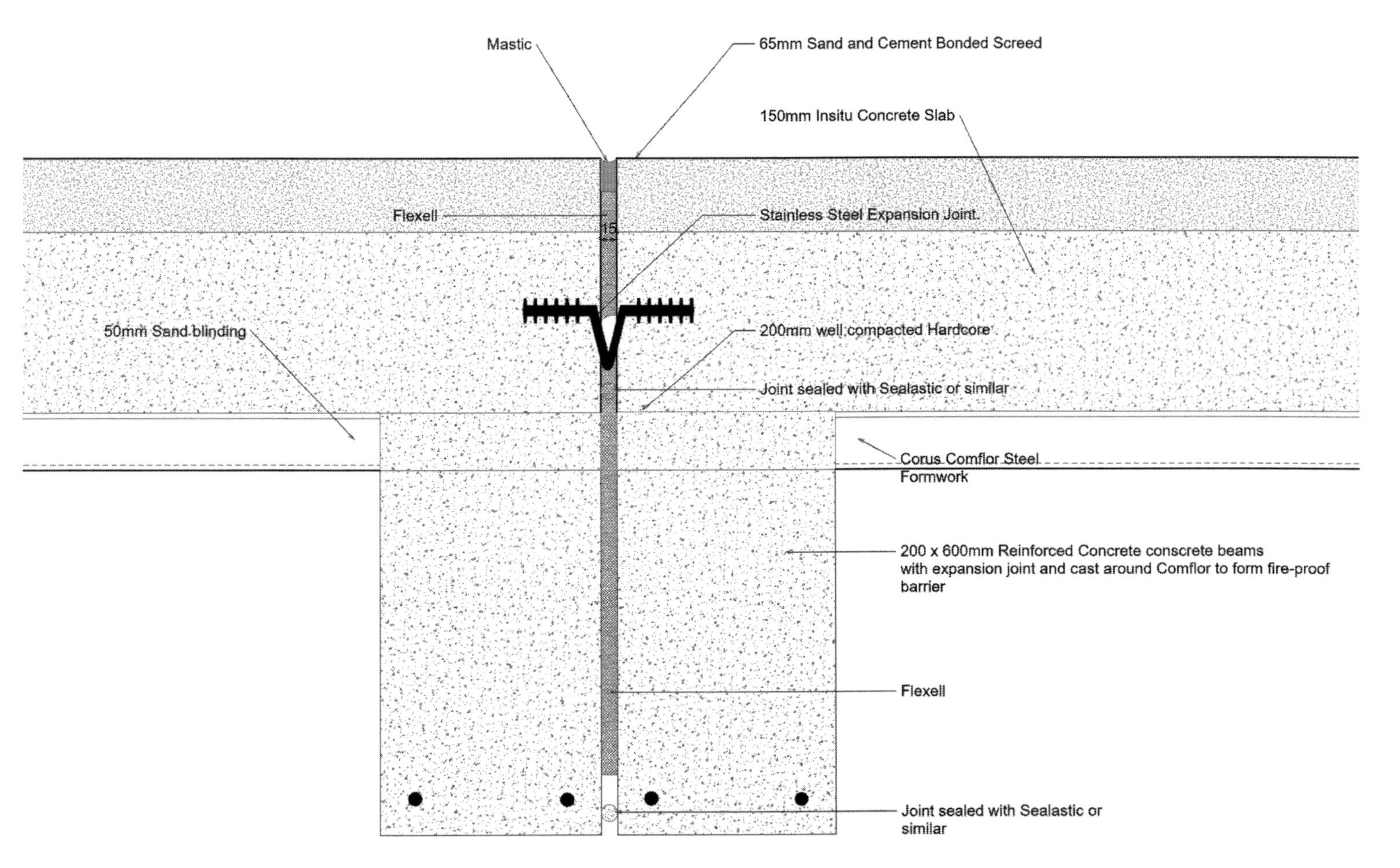

Figure 5.47 Typical movement joint in first floor slab of a framed building.

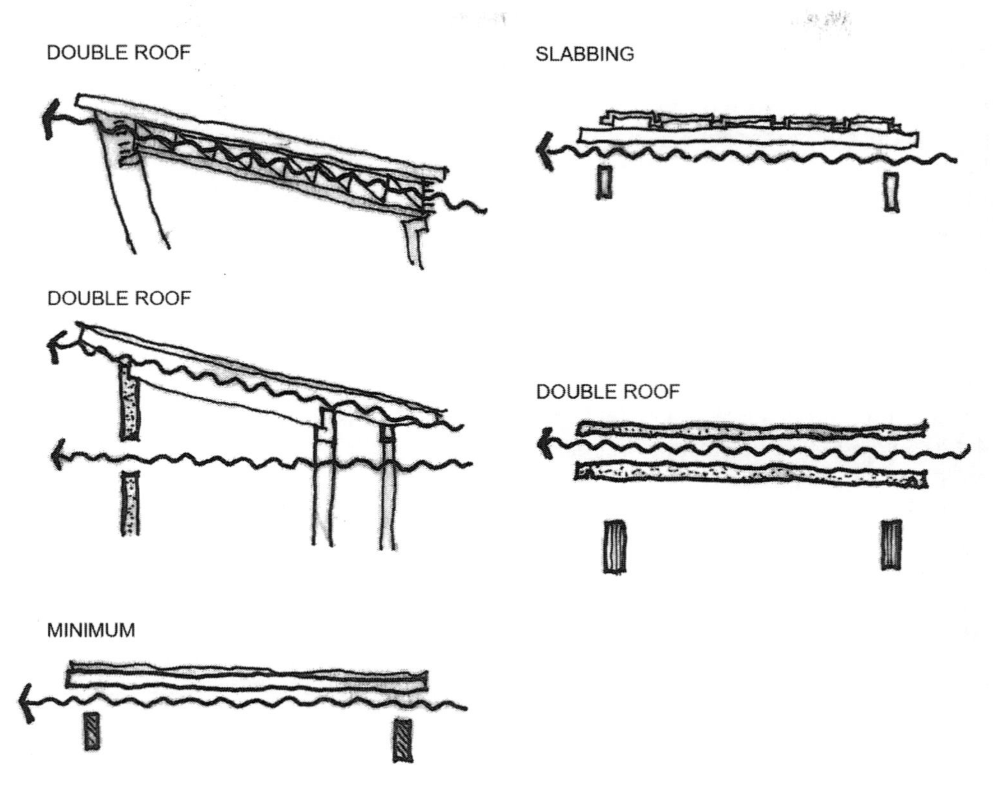

Figure 5.48 Ventilation in roofs.

the tropics: hot and humid and hot and dry conditions respectively. Good ventilation in hot and humid conditions is the key as the air temperature is fairly constant throughout day and night unless there is rain. The parasol roof or double roof provides a solution to the heat lag problem in ceiling or primary slabs in both climates as discussed earlier as exhibited in the houses of Le Corbusier in India. In the hot and dry tropics traditionally both double and single roof methods are used and constructed using a thick layer of mud to absorb heat, covered by a layer of impervious material which can be some form of asphalt or oil, or square or broken tiles, helping rain to run off quickly. Tiles are normally used where roofs are slept on. These roofs are heavy and only suitable for small spans and are notoriously difficult to construct correctly, tend to leak and cost a lot of money. In the hot and humid tropics Fry (1964) describes an experimental roof having a light reflecting quality of corrugated aluminium supported above a ceiling with an adequate air gap between. Fry (1964) advocated a monopitch roof with the lower side facing the direction of the sun so that heat could rise and escape at the higher peak. Being built of light materials, this roof has no reservoir of heat and an hour after sunset remains at the night time temperature. This is a low cost solution which can be associated with the *bumbung lima* of the Malay Traditional House where heat escapes through the roof tiles, although Fry's version has the benefit of the reflective material. The area of reflectivity is important in hot and dry climates where a certain quantity of foil backed insulation can be used to both repel heat transmission by virtue of low thermal conductivity and achieve a high emissivity so that heat can also radiate out from the material. Sing Saini (1980) points out that the greater the amount of insulation used does not necessarily provide an exponentially better result in repelling solar radiation. This matter concerns an offset of emissivity of longwave radiation to solar radiation reflectance. Emissivity is the ability of a surface to emit thermal radiation in comparison to a perfect emitter. The scale therefore runs from 0 to 1 with the lower value being better in reducing radiative transfers as the surface has a low emissivity (Nicholls 2008). If a large amount of insulation is used the roof will retain more heat and the emissivity will remain constant for a foil backed surface. An example of this is an experiment carried out by the Commonwealth Experimental Building Station in Sydney. The first 25 mm of vermiculite, an insulation product, reduced the maximum ceiling temperature by 4.4°C while the second 25 mm reduced it only by a further 1.1°C. A similar result is seen in the use of mineral wool insulation. U-values are a part of the analysis of roofs in tropical climates also. Drysdale (1959) recommends at least 25 mm of mineral wool insulation between roof and ceiling, giving a U-value of 1.08 $W/m^2 {}^\circ K$ which also corresponds to the findings of Lotz and Richards (1964). Optimum amounts of insulation in more extreme climates such as the Middle East and North Africa can have even lower U-values such as 0.28 $W/m^2 {}^\circ K$ for roofs constructed with air spaces. Koeningsberger and Lynn (1965) recommend that the temperature for the underside of the ceiling should never be more than 4.4°C higher than the dry bulb temperature after recording a series of roofs in Accra, Bombay, Colombo, Bangkok, Singapore and Kuala Lumpur.

In either case reiteration should be given to the fact that parasol roofs are the most efficient at keeping the interiors cool although a heavy parasol performs worse than a lightweight shading device such as painted aluminium or mill finished aluminium with a highly reflective shine. White surfaces have a higher emissivity (about 0.85–0.95) for longwave radiation such as occurs in the temperature range of 10–37.8°C whereas the corresponding emissivity of metals is very low. So although white surfaces absorb a slightly greater proportion of short wave radiation than shiny metallic surfaces, they are able to lose heat due to the surroundings by longwave re-radiation.

Roofs are especially critical in their performance in the event of tropical rain storms, cyclones and hurricanes, and there is some simple detailing that could help in the survival of the structure. The nature of wind and roofs is that they act as a sail if they are lightweight and overhang the property to guard against heavy rain and therefore must be tied down correctly. A reinforced concrete ring beam at eaves level, which itself is tied to the masonry wall below, is a useful method of strengthening the attachment of the roof to wall.

Similar details can be developed in timber frame but care must be taken to ensure that the eaves of the building have a timber ladder frame of sturdy construction so that the wallplate can be adequately fixed to it to prevent roof uplift. Other systems such as intensive and extensive sedum roofs can be used. Sedum roofs offer a good alternative in that they insulate the building well from heat gains and hold water for cooling and reduction of impact on the drainage system. They also prevent the build up of heat islands in developed areas by the absorption and storage of heat.

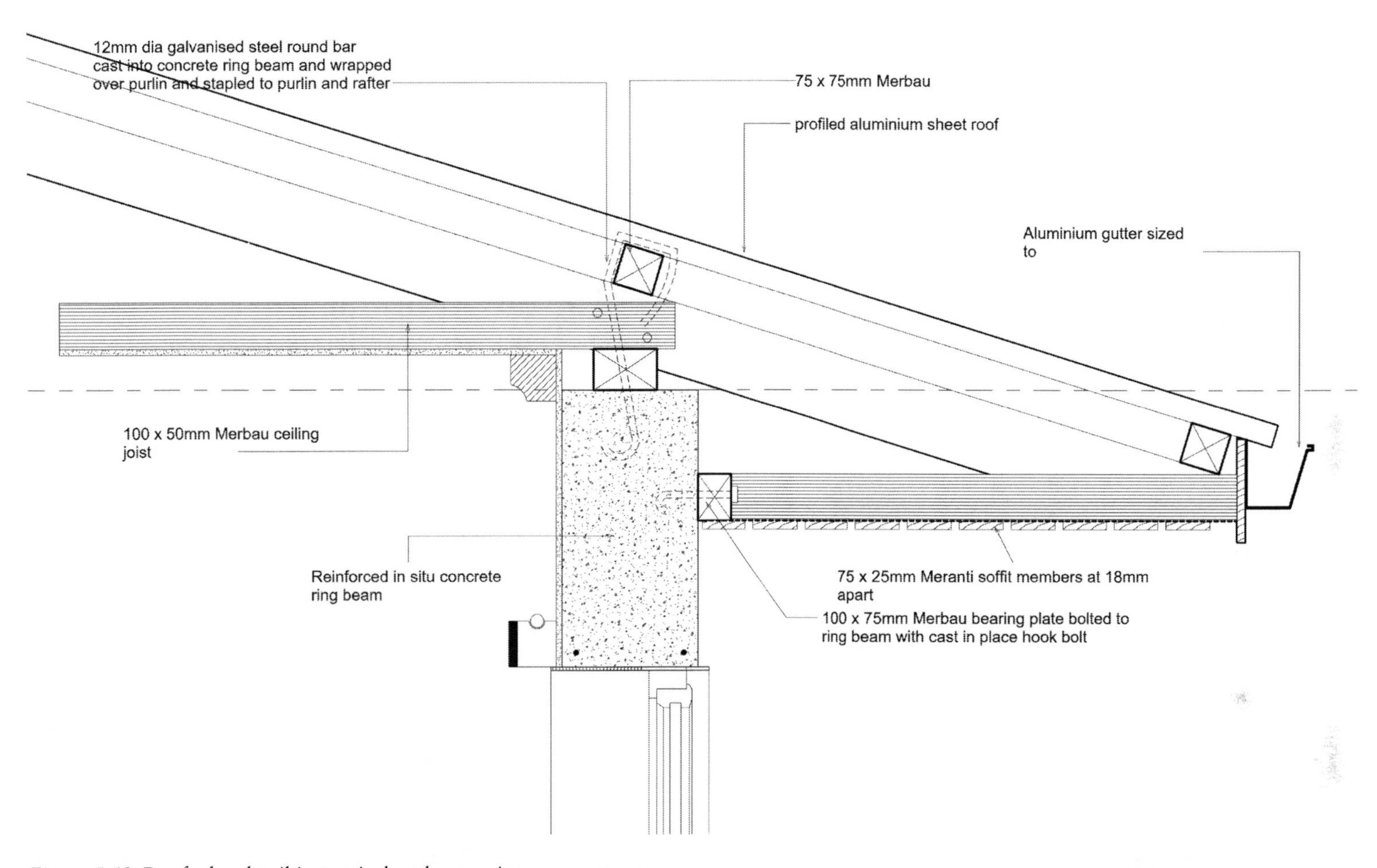

Figure 5.49 Roof edge detail in tropical cyclone region.

External enclosures

Other than the shading of the external wall enclosures to prevent solar radiation resting on the surfaces, the type of wall construction and size of openings is critical in all types of tropical regions. The main functions of the external enclosures is to resist the passage of solar radiation, control the passage of light, air and sound both inwards and outwards. Where the percentage of opaque wall/curtain walling to window is carefully controlled it can be seen that there are different values for wall types in hot and dry climates. In Figure 5.50, Sing Saini (1980) shows the temperature ranges for both timber wall/insulated roof and 9 inch (215 mm) brickwork/insulated roof constructions in a hot and dry tropical region. It is clear that the brick structure has a more constant temperature throughout the day where the enclosure is protecting against heat gain due to the time lag in dissipating heat. Furthermore the structure emits heat in the night time when the outdoor temperature gets colder. It is therefore likely that a brick, concrete block or stone wall would be used in a modern house in a hot and dry tropical region. Also Adobe, fired mudbrick or Pisé (rammed earth) walls can be used as they provide good thermal resistance and time lag. Time lag is the delay in a material to emit heat that it has absorbed from the sun. Ideally the longer the lag the better so that heat can be dissipated away during the night when outdoor temperatures get colder. Values for other materials are shown in Table 5.5.

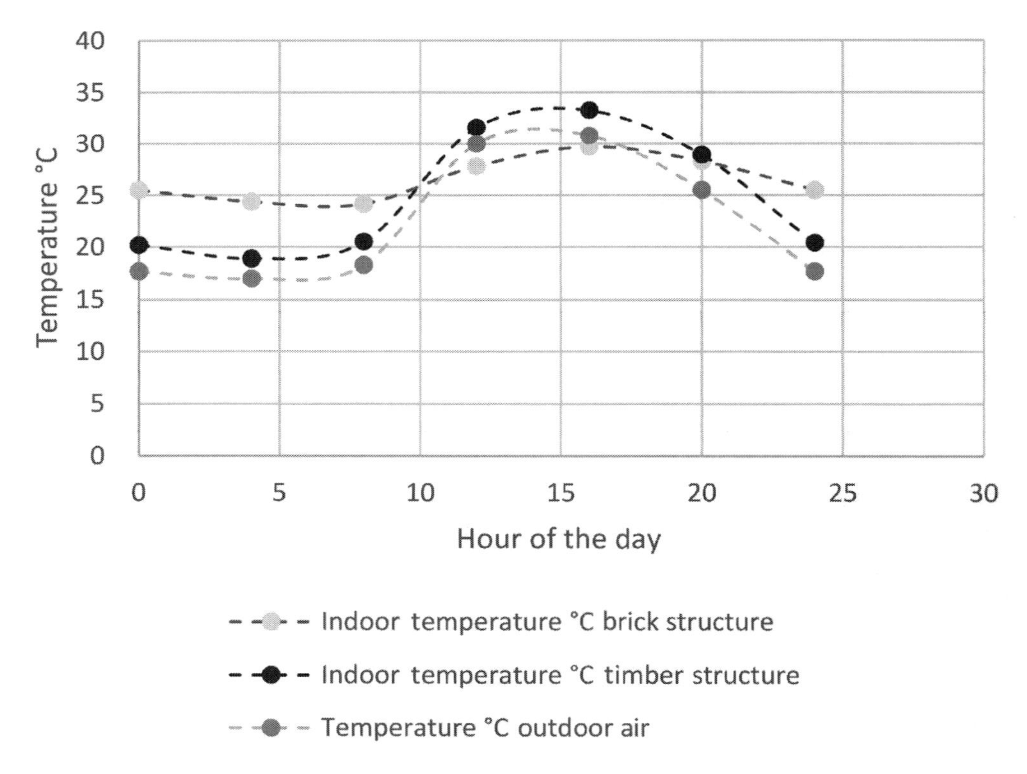

Figure 5.50 Hot and dry tropics – temperature ranges.

Table 5.5 U-values and time lags for some building materials

Material	*U-value (W/m²°K)*	*Time lag (hours)*
Stone 12 (305 mm thick)	3.1	8.0
Solid concrete (152 mm thick)	4.2	3.8
Solid concrete (305 mm thick)	3.0	7.8
Common brick (102 mm thick)	3.4	2.3
Wood (50 mm thick)	1.7	1.3

The orientation of the wall is an important factor and has been discussed previously, but the main aim is to keep the external walls in shade as much as possible where it will remain at shade temperature. Domestically shade can be created by projecting eaves, verandahs, sun breakers or other inventions. Traditionally colonial buildings in tropical regions had verandahs on all four sides which worked to shade the external wall enclosures and keep the rain off facades, but this rendered the interiors gloomy. More recently modern tropical buildings have incorporated more glass but this material absorbs heat completely and is reluctant to emit heat once it has been trapped. In humid and wet tropical regions glass is used regularly with shading devices as a glass opening has a good ability to provide porosity to a facade by the large opening size it can achieve or by including a solar screen or louvre with the panel.

Garde-Bentaleb *et al.* (2002) write about the importance of the porosity of external walls in the design of houses for the humid tropics. They introduce the ECODOM project in a paper about the improvement of thermal qualities of tropical housing which was an energy standard initiated by the French Government for their colonies in the tropics (Reunion and Guadeloupe). The ECODOM standard in a typical house type requires a porosity on each facade of 2.82 m^2 which is quite difficult to achieve with standard sized windows. This is achieved by the use of French windows at ground and first levels and also by slatted ventilation openings. If the problem of achieving porosity values proves irksome, creativity has to be used to develop windows and doors with ventilation panels above. Prianto and Depecker (2002) analyse the size, arrangement and positioning of ventilation and windows in a project for French Guiana, a humid tropical region where the wind speeds are quite low. The article cites that porosity has to go up to 50 per cent where this occurs so some meteorological checks will need to be in place before arriving at an optimum porosity level. The Venturi effect also comes into play where small openings create a faster wind speed which is beneficial for thermal comfort. There needs therefore to be a balance between opening size and porosity which can also be affected by the angle of the ventilation louvres and windows. An even faster wind speed can be achieved if smaller openings on the breeze side of the house are created with large openings on the opposite.

With regard to the solar louvres, these can be combined with a shutter system so that the louvre folds down when a tropical storm is on the way and is bolted to form a secure barrier against violent wind and rain. It is important that with any openable screens there is a continuity between opposing facades so that there can be a through draft. These are generally situated on the north and south edifices in order to maximise the coolest pockets of shaded air. In developing regions bamboo, meranti and eucalyptus screens are used for external walls. The importance of eucalyptus is in its

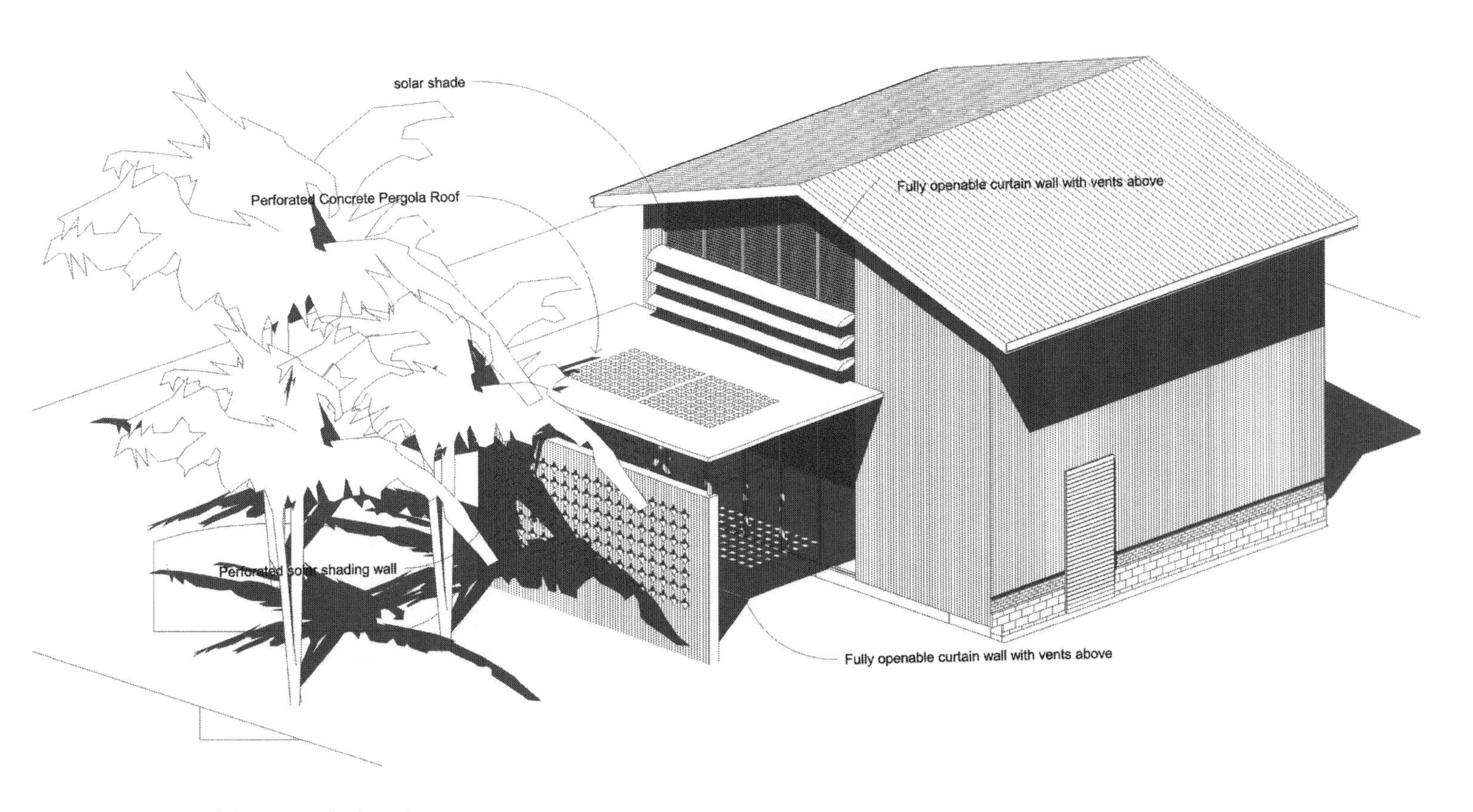

Figure 5.51 Typical domestic shading devices.

Figure 5.52 Example of filigree shading walls that also provide ventilation to circulation spaces.

ability to resist wood-boring termites. The timber species was introduced to Africa by the Australians and is used in unsawn and PAR conditions. The Forestry Stewardship Council (FSC) should verify that any tropical timber species are certified and come from renewable plantations. Eucalyptus and bamboo are used in many traditional structures in humid parts of Africa coupled with fern leaves and grasses; external enclosures that are breathable can be developed with the addition of mud or wattle and daub to finish.

Traditionally in the dry tropics external walls with a greater mass are adopted so that heat can be stored and slowly dissipate. The thicker the cross section of the wall the better as this will provide an inherent shading device. This alludes to fin walls or walls with ventilation spaces within their mass. The whole essence of what needs to be achieved is aimed at the quality of thermal comfort in the interiors in the tropics. Due to the stifling heat and air quality in the dry tropics the cross ventilated enclosure is not generally adopted. Heat rises and therefore the upper floors are likely to be hotter than below. The use of ventilated cavity walls with lightweight terracotta tiles can be used in these regions such as seen in the post office building in Ghardaia in the Mzab Valley in Algeria. Solid limestone is used for the groundbearing walls but the upper walls are of cavity construction with an inner leaf of cement bricks and an outer screen of terracotta. The vents at first floor level also provide an interesting element to the facades.

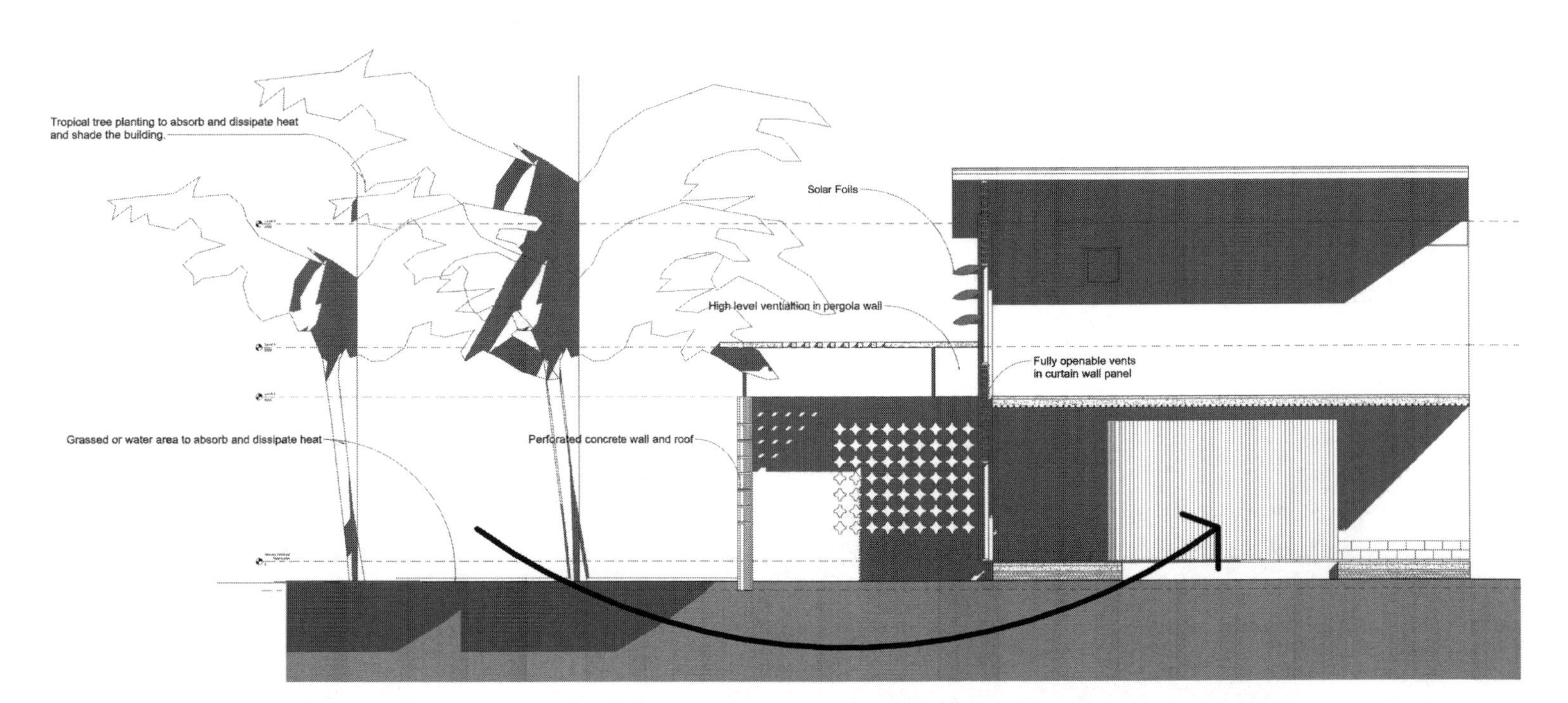

Figure 5.53 Typical cross section through an external solar screen courtyard showing position of grass and trees to reduce heat of adjacent land. This heat can transfer to the ground slab if not shaded. Grass absorbs and retains the heat.

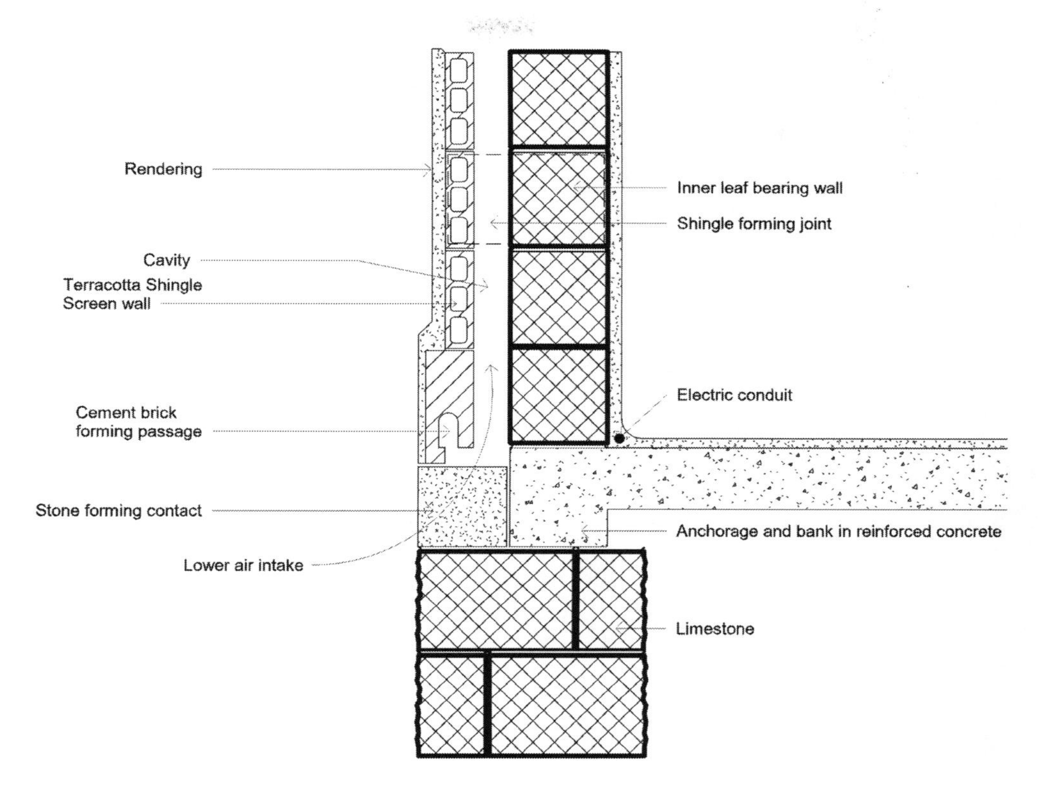

Figure 5.54 Ventilated external wall, Post Office, Algeria.

Lightning protection needs to be used domestically in areas where there is likelihood of tropical storms, whereas in the UK this is reserved for large buildings. The cable needs to run up the outside or through a wall, ensuring that there are no bends that are more than 90°.

Internal division

The internal division in houses is crucial in that there needs to be a cross flow of air from north to south facade depending on the prevailing wind direction. This means that internal walls and doors need to be porous or contain vents within at high level. The Jalousie type screen or window has the issue of poor acoustic insulation although acoustically attenuated vents can be used which are generally likely to be more expensive.

Porous materials in humid climates can be bamboo and hardwood lapped panels with glass or metal louvres overhead. Screens are useful in delineating spaces and allowing air to move freely without the use of un-passive electric fans. These can be used with a mechanism of mosquito screens, as can external walls and windows. Care needs to be taken in specifying internal timber in tropical regions susceptible to wood-boring termites. Usually a cut porcelain tile is used at low level (mud-boring termites can attack some terracotta tiles) for skirting boards instead of timber. The detail at the base of door frames and architraves can also be attacked. Usually concrete stools at low level are fitted to door frames to prevent infestation or metal legs used to timber screens. Internal loadbearing walls and timber floors should contain galvanised or stainless steel ant caps as shown in Figure 5.55. Internal screens can be a useful mechanism for venting the centre of the house vertically. This would include a passive venting chimney in the centre or even on the external facade. The name given to such a mechanism is a solar chimney. More studies on this subject can be pursued by reading Yong Kwang Tan and Nyuk (2013) in which a parametric model of a house with a solar chimney was completed to evaluate the levels of air speed and ventilation achieved.

Industrial and commercial building in the tropical regions

Building form and shape

Industrial buildings tend to have a larger wall area in ratio to roof area which can be a problem for overheating in the tropics. With height there is more surface area on which the sun's rays can rest. Large deep plan configurations tend to also overheat due to the large amount of roof area and care needs to be taken that openings to a roof intended to allow natural light in do not cause a glare and excessive solar irradiance. Generally saw-tooth roofs should have glazing angled away from the sun.

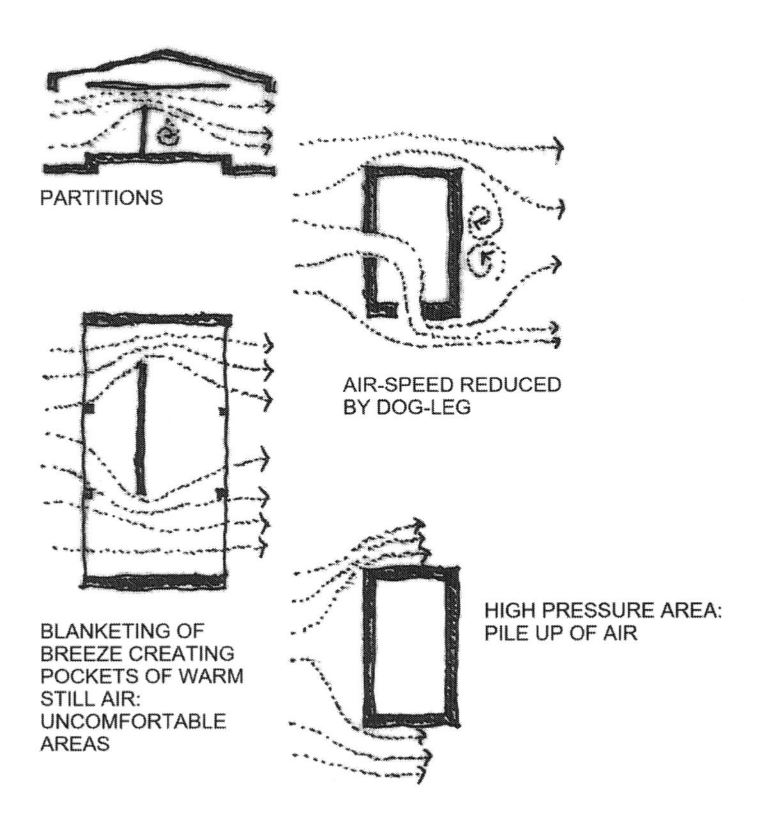

Figure 5.55 Internal shading devices.

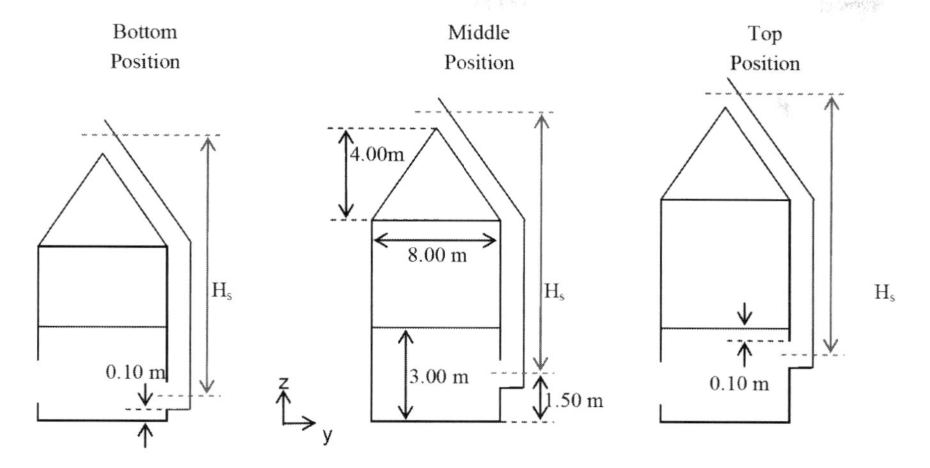

Figure 5.56 Solar chimneys. Yong Kwong Tan and Nyuk (2013).

In the northern hemisphere this is north facing but in the southern hemisphere this will be south facing. As the sun is overhead in equatorial areas for most of the year, then thought should be given to the use of opaque, light coloured glass. This prevents glare and also increases the emissivity of the glass. Within deep plan buildings thought should be given to the adoption of internal courtyards interspersed to provide through ventilation. Although industrial buildings in the tropics tend to be meagre in fenestration and architectural detail, in searching for form and function in architecture for this typology reference can be sought in the work of Renzo Piano at the Menil Museum, Houston, Texas (Brookes and Grech 1990). Although this

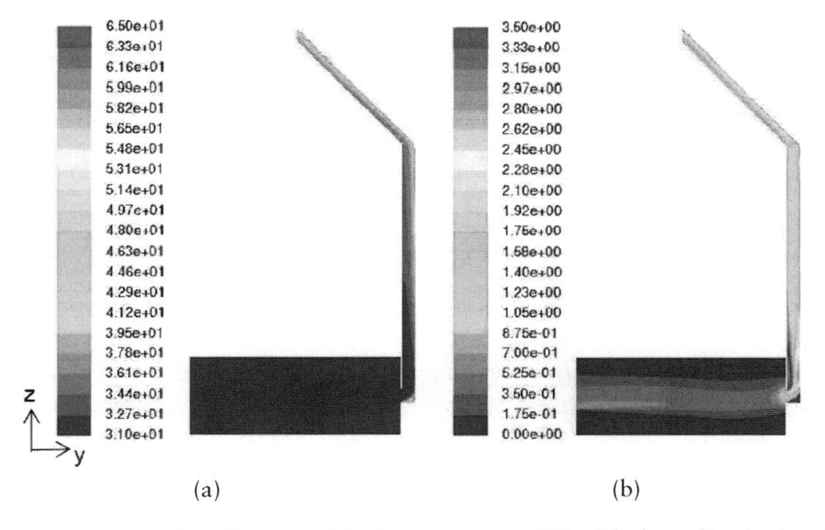

Figure 5.57 Solar chimneys. (a) air temperature (°C mid-plane distribution for base simulation); (b) air speed ((m/s) mid-plane distribution for base case simulation). Yong Kwong Tan and Nyuk (2013).

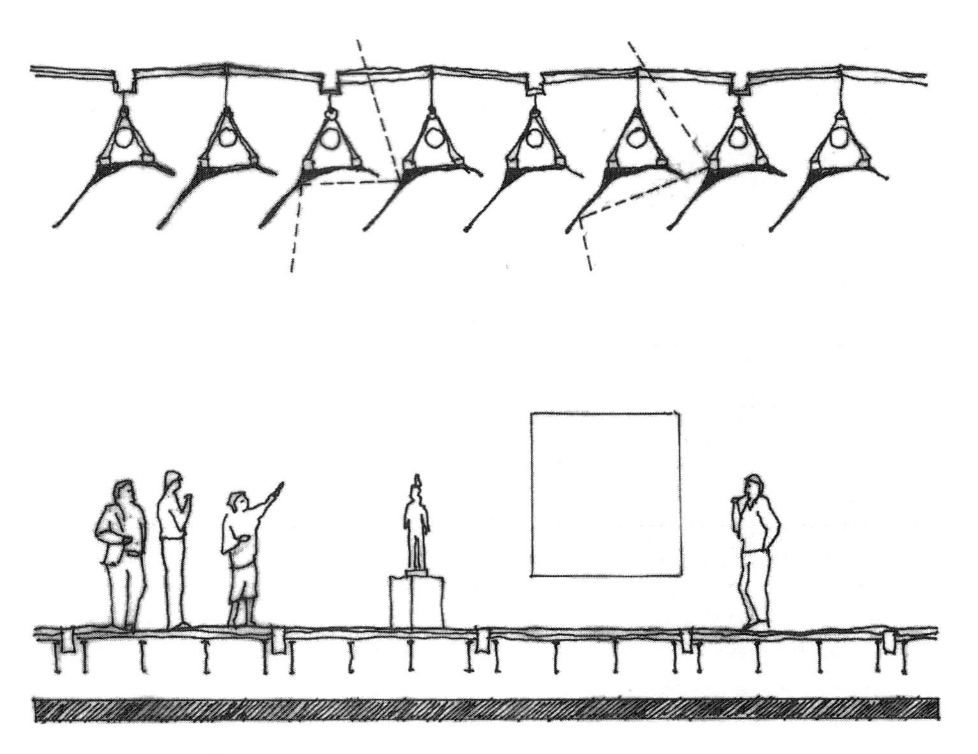

Figure 5.58 Menil Museum, Renzo Piano.

building is not in the tropics, the overriding concept has characteristics that can be adapted to humid and dry climates.

A mechanism of ferrocement leaves allows indirect natural light from above but also allows shading by the overlapping of the screens. Although the shades are below the level of glazing and would need to be above, this mechanism could take the place of a parasol roof in the tropics, albeit it is an expensive solution. Similar shading devices can be seen in much Hi-Tech architecture by the aesthetic of putting the building services on the outside. Buildings such as Lloyd's of London and the Pats-center, New Jersey by Sir Richard Rogers are examples of this building type. Industrial buildings usually appear with a water tower in the tropics and these elements could provide welcome shade to the factory.

In reflecting on commercial architecture in the tropics, there should be mention made of the great ancient souks (markets) of cities such as United Damascus, Dubai and Marrakesh. These building forms are great linear markets that are a direct response to climate. Although unsanitary, these structures have buildings close together in a hot arid climate which form an innate shade and protection from dust.

As discussed earlier, in equatorial climates where the sun is overhead, these souks allow a linear arrangement in the east-west direction, making use of breezes from the north and south. Larger cross sectional space in the north-south street axes can provide resistance to the heat sink effect. Form would therefore follow a set-back approach as the building height increases.

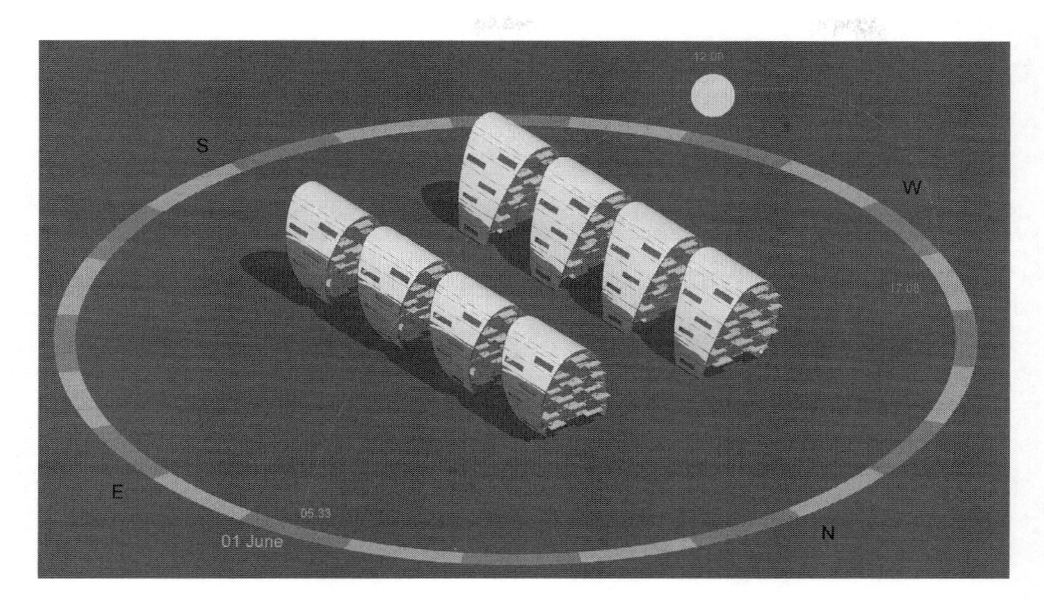

Figure 5.59 Office project for Uganda.

There is also an inherent problem with stepping back buildings in the tropics as the enlarged surface area increases heat gain. Smoother surfaces are better so that the shape factor value is kept low. Building forms similar to the Greater London Authority building by Lord Foster would be suitable.

In terms of energy use, recently there have been examples in construction of commercial buildings that have been designed to function in a semi-passive mode. Most of the best examples of these are built in temperate regions although the theory behind them can contribute to the 'new age' of commercial buildings in recently

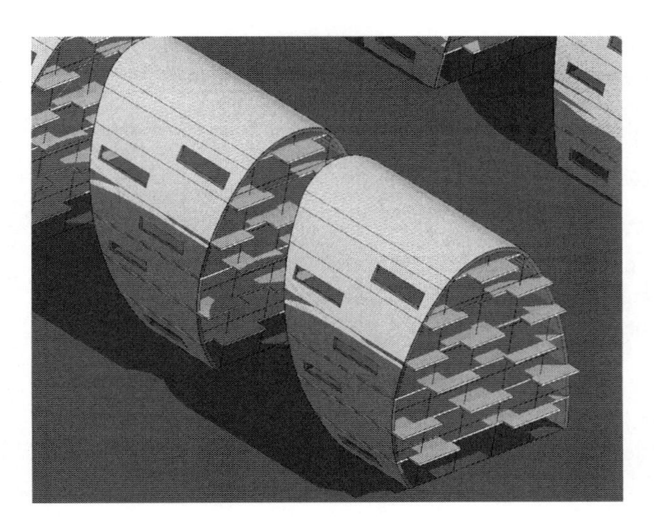

Figure 5.60 Office project for Uganda.

developed countries. It is no coincidence that regions in Asia, that have undergone massive amounts of new development, and the African continent, forecast to be the next major growth area, are situated in or close to the tropics. Although they are beautiful iconic buildings, developments such as the Burj Kalifa or Shanghai World Financial Centre are in essence poor examples of sustainable development in the tropics. Although designed with high density, which aids energy efficiency, the fact that they contain large amounts of glass and that they are HVAC cooled renders them inefficient thermally. Although there are many types of glass that are designed for use in hot regions (which will be discussed later), there is scepticism that these still increase heat gains compared to other dense cladding materials with higher emissivity and heat lag, such as stone, concrete and glass reinforced cement.

The Shanghai Tower (Figure 5.67) is a 124 storey skyscraper in the Luijiazui district of Pudong, Shanghai, designed by Jun Xia at Gensler. This building is a proponent of the use of high level atriums interspersed throughout its height which act as heat sinks. The double skin glass facade is also used; both of these characteristics help to reduce cooling loads. The risk of seismic shock in Asia is considerable as Shanghai, Hong Kong, Taiwan, etc. lie in 'the Ring of Fire' which is a ring-like belt of tectonic plate movement which stretches as far as Australasia.

The Shanghai Tower has a total architectural height of 632 m while the total structural height is 580 m, adopting the steel-concrete hybrid mega frame-core tube outrigger structural system. Since the total structural height and irregularity exceed the limit specified in the Chinese seismic design code, Shanghai Tower is classified as a code-exceeding tall building. According to the advice provided by the seismic peer-review panel, the building reacts in two modes: under the basic seismic model the building will act under its elastic properties; under the rare seismic model it acts with an increased method of damping where temporary mechanisms allow the structure to move.

The overall form of the mega-structure also has a twist which allows it to react better in the high wind-loads of a tropical cyclone. This is aided by huge radial trusses that transfer loads from the curtain walling via belt trusses to the hybrid concrete and steel core (Jiang 2014). In terms of form, however, 30 St Mary Axe, otherwise known as 'the Gherkin', was designed by Lord Foster to incorporate six gaps in the floorplates to act as kind of giant solar chimneys which draw warm air from each floor. The warm air then spirals upwards to escape through ducts at every sixth floor which are then fire stopped (Figure 5.61).

Building materials

Glass is obviously a first choice for any commercial building although it doesn't perform well in terms of resisting the passage of heat in tropical climates. Generally the type of glass that needs to be specified is one that has a low solar radiant transmittance which is termed as the *g value*. Most manufacturers will give a value as a percentage figure of 0.00 to 1.00 for this value. The lower the solar radiant transmittance there is in a product, the lower the light transmittance. So with a glass with good protection in the tropics there will be some compromise on the amount of light that reaches the interiors.

Solar shading will be required also especially to the east and west facades. Pilkington produces a range called Suncool™ which has glass types with light and

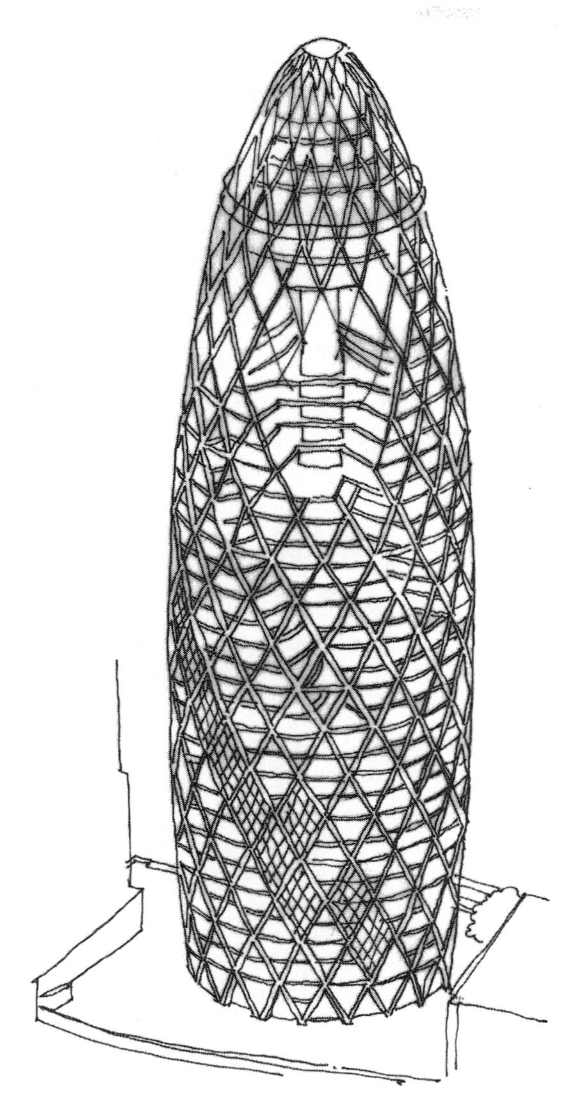

Figure 5.61 30 St Mary Axe, Sir Norman Foster.

solar transmittance of varying values. Other coatings that can be added to glasses in commercial buildings can have self-cleaning properties. These work by the chemical reaction of the coating with dust that then gets washed away by rain. Care needs to be taken in hot and dry tropical regions where rain is scarce and in these areas alternative methods for cleaning should be sought. Pilkington has a specifying online toolkit which shows the properties of glass assemblies of the specifier's choice. Figures 5.62 and 5.63 show two different options; the first is for a simple double glazed unit and the second is a twin wall situation. Twin walls benefit from a ventilated air space which can allow residual heat to escape. *G values* in these types of

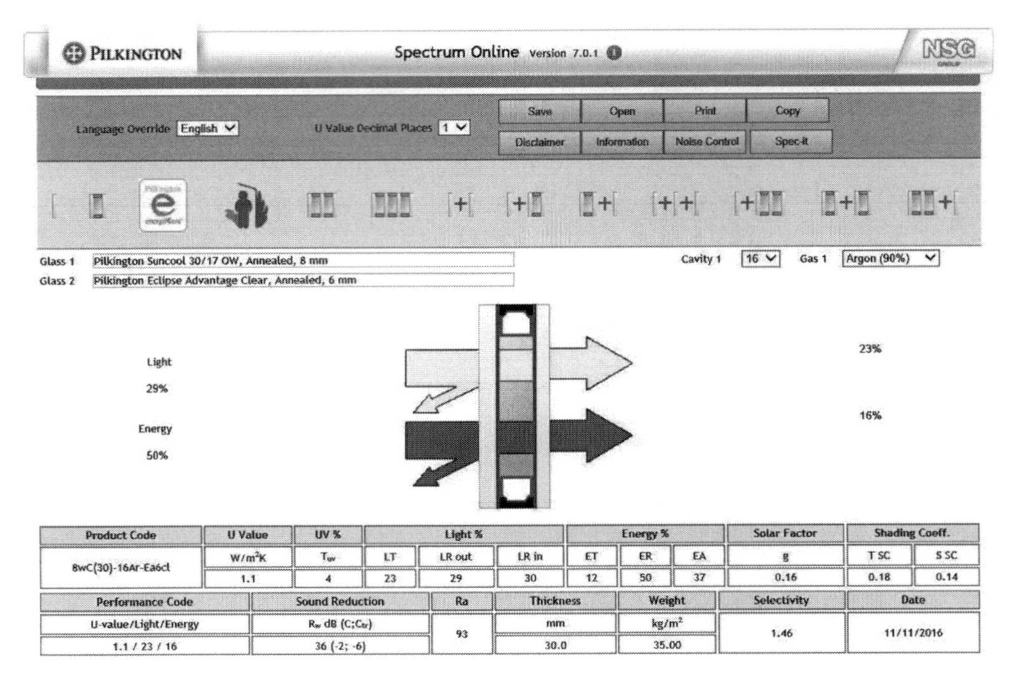

Product Code	U Value	UV %	Light %			Energy %			Solar Factor	Shading Coeff.	
8wC(30)-16Ar-Ea6cl	W/m²K	T_{uv}	LT	LR out	LR in	ET	ER	EA	g	T SC	S SC
	1.1	4	23	29	30	12	50	37	0.16	0.18	0.14

Performance Code	Sound Reduction	Ra	Thickness	Weight	Selectivity	Date
U-value/Light/Energy	R_w dB (C;C_{tr})	93	mm	kg/m²	1.46	11/11/2016
1.1 / 23 / 16	36 (-2; -6)		30.0	35.00		

Figure 5.62 Spectrum online readouts.

Source: courtesy of Pilkington.

assemblies can get as low as 0.09 but the light transmittance reduces to 8 per cent which is the *LT* (Pilkington 2016). This kind of glass absorbs more heat than standard annealed glass and will be subject to more expansion and contraction under tropical conditions. Therefore care should be taken in specifying and detailing curtain walling so that there is adequate tolerance in the frame section capping and gaskets.

More traditional materials are generally better suited to tropical climates. Vernacular architecture has been developed over centuries and inhabitants of these regions have found successful solutions by both scientific and logistical processes of elimination. Generally in hot, dry climates materials with a high mass and density need to be used that are also light in colour for reflectance of short wavelength radiation. Although these materials keep the long wavelength radiation inside the property they have larger time lags and dissipate heat internally more slowly. Stone was traditionally used in these regions which now has many similar modern equivalents such as precast concrete panels, glass reinforced cement panels, concrete block and brickwork and also many forms of industrial metal claddings. Generally metal claddings such as aluminium and steel have a lower emissivity than denser materials, a higher time lag and a higher rate of conductivity. This results in a high proportion of heat radiating through the panels. The best material to use is white painted aluminium cladding which has a reflection coefficient of 0.78 in comparison to new asphalt (black) which is 0.09. Dark roof coverings such as felt and asphalt should be avoided unless white solar reflective

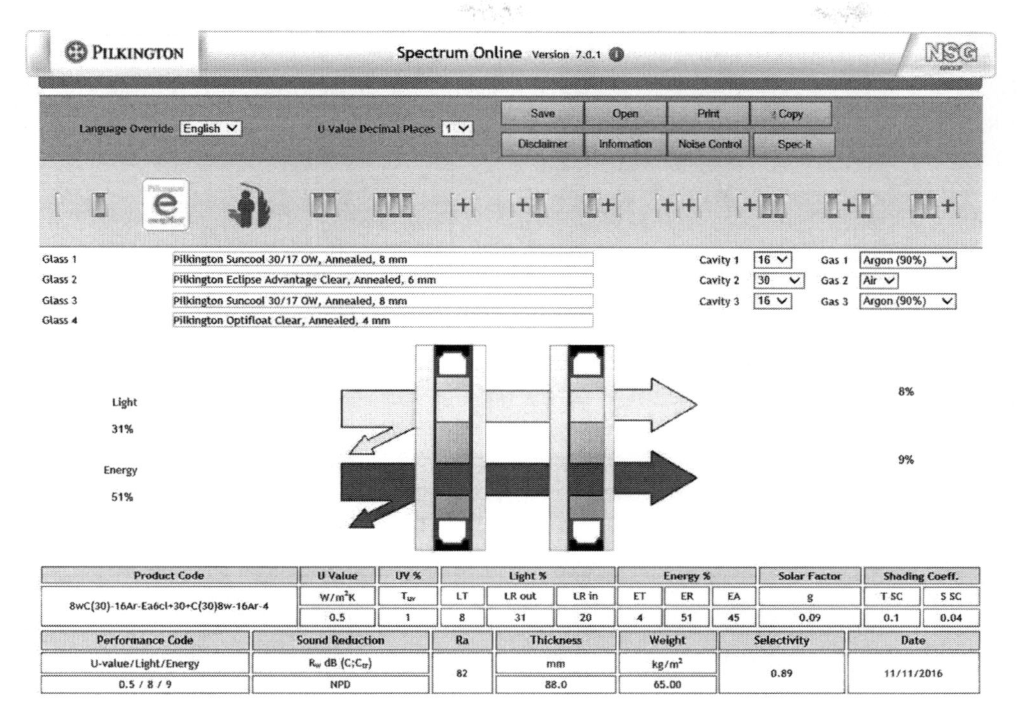

Figure 5.63 Spectrum online readouts.

Source: courtesy of Pilkington.

chippings or precast concrete paving slabs are used as a protective covering. White-washing would have the same effect but is quite high in maintenance. In the UK 18 mm thick sanded asphalt can reach temperatures up to 43°C whereas the same material with reflective white chippings has been tested at 27°C.

Roofing materials react differently when they are associated with a ventilated air space. The efficacy of these materials has been catalogued by Fry (1964) with regard to the combination of reflectance with ventilated air spaces:

- Ferrous metals and zinc corrode rapidly in hot wet tropics unless they are prominently exposed and free from the ground. Zinc is unsuitable due to corrosion unless terne coated. These characteristics are not applicable to hot dry conditions.
- Aluminium and alloys are light and transportable and suitable for all tropical climates although they can be expensive.
- Copper, brass and bronze are excellent in hot and humid conditions.
- Lead suffers from fatigue cracking due to thermal expansion and contraction.
- Concrete and other cement products have difficulty with premature hydration of cement in humid conditions and there is usually a lack of adequate water supply in hot dry areas. River water also can be acidic. The performance of

Table 5.6 Reflectance coefficients of a variety of materials

Material	*Colour, finish or condition*	*Reflection coefficient*
Slates	Silver grey	0.21
	Dark grey (smooth)	0.11
Clay tiles	Machine-made red	0.38
Concrete tiles	Uncoloured	0.35
	Brown	0.15
	Black	0.09
Galvanised iron	New	0.36
	Very dirty	0.08
	Whitewashed	0.78
Copper	Polished	0.82
	Tarnished by exposure	0.36
Lead sheeting	Old	0.21
Bituminous felt	–	0.12
	With aluminium surface	0.60
Asphalt	New	0.09
	Weathered	0.18
Mortar screed	–	0.27
Steel sheet	Vitreous enamelled/polyester powder coated white	0.57
	Green	0.24
	Dark red	0.19
Bricks	Gault – cream	0.64
	Stock – light fawn	0.44
	Stafford – blue	0.11

high alumina cement is affected by heat. Exposed cement products can blacken in hot wet tropics. The transportation of cement and steel rods is achievable to most sites. Rural Africa can have difficult road conditions. Once cured correctly and if sufficient concrete cover is provided, these products are adaptable and durable in extreme conditions.

- Steel reinforcement can encounter corrosion problems in hot wet areas due to high rainfall, atmospheric humidity, unwashed and saline aggregate, improper curing and the use of salty water.
- Concrete blocks and cement mortar renderings become brittle in hot dry climates although this is similar in temperate regions. Concrete blocks and renderings are economic and simple to manufacture without the use of too much machinery and remain uncracked in hot dry climates.
- Stone in high temperatures may crack in masonry and there is usually a low skilled local labour force which makes it difficult to detail. On the whole stones of low quality can be used with satisfactory results in all areas.
- Glass is known to deteriorate more rapidly in hot wet conditions than hot dry but only to a small extent. Water absorption with recrystallisation of alkaline constituents and growth of mould are two difficulties. Glass is not always suitable for tall buildings in zones suffering from tropical cyclones (Lee and Wills 2002).

Table 5.7 Efficacy of certain roofing materials for reflectance in a ventilated roof

Materials	*Surfaces of sheet*	*Efficacy*
Metal sheet	Black/black	0
Metal sheet	Black/aluminium paint	16
Two metal sheets with 25 mm unventilated air space	Black/black (upper) Black/black (lower)	22
Two metal sheets with 25 mm ventilated air space	Black/black (upper) Black/black (lower)	46
Metal sheet	White/aluminium paint	80
Metal sheet	Tinplate/tinplate	86
Metal sheet	White/tinplate	100
Two metal sheets with 25 mm unventilated air space	White/tinplate (upper) Tinplate/tinplate (lower)	102
Two metal sheets with 25 mm ventilated air space	White/tinplate (upper) Tinplate/tinplate (lower)	107

- Earth and stabilised earth walls are susceptible to termite depredation unless stabilised. Stabilised earth blocks that are made in hand operated compressed moulds have lower compressive strength than concrete blocks. Hand made blocks need to be kept to single storey constructions whereas machine made earthen blocks can be used in up to three storeys subject to earth quality. Certain earths are durable in hot wet areas and very durable in hot dry zones. Their use is greatly improved by modern techniques in rendering and soil stabilisation. They can be constructed on site which reduces transportation costs and skill levels of operands.
- Gypsum products in hot dry climates can sometimes be used successfully externally.
- Untreated hardboard and plywood are liable to wood-boring termite attack. Hardboard and resin-bonded plywood clear of termites are used with satisfactory results in hot wet climates but should be painted. Durability tests on various cellulesic wall boards show that under hot wet conditions, they suffer from algal and mould growth, although results are similar to that of a temperate climate.
- Timber in hot dry climates suffers from deterioration due to continual expansion and contraction. Paint therefore can suffer from sequential damage. These characteristics are small when compared to difficulties in depredations of fungi and insects in hot humid zones. Deterioration due to rotting is not as great in hot wet areas as might be expected. There is an abundance of supply in tropical regions of reasonable and excellent quality hardwoods.
- Thatch becomes deteriorated after 18 months to two years and suffers from fire risk and vermin. Thatch is simply and speedily erected and repaired, and has good ventilation qualities.
- Plastic materials do not suffer from weathering although in hot desert conditions thermo plastic materials warp and crack. Most plastics are susceptible to termite attack and must be kept free of the ground. Thermo plastics can be used for plumbing although they should not be used underground.

- Paints break down due to photo-chemical action, thermal radiation, cycles of heating and cooling, wetting and drying, moisture prevention and occasionally invasion by micro-organisms. Poor workmanship is an important factor. Storage at high temperature promotes gelation. White pigmented paints fail more rapidly than darker ones. Repainting is more frequent in hot wet (every two to three years) than hot dry (four to five years). Paint is not regarded as an adequate protection for either timber or steel in humid conditions.
- Bituminous materials tend to blister, creep and become brittle in hot conditions, unless treated with whitewash, aluminium, paint, etc. They have great properties in protecting timber against rot, fungi and insects.
- Burned clay products are not always satisfactory where there is a lack of technical skill in manufacture, which may result in under burning. They are subject to no abnormal deterioration due to tropical conditions.
- Fibreglass is ageless, unattractive to vermin, will not rot and is highly fire resistant. It can be used in loose infill through flexible mats to boards of fair rigidity. It is incapable of being used as partitions and walls in the absence of a surfacing material such as hardboard or plasterboard.

(Fry 1964; BRE 1981; Slessor and Halnan 1990)

Structural forms

The overriding characteristics of structural forms in tropical regions need to cope with heat, heavy rains, tropical storms and cyclones, and earthquakes. In three of these phenomena the plastic state of the structure needs to be coded in terms of adequate deflection of the components and connections. Traditionally in hot and dry climates, loadbearing structures have been adopted because of their inert capabilities

Figure 5.64 Mosque at Djenné, Sudan.

under extremes of hot and cold temperatures. Historically, the ziggurat and Ur in Iraq and the Egyptian pyramids are obvious massive monolithic structures that defied the structural techniques and labour skills of the age. Loadbearing structures are evident in Africa in the Adobe block idiom at Mali and Djenné, Sudan, and include intricate methods of wall/roof support and in-built decorative rodier palm (*Borassus aethiopum*) sticks, called *toron* that project 600 mm from the edifices and provide permanent access to maintain the earthen structure. Towers and pillasters act as buttresses to the massive Adobe block walls which are mortared with earth and sand and rendered with mud.

More recently parabolic arch and domed structures have been used to good effect in burnt clay masonry constructions due to the lightness and excellent span capabilities for commercial buildings. Use of local labour skills to erect the timber formwork for these buildings is generally readily available. Malaysian civic buildings are notably built from clay bricks due to the readily available raw materials in the Klang river delta. On larger span buildings reinforced concrete frames and shell structures are more common as the considerable cost of steel in developing countries renders this material unfeasible. Due to the advent of parametric computer software the traditional masonry structural forms may evolve to cope with the climatic conditions in a responsive way. These will inevitably be constructed in reinforced or precast concrete due to the unusual forms that can be analysed and more easily detailed. Examples of what may lie ahead in the future are the parabolic and folded plate structures of the 1960s in Africa, Asia and South America such as the Ibandan University College buildings, Southern Nigeria by Fry (1964), the Masjid Negara in Kuala Lumpur by Howard Ashley, Hisham Albakri and Dato Baharuddin Abu Kassim,

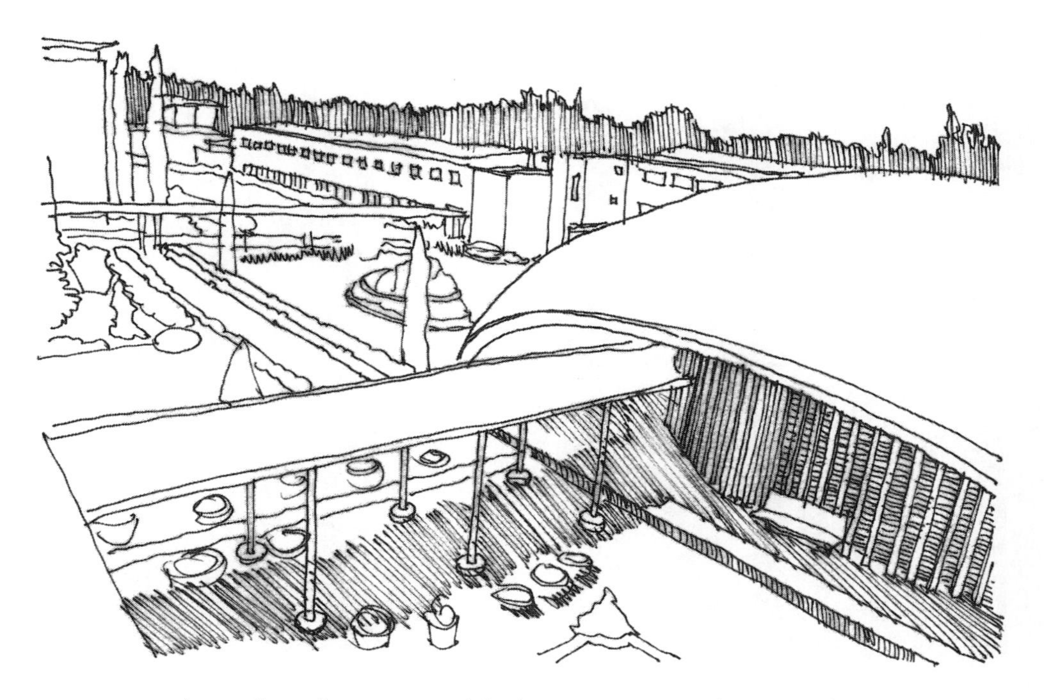

Figure 5.65 Sultan Bello Hall, University of Ibadan, Nigeria. Fry and Drew Architects.

and the Meeting Place of the Senate, Brasilia by Oscar Niemeyer respectively. These structures are synonymous with the tropics in terms of their parasol roofs.

In the Sultan Bello Hall at the University of Ibadan we see a parabolic shell roof used which is not reliant on the walls below for structure. Instead it rests on the point loads of the four corners of a parabolic dome. The walls then reach up but do not touch the roof to provide continuous ventilation at high level.

Furthermore structural frames in reinforced concrete or steel perform well in the tropics as they allow a variety of infill panel forms and free facades for different climates. Frames also allow long continuous openings at high level for ventilation. Cantilevers are used successfully to provide balconies that shade facades in hot climates with infill panels of burnt clay bricks for mass. Eucalyptus is generally used in an unsawn state for temporary structural support and scaffolding.

Glulam can be used in the tropics due to its excellent plastic state properties but should only be used for roof structures and above ground in order to resist termite attack.

In larger commercial buildings we see hybrid steel and concrete cores which act well in the event of an earthquake or tropical cyclone. The concrete retains the characteristics of a dense material whereas the steel columns allow the overall structure to flex. The Shanghai Tower is a good example of the design of a megastructure which falls outside the standard requirements of local building codes.

Figure 5.66 Example of construction project with cantilevered balconies, Northern Uganda.

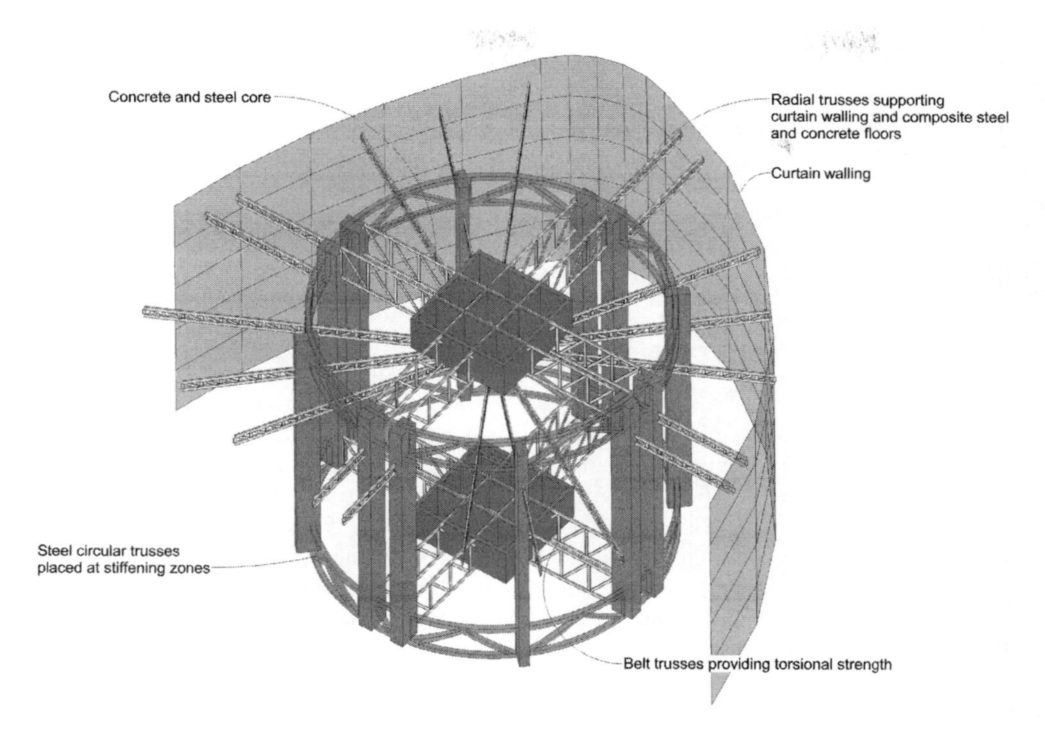

Figure 5.67 Floor section of the Shanghai Tower by Gensler (original).

According to the current Chinese code, the following items are identified as code-exceeding: (1) the total structural height of 580 m exceeds the limit of 190 m stipulated for steel-concrete composite frame-core tube structures; (2) the elevation irregularity including several stiffened and transfer stories exceeds the code limit; (3) the overhanging length of 14 m of the radial truss in the stiffened stories exceeds the limit of 4 m and 10 per cent of the total span.

It is for these reasons that the building was designed not under code but by a seismic peer review panel. The structure was designed with eight megacolumns which vary from 5.3 × 3.7 m in cross section at their base to 2.4 × 1.9 m at the top, rising to the top of zone 8 while the four corner columns stop at zone 5. The core is made from hybrid steel and concrete shear walls and rise the full height of the building and take up the lateral forces acting upon the building. Eight two storey high steel belt trusses interconnect with the core and are positioned at regular intervals throughout the structure and are termed as stiffening storeys. In each zone two parallel circular trusses are set to enhance the torsional resistance. Twenty-one one storey high radial trusses are installed in the stiffened storeys to transmit the vertical loads from outer curtain walls and floors to belt trusses.

In hot and dry climates it is imperative that dust storms and desert winds are dealt with in designing structures. In Le Corbusier's work for a new urban district in Algiers in the 1930s, he designed large concrete structures to protect the smaller buildings from desert climatic conditions. The structures included an enormous sea wall housing project flanked by a motorway that was to house 220,000 people. In poorer regions it is important in earthquake design that concrete and more rarely steel structures are designed to prevent disproportional collapse. We can see the difference in structural strategies by comparing the loss of life analysis between the Los Angeles earthquake of 1994, which reached 6.7 on the Richter scale causing 57 deaths, and the Haiti earthquake in 2010 that reached 7.0 on the Richter scale and was attributed to 230,000 deaths. This contrast was caused by the fact of simple stiffening to the concrete frames in the structures of Haiti whereas Los Angeles had buildings that were fully earthquake proof. Simple additions to frames such as haunch details at column and beam connections can prevent floors from collapsing on each other when the building is in vibration mode.

Despite the high cost of steel, the material is still used widely in the tropics although it is important to be aware that its use depends on the availability of a comprehensive selection of hot rolled section sizes; for a variance in structures, all section sizes are not always available without extortionate transportation costs, especially in equatorial Africa. A good example of the adaptability of designers in areas where this situation occurs is the Basilica of the Ugandan Martyrs in Namugongo, Uganda, which was consecrated in 1993. The building is a conical steel structure that is similar in form to the Cathedral Church of Christ the King in Liverpool. Due to the expense of casting or hot rolling tapering raking columns in rural Uganda, the architect decided upon the idea of using hot rolled steel plates, spiralling to form the components. Figure 5.68 shows this configuration together with steel ring beam and wind bracing.

Figure 5.68 Basilica of the Ugandan Martyrs, Namugongo – column details.

In Figure 5.69 the spirals of the edges of the plates can be seen on the columns which were rolled into position offsite and butt welded. The conical form of the structure allows heat to rise and be ventilated at high level.

Building fabric and enclosure

In the tropics the use of external screens is prevalent. Filigree patterned concrete or stone fabrications are common. These allow a secondary external screen from which inner windows can obtain borrowed light. Secondary external screens can also provide much needed shaded outdoor space where people can congregate and interact. Usually walls of heavy mass are used either perforated or unperforated. The ingress of large amounts of light is culturally not seen as a necessity in the tropics; in fact it infrequently becomes a phenomena that is guarded against as much as snow and ice in temperate climates. Figure 5.71 shows an example of this in an arts centre in Kampala, Uganda. Figure 5.70 shows examples of filigree screening in the Putra Mosque in Putrajaya, Malaysia.

Figure 5.69 Basilica of the Ugandan Martyrs, Namugongo.

Figure 5.70 Putra Mosque, Putrajaya, Malaysia.

In commercial buildings, these can also take the form of brise soleil that create both vertical and horizontal shading techniques. The benefit of vertical solar shades is that they can shade against solar radiation but also funnel wind onto the facades to either cool the fabric or allow improved ventilation to openable windows and shutters. With the use of parametric software and building monitoring systems (BMS), these brise soleil can be analysed and animated to follow the altitude (horizontal type) and azimuth (vertical type) of the sun.

Cantilevered balconies are inherently good shading devices and Prianto and Depecker (2002) produced research into their effectiveness in an urban situation in Cayenne, Guyana and Semarang, Indonesia.

Figure 5.71 Ndere Arts Centre, Kampala, FBW Architects.

Increasingly twin wall glass facades are being adopted in the tropics and can be seen in such buildings as the Shanghai Tower and 30 St Mary Axe, London. Although the latter is in a temperate region it can be seen from Figure 5.72 that the twin wall construction can allow hot air to travel vertically in the ventilated airspace and escape at high level.

The area of the tropics where building fabric has evolved due to the culture of trade is in Malaysia where the *Baba Nonya* developed large areas of commercial properties; namely Penang, Melaka, Singapore and Kuala Lumpur. The *Baba Nonya* were the Chinese who had travelled from southern China to dwell in the regions of South-East Asia around the Straits of Melaka. The commercial properties, named the *Shophouse*, became a building typology that was designed and adapted successfully for the hot and humid environment of this region. The exterior was shaded by the five foot street which was introduced by Sir Francis Light as a planning condition. The overhanging archways were linked and also provided a continuous covered street where people could shelter from the tropical heat and rain storms. The Chinese introduced a variety of ventilation openings in the external enclosure which relate to Taoist and Buddhist principles and provided a variance in architectural detail. A courtyard created an indispensable internal ventilation shaft that was enclosed with a variety of timber jalousie screens. Stone ground floors introduced a

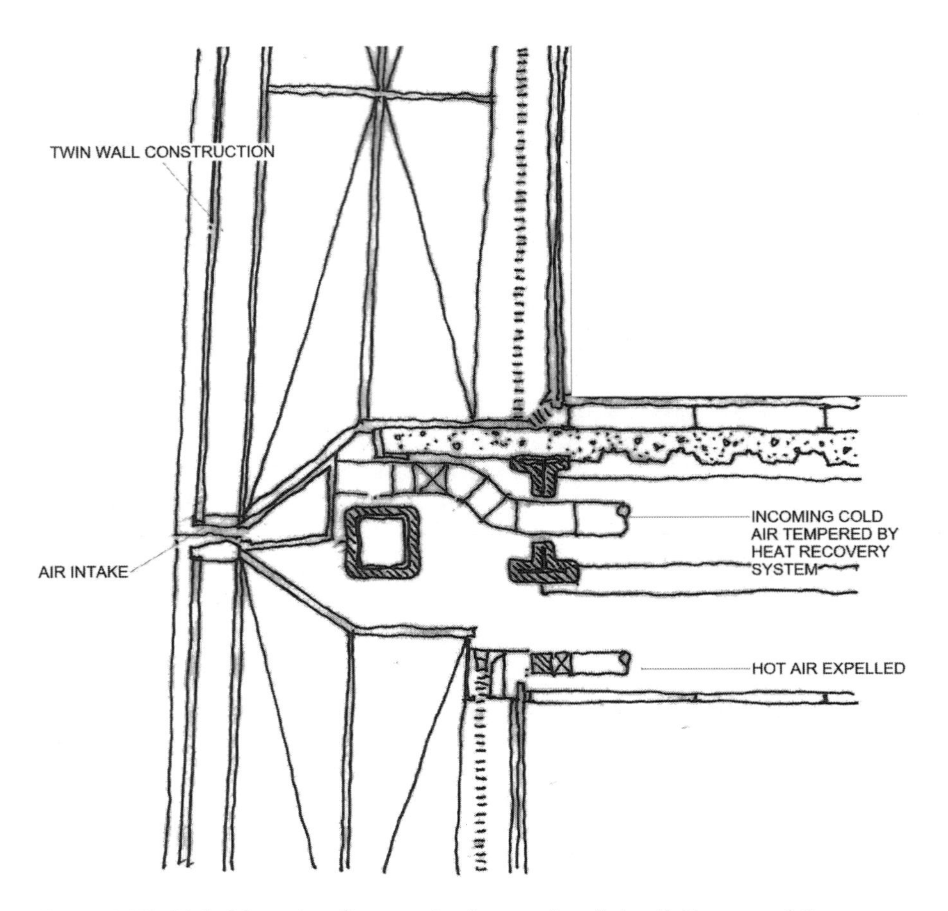

Figure 5.72 30 St Mary Axe floor section/external wall detail. Foster and Partners.

cooling fabric to the whole unit: granite that was transported from China as ballast for the returning ships trading in spices and textiles from the Straits. This showed Chinese ingenuity in using all available materials for construction.

Environmental services

Most of the discourse in this chapter has involved the passive use of tropical architecture, whereas it is a fact that predominantly buildings will overheat in these regions. It is common in Africa to long for a nice car, plasma TV screen and air-conditioning unit before owning a nice house. There are two main issues with air-conditioning in that (1) a system uses huge amounts of electricity to operate and (2) the systems are designed based on the presumption that the building will be sealed from ventilation leaks. If air-conditioning is being used and there is passive ventilation, it is very wasteful of energy as some cold air will be escaping through the external envelope. The dilemma is evident in Hong Kong where high rise developments have individual air-conditioning units to each property, even each room. To have individual units cooling single rooms is inherently inefficient and there should be some design effort attributed to whole building systems in the commercial context. In order to design an air-conditioning system efficiently in the current climate change agenda, thought should be given to using a combined heat and power system to run a whole building cooling system. Also in order to save energy, a zonal system should be designed to operate in only occupied areas; to achieve this a BMS should control this.

Figure 5.73 Air-conditioning in Hong Kong.

To design a system accurately, the design engineer needs to know the following information:

a Records averaged for several years of the mean monthly maximum and minimum dry-bulb temperatures.
b Similar records showing the absolute maximum and minimum dry-bulb temperatures recorded for each month.
c The relative humidity or wet-bulb temperature which obtains at times of the maximum and minimum dry-bulb temperatures described in (a) above.
d Some indication of the diurnal variation of wet- and dry-bulb temperatures during typical hot and cold days. Occasionally hourly records are available for typical days.
e General description of climate giving details of monsoon conditions, prevailing wind, rainfall, existence of sandstorms, etc.

In the humid tropics where design conditions may be from 32°C to 38°C dry-bulb with about 27°C wet-bulb, an effective temperature of 23°C is recommended for maximum thermal comfort in continuously occupied spaces in commercial buildings, 24°C in the humid tropics and 25°C in the arid tropics (Fry 1964). It is ideal to include relative humidity values with a fully responsive HVAC system and in the humid tropics this value should be 55 per cent, whereas in the arid zones a lower RH of only 45 per cent is achievable. The cost of adding this technology to HVAC systems is considerable in what are often poorer countries with lower budgets.

Bunning and Crawford (2016) advocate the use of directionally selective solar shading devices linked to a BMS controlled HVAC system. They catalogue the energy use comparatives for using internally as opposed to externally placed venetian blinds in terms of the life cycle energy use in Terajoules for Melbourne and Brisbane over 25 years. It is clear from his analysis that externally placed solar shading has a considerable saving due to the reflection of solar radiation.

Artificial lighting is another dilemma in buildings that have a low light transmittance glazing system. The lighting systems should be fitted to a PIR BMS to only be used in occupied areas.

Another element of environmental services that needs to be considered in tropical regions is adequate sanitation systems and access to a clean water supply.

In sub-Saharan Africa clean water is an issue. In urban areas wells and water towers are commonplace as there is not an adequate supply of clean water for washing and boiling for potable purposes. In Kampala as an example most of the city wells have a charge for personal collection of water from communal taps. In rural Africa community wells are mainly used and it is a luxury if there is a mechanically drilled well. Mechanically drilled wells reach to about 100 m deep and have a better quality of water from deep underground water courses. More common are hand-drilled wells which are very difficult to complete to depths of approximately 20 m. Figure 5.75 shows a hand-held auger-well being dug in Gulu, Northern Uganda. Figure 5.74 shows a mechanically drilled well that is 80 m deep and has a water yield of 0.93 m^3.

Sanitation is another key issue in African cities and territories that have only pit latrines for the removal of waste. Only 10 per cent of the two million people that live in Kampala have access to latrines that run to a government sewerage system.

Figure 5.74 Machine drilled well, Gulu, Uganda.

Figure 5.75 Hand drilling a well, Gulu, Uganda.

The result of this is that cholera and diarrheal diseases are the biggest cause of premature death in this region. Pit latrines generally run liquid waste into soakaways and as this has contact with faeces, it pollutes the water courses from which the wells are sourced. New technologies such as composting units have been promoted by the Bill Gates Foundation for Africa and are the future, as to renew all African cities with a traditional sewerage system would be a gargantuan effort in terms of labour and cost. In comparison to composting units, traditional systems are energy inefficient as it takes considerable amounts of fossil fuels to purify waste. The latest techniques in these units use deflectors that separate urine from faeces and allow the solid matter to compost (Hoglund *et al.* 1998). Figure 5.77 shows a cross section through a Urine Separation unit.

Sustainable construction materials in the tropical regions

The most sustainable construction materials in tropical regions are the ones that are readily available in the locality of the site. The obvious choice is tropical hardwoods which are in abundance, although they hold some climatic reservations into their renewable use due to the fact that they are taken from rainforests that are crucial to the balance of the world's ecosystems. There are also many endangered species of timber in the world. Figure 5.78 illustrates this based on a sample of two countries per continent. Senior management commitment is essential for the successful implementation of a sustainable timber policy. A Timber Procurement Policy must be produced that is endorsed by the Board and the Managing Director or in particular

Figure 5.76 Perforated well pipe, Uganda.

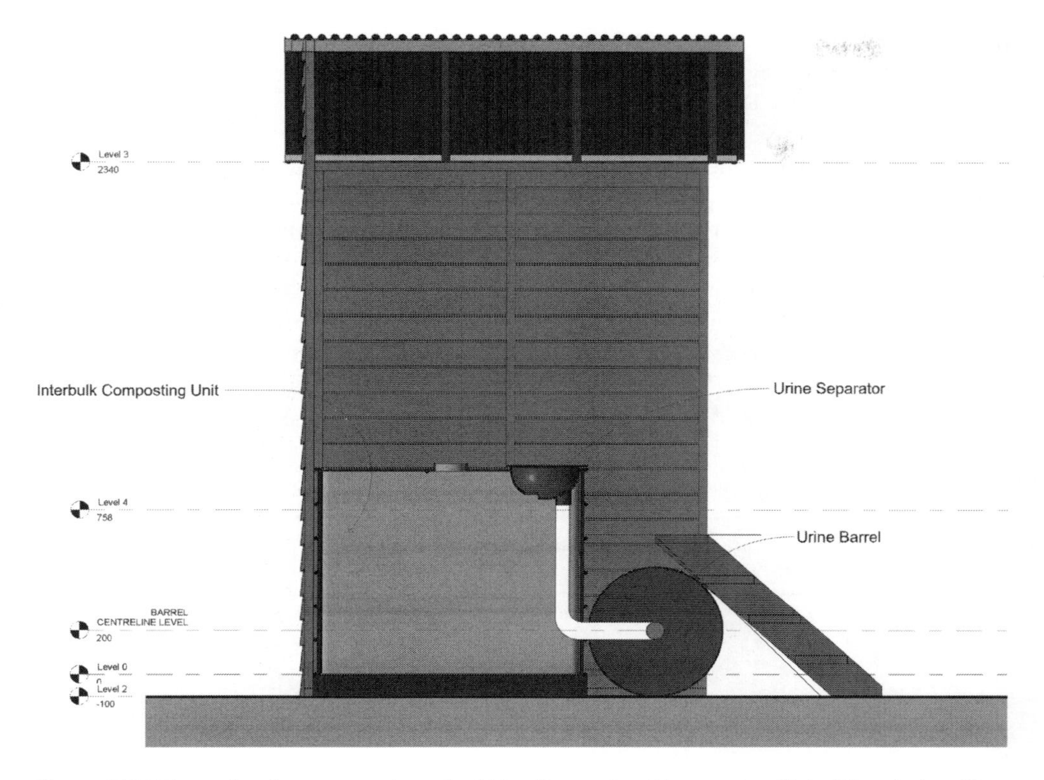

Figure 5.77 Example of a cross section of a Urine Separating Composting Unit. Thunderbox2go.

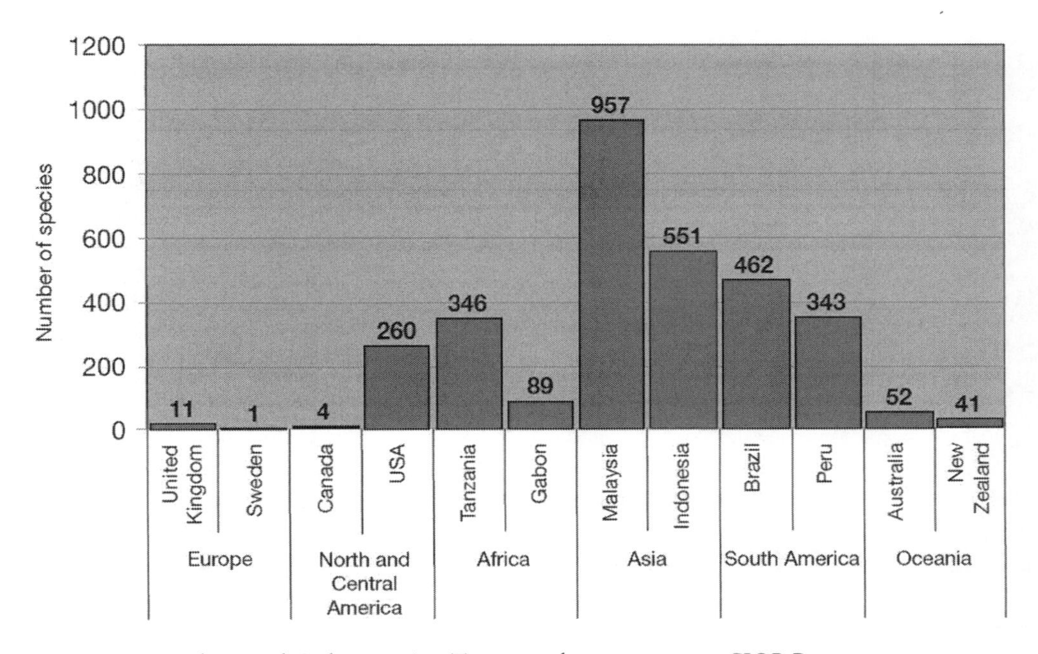

Figure 5.78 Endangered timber species. Two samples per country. CIOB Paper.

the Chief Executive Officer. The aim of the policy is to ensure that all timber procured by the company is sustainable. One solution can be to insist that all timber must be from a reclaimed, recycled or a fully certified source (Forest Stewardship Council or Pan European Forest Certification – FSC or PEFC).

At the global level, the proportion of the world's wood products sourced from illegal logging is estimated at 20 per cent. In Asia, the rainforest is being cleared at a rate of five million hectares per year, and once an intact rainforest has been clear-fell-logged, it will not recover for many hundreds of years (Cairns 2011). Greenpeace also gives advice on what timbers to avoid and what to specify.

Holistically, there should be reference to the import of timber materials from tropical countries to regions around the world as the depletion of the rainforests affects the microclimatic conditions in continents like Africa, South America and Asia. Tables 5.8, 5.9 and 5.10 show the quantities of illegal timber products from

Table 5.8 Imports of illegal timber from Africa into the EU

Country	*Estimated illegal or suspicious quantity of wood*	*Imported products*	*Customers in the EU*
Cameroon	645,000 m^3 (RWE) illegal	81% sawnwood 10% roundwood 6% veneer	Italy (24%) Spain (19%) Netherlands (16%) France (10%) Belgium (9%) Germany (5%)
Gabon	590,000 m^3 (RWE) illegal	45% roundwood 28% veneer 18% sawnwood	France (52%) Italy (24%) Germany (5%)
Cote d'Ivoire	530,000 m^3 (RWE) suspicious	56% sawnwood 25% veneer	Italy (33%) Spain (23%) France (12%) Germany (8%)
Nigeria	380,000 m^3 (RWE) illegal	90% charcoal 4% sawnwood	Belgium (34%) Netherlands (33%) Germany (13%) Italy (11%)
Democratic Republic of Congo	235,000 m^3 (RWE) suspicious	61% roundwood 33% sawnwood	Italy (25%) France (21%) Netherlands (20%) Portugal (18%) Belgium (14%)
Congo	180,000 m^3 (RWE) suspicious	68% roundwood 27% sawnwood	Italy (36%) France (26%) Spain (11%) Portugal (10%)
Ghana	170,000 m^3 (RWE) illegal	53% sawnwood 19% veneer	Italy (27%) Germany (15%) France (15%) Netherlands (11%) Belgium (10%)

Source: WWF (2008).

Table 5.9 Imports of illegal timber from South America into the EU

Country	*Estimated illegal or suspicious quantity of wood*	*Imported products*	*Customers in the EU*
Brazil	2.8 million m^3 (RWE)	37% plywood 20% sawnwood 19% furniture or other finished wood products 11% parquet	Great Britain (19%) France (14%) Belgium (13%) Italy (12%) Netherlands (11%) Spain (11%)
Bolivia	46,000 m^3(RWE)	37% sawnwood 28% parquet 27% furniture and other finished wood products	Netherlands (52%) Italy (14%) Spain (12%) France (9%)
Ecuador	21,000 m^3 (RWE)	74% furniture and other finished wood products 20% sawnwood	Denmark (33%) Spain (22%) Germany (15%) France (13%)
Honduras	15,000 m^3 (RWE)	61% sawnwood 27% parquet	Germany (27%) Spain (23%) Great Britain (18%) France (12%)

Source: WWF (2008).

these continents. The scale of the problem reaches from solid hardwoods to wood pulp and plywood boards. Certificates should be obtained that the respective batches of products are controlled by FSC or PEFC certification. Tropical hardwoods from the South-East Asian rainforest have the most species in danger. Timbers without FSC certification include:

- meranti – used for all mouldings, dowels and architraves;
- merbau – used for skirting and joinery;
- ramin – mostly used for picture frames and fine joinery;
- Pacific maple – used for all mouldings, dowels and architraves;
- Philippine mahogany, Calantas Pretend red cedar – used for fireplaces, stairs and furniture;
- Keruing, naytoh, narra, kapur – used for joinery;
- Teak – used for outdoor furniture, carved beams and cabinet work;
- Jelutong – used for joinery, carved work and toys;
- Motoa, merawan, batu – used for house posts.

(Cairns 2011)

The embodied energy of construction materials has a high impact on their sustainability. The embodied energy values relate to the fossil fuel depletion in extracting the raw materials, making the product and the transportation and processing of the materials in question. These values vary; concrete has a value of approximately two gigajoules per ton whereas aluminium has a value of hundreds of gigajoules per tonne. Recycled materials have much lower embodied energy levels. Recycled aluminium has less than 10 percent of the embodied energy of aluminium manufactured from raw materials. Table 5.11 shows the embodied energy values for the most common materials.

The use of concrete is a vital component in the modern construction process as the material is durable and has good longevity. The emission of greenhouse gases in its production is a major problem as harmful gases are emitted in burning the raw materials to make cement and also in the curing process. Use of cement that contains a proportion of agents such as fly ash or ground blast furnace slag reduces the emission of greenhouse gases and helps deal with industrial waste. The use of these ecologically sourced agents also raises the cost of the product. The production of steel has a high embodied energy due to the mining process of the raw materials and the refining process. However, it is a very durable material and can increase the life of a building. Using steel sparingly and using recycled sections is the key to responsible specifying.

Table 5.10 Imports of illegal timber from South East Asia into the EU

Country	*Estimated illegal quantity of wood*	*Imported products*	*Customers in the EU*
Indonesia	4.2 million m^3 (RWE)	38% furniture and other finished wood products 17% pulp 16% parquet 10% plywood	Netherlands (20%) Belgium (15%) Italy (14%) Germany (13%) France (13%) Great Britain (12%)
China	3.7 million m^3 (RWE)	52% furniture and other finished wood products 19% plywood 13% paper	Great Britain (30%) Germany (14%) Spain (8%) France (8%) Netherlands (7%) Italy (7%)
Malaysia	280,000 m^3 (RWE)	33% furniture and other finished wood products 30% sawnwood 21% plywood 6% parquet 4% charcoal	Great Britain (37%) Netherlands (20%) Belgium (13%) Germany (7%) France (5%) Italy (5%)
Vietnam	250,000 m^3 (RWE)	89% furniture 9% finished processed wood products	Great Britain (22%) France (18%) Germany (17%)
Thailand	250,000 m^3 (RWE)	30% furniture 30% finished wood products 21% paper	Great Britain (26%) Italy (13%) Netherlands (12%) Belgium (12%) Germany (11%)

Source: WWF (2008).

Table 5.11 Table of embodied energy of certain material

Material	*MJ/kg*
Kiln dried sawn softwood	3.4
Kiln dried sawn hardwood	2.0
Air dried sawn hardwood	0.5
Hardboard	24.2
Particleboard	8.0
MDF	11.3
Plywood	10.4
Glue-laminated timber	11.0
Laminated veneer lumber.	11.0
Plastics – general	90.0
PVC	80.0
Synthetic rubber	110.0
Acrylic paint	61.5
Stabilised earth	0.7
Imported dimensional granite	13.9
Local dimensional granite	5.9
Gypsum plaster	2.9
Plasterboard	4.4
Fibre cement	4.8
Cement	5.6
In situ concrete	1.9
Precast steam-cured concrete	2.0
Precast tilt-up concrete	1.9
Clay bricks	2.5
Concrete blocks	1.5
AAC	3.6
Glass	12.7
Aluminium	170.0
Copper	100.0
Galvanised steel	38.0

Source: Cairns (2011).

References

Agra de Lemos Martins, T., Adolphe, L., Gonçalves Bastos, L. E. and Agra de Lemos Martins, M. (2016). 'Sensitivity analysis of urban morphology factors regarding solar energy potential of buildings in a Brazilian tropical context'. *Solar Energy* 137: 11–24.

Aia, Z., Mak, C. M., Niu, J. L. *et al.* (2011). 'Effect of balconies on thermal comfort in wind-induced, naturally ventilated low-rise buildings'. *Building Services Engineering Research Technology* 32(3): 277–292.

Bagnold, R. A. (1941). *The Physics of Blown Sand and Desert Dunes*. London, Methuen.

Batty, M. (2008). 'The size, scale, and shape of cities'. *Science* 319: 769–771.

Bose, S. S. S. (2015). 'Top floors of low-rise modern residences in Kolkata: preliminary exploration towards a sustainable solution'. *Current Science* 109(9): 1581–1589.

BRE (1981). 'The termite resistance of board materials'. *Information Paper*. B. Press. Watford, UK, Building Research Establishment.

Breheny, M. (1992). *Sustainable Development and Urban Form*. London, Pion.

Brookes, A. J. and Grech, C. (1990). *The Building Envelope*. London, Butterworth Architecture, 61–63.

Bunning, M. E. and Crawford, R. H. (2016). 'Directionally selective shading control in maritime sub-tropical and temperate climates: life cycle energy implications for office buildings'. *Building and Environment* 104: 275–285.

CSIRO (1968). 'Hot weather concrete'. *Tropical Building Research Notes*, Commonwealth Scientific and Industrial Research Unit, 30.

Cairns (2011). *Sustainable Tropical Building Design – Guidelines for Commercial Buildings.* Cairns Regional Council, Australia.

Droege, P. (2007). *The Renewable City: A Comprehensive Guide to an Urban Revolution.* Chichester, UK, Wiley.

Drysdale, J. W. (1959). *Designing Houses for Australian Climates.* Commonwealth Experimental Building Station, Sydney.

Fry, M. D. J. (1964). *Tropical Architecture in the Dry and Humid Zones.* London, BT Batsford Limited.

Garde-Bentaleb, F., Miranville, F., Boyer, H. and Depeker, P. (2002). 'Bringing scientific knowledge from research to the professional fields: the case of the thermal and airflow design of buildings in tropical climates'. *Energy and Buildings* 34: 511–521.

Grosso, M. (1998). 'Urban form and renewable energy potential'. *Renewable Energy* 15: 331–336.

Hoglund, C. S. T., Jonsson, H. and Sundin, A. (1998). 'Evaluation of faecal contamination and microbial die-off in urine separating sewage systems'. *Water Science & Technology* 38(6): 17–25.

Jiang, H. J., Lu, X. L., Liu X. J. and He, L. S. (2014). 'Performance-based seismic design principles and structural analysis of Shanghai Tower'. *Advances in Structural Engineering* 17(4).

Kéré, D. F. (2013). 'Gallery: a school and a clinic, built from compressed clay'. Tedtalk's Soup.

Koeningsberger, O. and Lynn, R. (1965). 'Roofs in the warm humid tropics'. *Paper No. 1.* London.

Lacerda, W. A. (2007). 'Landslide initiation in saprolite and colluvium in southern Brazil: field and laboratory observations'. *Geomorphology* 87: 104–119.

Lee, B. E. and Wills, J. (2002). 'Vulnerability of fully glazed high-rise buildings in tropical cyclones'. *Journal of Architectural Engineering* 8(2).

Lotz, F. J. and Richards, S. J. (1964). 'The influence of ceiling insulation on indoor thermal conditions in dwellings of heavy-weight construction under South African conditions'. *Research Report no. 214.* Pretoria.

Marshak, S. (2008). *Earth: Portrait of a Planet.* New York, W.W. Norton and Company Inc.

Martins, T. A. L., Adolphe, L. and Bastos, L. (2014a). 'DOE sensitivity analysis of urban morphology factors regarding solar irradiation on buildings envelope in the Brazilian tropical context'. *Passive and Low Energy Architecture.*

Martins, T. A. L., Adolphe, L. and Bastos, L. (2014b). 'From solar constraints to urban design opportunities: optimization of built form typologies in a Brazilian tropical city'. *Energy Build* 76: 43–56.

McCabe, L. C. (1961). *The Identification of Air Pollution Problem.* UN Organisation. Geneva, World Health Organisation, note 46: 39.

Nicholls, R. (2008). *The Green Building Bible Volume 2.* Llandysul, Green Building Press.

Oliver, P. (1971). *Shelter in Africa.* London, Barrie and Jenkins, 143–152.

Owens, S. E. (1986). *Energy, Planning and Urban Form.* London, Pion.

Pilkington (2016). 'Spectrum'. *Spectrum Online.* Retrieved 11 November 2016 from http://spectrum.pilkington.com/Main.aspx.

Prianto, E. and Depecker, P. (2002). 'Characteristic of airflow as the effect of balcony, opening design and internal division on indoor velocity. A case study of traditional dwelling in urban living quarter in tropical humid region'. *Energy and Buildings* 34: 401–409.

Ratti, C., Raydan, D. and Steemers, K. (2003). 'Building form and environmental performance: archetypes, analysis and arid climate'. *Energy Build* 35: 49–59.

Riley, M. and Cotgrave, A. (2008). *Construction Technology 1: House Construction*. Basingstoke, UK, Palgrave Macmillan, 92–144.

Roux, A. J. A. (1948). *Heat Interchange between a Roof and its Surroundings*. L. National Botanical Research Institute, India. Pretoria, Council for Scientific and Industrial Research.

Roux, A. J. A. and van Straaten, J. F. (1951). *Some Practical Aspects of the Thermal and Ventilation Condition in Dwellings*. June. N. B. R. Institute. Pretoria.

Saini, B. S. and Borrack, G. C. (1967). 'Control of wind-blown sand and dust by building design and town planning in the arid-zone'. *Proceedings of the Third Australian Building Research Congress*.

Sanaieian, H., Tenpierik, M., Linden, K. V. D., Seraj, F. M. and Shemrani, S. M. M. (2014). 'Review of the impact of urban block form on thermal performance, solar access and ventilation'. *Renewable & Sustainable Energy Reviews* 38(1): 551–560.

Serralde, J., James, D., Wiesmann, D. and Steemers, K. (2015). 'Solar energy and urban morphology: scenarios for increasing the renewable energy potential of neighbourhoods in London'. *Renewable Energy* 73: 10–17.

Sing Saini, B. (1980). *Building in Hot Dry Climates*. London, John Wiley and Sons Ltd, 36–68.

Slessor, C. and Halnan, M. (1990). 'Fruits of the forest'. *Architect's Journal*, 8 August, 45–48.

Snell, M. B. (2005). 'Better homes and garbage'. *Sierra Club* 90: 28–110.

Sobel, A. H. (2012). 'Tropical weather'. Retrieved 30 October 2016 from www.nature.com/scitable/knowledge/library/tropical-weather-84224797.

Steemers, K. (2003). 'Energy and the city: density, buildings and transport'. *Energy and Buildings* 1(35): 3–14.

van Straaten, J. F. (1961). 'Some thermal aspects of curtain wall construction'. *South African Architectural Record* 46(3).

WHO (2016). WHO's Urban Ambient Air Pollution database – Update 2016. S. a. E. D. o. H. D. Public Health. 1211 Geneva 27, Switzerland.

Williams, K., Burton, E. and Jenks, M. (2000). *Achieving Sustainable Urban Form*. London, E&FN Spon.

WWF – World Wide Fund For Nature (2008). 'Illegal wood for the European market. An analysis of the EU import and export of illegal wood and related products'. July. WWF.

Yong Kwang Tan, A. and Nyuk, H. W. (2013). 'Parameterization studies of solar chimneys in the tropics'. *Energies* 6: 145–163.

6 Building pathology, maintenance and refurbishment

Zahiruddin Fitri Abu Hassan, Azlan Shah Ali, Shirley Jin Lin Chua and Mohd Rizal Baharum

After studying this chapter you should be able to:

- appreciate the tropical context of construction defects;
- understand the unique family of defects that afflict buildings specific to tropical regions;
- relate the visible aspect of the defect to the defect mechanism;
- discuss the link between building physics and how defects occur;
- recognise the importance of sympathetic repair especially on historical structures;
- reflect on the changing trend and growth of refurbishment work in tropical climates;
- identify the refurbishment options, complexities and uncertainties in tropical climates;
- understand the performance of refurbishment work in tropical climates and;
- appreciate the potential of sustainable refurbishment in tropical climates.

Introduction

Tropical and equatorial construction is subject to its own unique set of defect characteristics. This is because of the intensity of the sun, which sits at a higher azimuth from the horizon and is constant throughout the year, influences the weather and humidity of this area. The tropical and equatorial belt has high humidity due to the downdraft of the jet stream pushing warm and humid air downwards. This climatic feature is probably one of the main distinguishing characteristics of the tropical countries and it is significant in the manner in which building defects occur and propagate. In this context humidity is the main factor in the tropical regions that makes them different from the climate in other parts of the world. As such it is a factor that forces distinction in the nature and extent of defects in building structure, fabric and components.

For example, unlike temperate regions, decay mechanisms influenced by water and water vapour work constantly throughout the year with minimal variation. In addition the exposure to consistently elevated temperature works to accelerate chemical reactions and the extent of degradation of materials and components. The tropical regions are also distinguished, in part, by high levels of rainfall and in some instances monsoon climates. This also impacts upon the type of defects experienced and the implications of failure of building components and finishes. A minor defect in the external building enclosure that might result in a modest leakage in a temperate climate might be the cause of catastrophic water ingress in a monsoon climate.

As a result, local (especially vernacular) building practice has responded and has been good at playing to the nuances of the climate, with specific adaptation to deal with the threat of defects. For example, in possible flood zones, it is normal practice in vernacular construction for dwellings to be built on elevated piers, depending on the height of the flood water history in the local living memory.

For building materials, a good example is the process of painting and protecting timber structures with a mixture of diesel oil and used-up lubricant as a very low tech, local response to the risk of decay and insect infestation. This process serves both as a deterrent to termites and other insects and repels rainwater from affecting the moisture of the timber. There are many such examples across the tropical regions that are manifested in traditional, vernacular construction practice.

However, as buildings and construction practice have homogenised internationally to reflect a 'global' construction form for modern commercial buildings, the approach to dealing with pathology and defects has changed. Construction processes and materials selection have shifted to less climatically specific forms using steel, concrete, bricks, mortar and plaster. As such the typography of defects in tropical areas has changed with these materials. The construction of buildings in these areas does take this into account to some extent. The nuances of the tropical climate mean that the defects that occur in these modern buildings are different to the types of defect that befall similar buildings in other climates. The truth is that the local experience has to be given renewed prominence to deal with materials and processes that may not be entirely adaptable to the local climate.

Overview

Along the equatorial line, many parts of the world are developing countries which have been, at some point in their history, influenced by colonialisation of foreign powers. Apart from administrating these countries, the colonial powers brought with them their building materials and introduced new construction methods and styles. However, evidence suggests that they were largely successful in adapting these imported materials to the local climate through detail changes to take into account, for example, more rainfall/higher humidity/effects of solar radiation that are typical of the local climate. As an example the manufacture of UK style bricks using local clays was widespread and successful.

Fast forward into the future, and local adaptation seems to give way to the ad hoc construction trends imported from more temperate regions that characterise modern architecture such as designs from Scandinavia and Europe that are currently in fashion.

The purpose of this chapter is to introduce the defects that are typical of traditional and modern buildings across tropical climates caused by the multitude of factors alluded to above. Based on that premise, the chapter focuses on looking at the context and the causes of defects in the tropical setting. These will be unique to these regions due to the effects of climate on the existing modern(ish) buildings predominantly consisting of masonry construction. Understandably the detailing of buildings specific to the climate does play its part in avoiding problems but one has to address the consequence of introducing brick, mortar and plaster construction to the localities.

In saying that, it does not mean that local or vernacular construction has not had problems with defects but the local experience has had the advantage of several centuries of adaptation and acclimatisation compared to 50–100 years of the new materials and construction. We are now compounded with the ever faster shift in climate change that is currently posited.

Key elements in this chapter include discussions on the interplay between building physics and causes of defect and its mechanism. The chapter also considers the importance of matching the correct repair methodology to suit the climate conditions especially when dealing with older prewar structures that use different building materials and processes to the modern buildings.

Common building pathology

Water and moisture

Atmospheric moisture content in tropical and equatorial climates averages at around 80 per cent and dew point is at 20°C average. At this level of relative humidity, deterioration agents such as corrosion, insect infestation, degeneration and breakdown of timber cellulose are elevated as their rate of degradation is closely related to the moisture content of materials.

Water and moisture play a large role in affecting buildings in the tropical region. Building envelopes in the tropical regions are normally constructed as a single half brick skin, due to the practicality of not having to deal with keeping warm air in. Moisture penetration due to rain is reduced by rendering and painting the outer wall, and the moisture that penetrates the bricked up building envelope is generally evaporated off quickly enough due to the high temperature.

However, when the humidity is high, especially during heavy rain season, the risk of mould developing on these plastered surfaces is equally high. Any surface, especially sheltered faces that dry slowly, are susceptible to mould attack even though they are situated externally. Anti-fungal paint coatings are normally effective up to around five years, after which if repainting work is not carried out, mould can reappear.

Figure 6.1 illustrates examples from Malaysia and Sri Lanka of external mould growth on relatively well maintained buildings. This is a direct opposite of the interior condensation issues experienced in colder climates where warm moist air in the building condenses on cold external walls, causing dampness and mould growth internally.

Buildings are now increasingly air conditioned. However, if no allocation is provided for insulation, as is often the case in tropical areas, the risk of interstitial condensation is elevated. The lower the internal temperature against the outside temperature, the higher the risk for interstitial condensation.

Water deep in the brickwork that is not evaporated will collect over time – due to gravitational force – at the lower levels in external walls, particularly at the interface between brick and reinforced concrete beams. After several years, due to the effect of continual wetting and evaporation at this brick and beam interface, paint film will deteriorate and spall, sometimes giving an impression of rising dampness. The condition effect of this is similar to the concept of rising moisture in temperate climates, although in reverse.

Water evaporation can wreak havoc on older brick buildings when repair or re-patching of mortar does not take into account the original building construction

Figure 6.1 Sheltered surfaces on a high rise flat infected with mould, while surfaces that face direct sunlight remain clean.

methods. Some colonial buildings are built as facing brick and use lime mortar. Where re-patching work is done using the harder cement mortar, it has been found that the brick face starts to crumble at the edge. The crumbling of the brick is attributed to the mechanism of water evaporation that takes place preferentially through the brick instead of the harder mortar.

In timber construction, water related defects could present a problem where prolonged exposure to moisture causes an increase in wood moisture content. This results in the accelerated decay of the timber cellulose, especially on softwood or green timber. Even without insect or fungal attack, timber may disintegrate because of the excess moisture and subsequent repeated wetting and drying. Typically timber is susceptible to attack from wood-rotting fungi at moisture levels in excess of around 20 per cent. Attack from wood-boring insects is also often linked to high moisture levels. Hence the performance of timber in tropical areas needs careful material selection and construction detailing.

Other pathological problems may be water related and they will be described in detail in the following paragraphs.

Insect attack

Insect attack on timber can be devastating in tropical regions, ranging from major physical deterioration leading to building failure to a more cosmetic effect. Insect infestation generally results in the generation of a range of physical and visual symptoms including:

- presence of insect larvae;
- flight and boreholes within the body of the timber;
- loss of strength and visceration of the timber.

A major example of the issue of insect attack is illustrated with reference to Borer beetles. Borer beetles, although different from their wood-boring cousins from other climates, have the same infestation behaviour. *Hylotrupes bajulus* or longhorn beetle is a species of wood-borer beetle which exists in the UK and which belongs to the *Cerambycidae* family. In Malaysia, *Batocera rufomaculata* is the species of borer beetle whose behaviour is the same as the *Hylotrupes bajulus* and the same family of beetle (*Cerambycidae*). However, physically they are rather different from each other. Both are commonly known as the longhorn beetle due to their long antennae segment. The increased size of the tropical beetle makes the impact of its activity in the wood much greater than that of the UK version. Such an attack can be structurally very significant.

Infestation of wood-boring beetles normally takes a longer time to develop than the common termite attack which can occur rapidly; typically within five to six weeks. In contrast wood borers can take more than six months to develop and effect a major attack. Wood-boring beetles attack wood via two distinct mechanisms. The phytophagus group of insects feeds on vegetation including timber. The xylophagous group refers to insect larvae that bore into timber for food and protection. Since the tropical environment affects the timber moisture, where borer beetles like the high moisture content in the timber, proper drying and preservative treatment is essential in prolonging the life of timber. The nature of the impact of wood-boring insects in

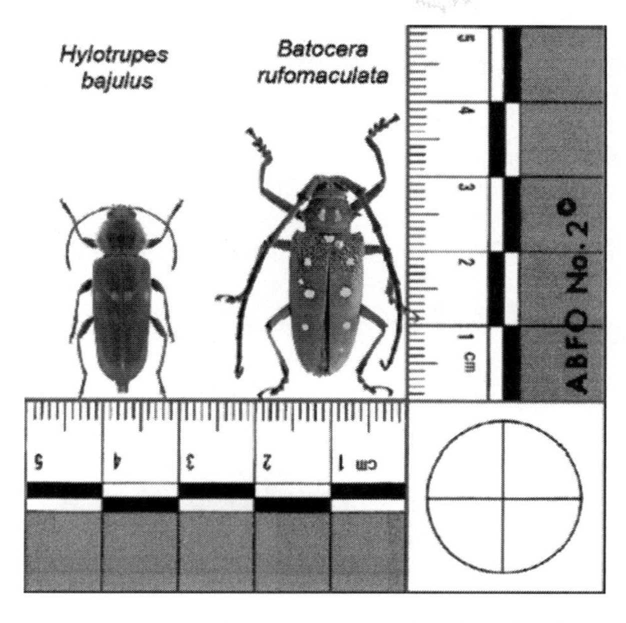

Figure 6.2 Tropical and temperature longhorn beetle.

timber is related to the different species involved and their generic life cycle as illustrated in Figure 6.3.

Termites are uncommon in temperate zones; although they do exist in Europe and North America they are much less prevalent than in tropical areas. There are numerous species in existence with more than a thousand species found in Africa alone. It is estimated that over one million active termite mounds are in existence in the Kruger National Park alone (Meyer 1999). In Asia and Australasia they are endemic and give rise to considerable damage to buildings through infestation if suitable measures are not effected to counter them. Termites fall into three broad categories or ecological groups. First, dampwood termites which are found only in coniferous forests; second, drywood termites which inhabit hardwood forests; and finally subterranean termites. Depending upon the locality and the species which are prevalent

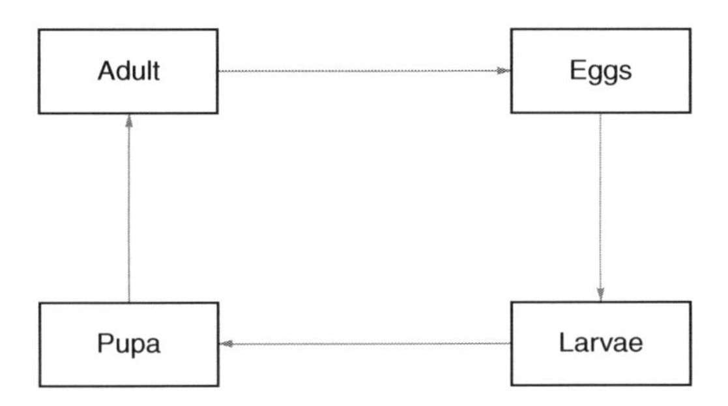

Figure 6.3 Insect life cycle.

in a particular region termites can be the source of considerable, even devastating damage to timber elements of buildings and other structures. In many instances the presence of the infestation may go undetected until damage to the timber is quite advanced. The nature of the action of the termite is such that the timber is consumed from the inside, leaving the insects largely concealed with a thin layer of material remaining on the external faces, leaving a void internally. Termites thrive in warm, tropical climates and as such are the source of concern in many traditional timber buildings. The traditional steps to avoid such damage include the use of eucalyptus, which is disliked by the termite, and the impregnation of timber with used fuel oil and other available preservatives.

Biological agents

Fungi can flourish well in climates where the relative humidity is high (around 70 per cent). Other mould contributing factors are availability of nutrients, temperature (5–50°C with the optimum of 22–35°C – perfect conditions exist in tropical climates), oxygen and spores. Fungi grow best on damp surfaces. Research has shown that fungal growth is viable on almost all building materials. Figure 6.4 shows the contributory factor for mould growth. Figure 6.5 illustrates how the effects can be manifested with the example of a building in India.

Fungi survive by dissolving food sourced from organic building materials in order to assist their growth. Different fungus types display differing physical characteristics but they tend to share a common life cycle, as illustrated in Figure 6.6. Spores are almost omnipresent in the atmosphere. When they alight on a suitable material, in suitable conditions, they germinate and create the basis for fungal growth. As this develops it seeks nutrients for survival and strands known as hyphae spread to seek moisture and food. They develop into thick clusters known as mycelium and allow the development of the fruiting body, which then emits spores to repeat the life cycle. The filament called hyphae seeks out this food by continually extending and breaking down organic material, spreading and infecting building material such as timber. These filaments will also flourish anywhere in the crevices of building cracks provided there is enough moisture presence. Apart from giving an unsightly appearance to buildings, certain fungal spores produce harmful toxins that could affect humans,

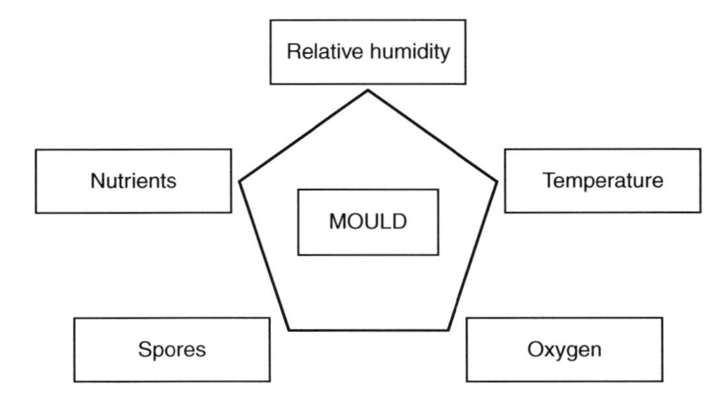

Figure 6.4 Fungal growth factors.

Figure 6.5 Mould growth on residential building.

especially those with compromised immune systems. Therefore mould and fungal outbreaks can be dangerous in hospital areas or in housing for the young, elderly and the very sick.

In the tropical climate, mould and fungus can be found both on the inside and also on the outside surfaces of buildings. Anywhere there is shelter from the sun is the most common place for a mould to grow. It is often found in courtyards and air-welled areas with minimal sun exposure. Mould seems to dislike the intense light or heat from the sun. Fungal attack tends to take place in warm, moist and dark areas.

Design, construction and maintenance failures

The homogenisation of building design for aesthetic reasons increasingly follows international trends. In many tropical regions buildings are increasingly designed with large windows and flat roofs, borrowing design trends from temperate climates where there is less available sunlight to heat the building and lesser rain drainage requirements.

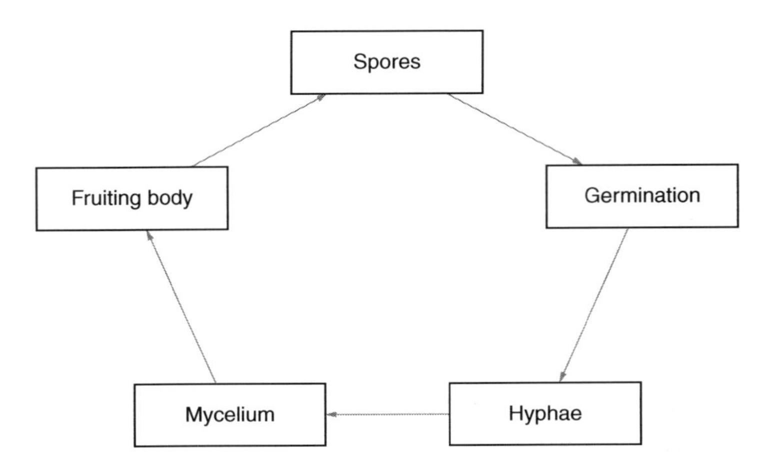

Figure 6.6 Fungal life cycle.

However pleasing the design, the compatibility of this practice with the local, tropical, climate could introduce issues of whole life performance. Large windows increase the air conditioning load of buildings, impacting on electricity use and energy conservation. Large windows will also need efficient water channelling ability which, increasingly, they are designed without, as frames are now flushed with the wall in rainscreen cladding.

Design detailing of construction work for efficient and effective water removal has been and will remain very important in this climate. During construction, close monitoring has to be exercised especially during the finishing work that paints and plaster/render does not bridge the drip throatings built into the window sills, eaves or other overhanging features in buildings, exposed to rain.

Flat roof and butterfly roofs have been largely unpopular in this climate in response to the amount of rain that this region experiences. This trend is reversing and more examples of this type of construction, especially in housing developments, are being experienced. This equates to an increasing risk of water related defects when the detailing for efficient water removal is not in place.

Flat roof membranes

Flat roof design solutions have been a common feature in commercial buildings, predominantly in high rise construction even in the tropical climate. However, due to the high temperature of the environment (27–40°C), combined with high occurrence of annual rain (ten year average – ~50cm annually), flat roof failure can be disastrous in such areas. Water, as highlighted before, is the main culprit in a lot of defect related problems.

Apart from clogging and ponding of water on the roof, the type of roofing membranes used has traditionally been based on standards specified in other countries that have no relation in terms of weather to the region (UK-BSI and US-ASTM). This takes the operational window of the materials outside that for which they are specified, and local standards are yet to be developed in many tropical countries. Empirical studies on roof membranes have found that the heat gain on these surfaces (normally dark in colour) can reach up to ~80°C on a hot midday (measured with thermal camera). The weather then suddenly turns and heavy rain downpour pushes the temperature of the surface down to ~20°C in a matter of minutes. This puts tremendous strain on the membrane as far as degradation is concerned. The primary cause of membrane failure is the inability to cope with repeated thermal and dimensional changes over time combined with the action of UV radiation.

Since the angle of the sun in these climates is at a more straight angle than it is in temperate climates, photo-degradation due to ultraviolet radiation is suggested to be at an elevated rate. The bituminous torch-on felts degrade by becoming hard and brittle over time, developing cracks that eventually allow water to seep through the felt onto the ceiling slab surface. Where enough water accumulates, blistering occurs pushing the bonded surface off the slab. Blistering problems occur not just through torch-on felts but can also be observed on other types of membrane materials such as PVC and KEE.

Structural movements

Defects arising due to movement vary according to construction methods. Generally, load bearing structures are more susceptible to soil and platform movement than reinforced concrete frame buildings. Shallow foundations are also more susceptible to movement than deep piled foundations provided that the work has been done correctly. In tropical zones there are often issues associated with land slip and high volumes of water in the ground.

Landslip or landslide is a result of movement of a body of soil material down a slope where the exerted pressure and kinetic energy in the soil is more than the friction keeping the soil in place. Because of the high amount of rainfall, slope areas are prone to landslips and landfalls as water acts as the lubricant, reducing the available friction force.

As population density increases and dwellings are increasingly being built at or near slope areas the risk of landslips affecting the general population also increases. Land clearing of slopes for building work in itself also increases the risk of a landslide as the roots of trees and vegetation contribute to soil stability and excess rain water absorption.

Earthquake and tsunami

Part of the tropical and equatorial regions sits within the Pacific Ring of Fire, an area with high volcanic activity and tectonic movement. A recent memory is the 2004 Aceh earthquake and tsunami.

In 2015 an ancient faultline near Borneo was suspected to be reactivated, causing a 6.0 magnitude earthquake to hit Mount Kinabalu, the highest peak in Southeast Asia (4095 m). Buildings built with modern rigid materials, for example, brick and mortar construction, are less durable in absorbing and dissipating vibrations caused by tremors, compared to traditional materials like wood, which is able to tolerate movements.

Adding to that, in countries where earthquake occurrence is very rare, such as Malaysia, earthquake building codes are non-existent. Even in Indonesia, where an earthquake code is in place, apart from engineered structures, occupier self-built masonry dwelling structures rarely conform to the earthquake code and are at the highest risk of earthquake. Tsunami is a direct consequence of earthquake and results in a devastating tidal movement of water which causes huge damage to buildings. Following the Aceh 2004 tsunami, the devastation of most buildings derived from the effect of the tsunami rather than movement from the earthquake.

Due to the massive nature of movement and damages caused by natural disasters such as landslips, sinkholes, earthquakes and tsunamis, the impact on building structure and fabric is often catastrophic in tropical areas.

Elemental defects

Single building elements such as plaster, tiling, ceiling panels and roof covering encounter different external and internal pressures that could cause defects to occur. These factors don't normally happen in isolation but often as a result of interaction with either other building elements or the environment.

Cracks

Due to the hot climate and direct exposure to sun, the risk of plaster and render cracks caused by rapid evaporation of mix water, especially on unsheltered sides, is high and care must be taken by the builders not to overwork their trowels, drawing mix water to the surface of the plaster work.

Cracks can appear and become unsightly or at the worst result from structural movement as a precursor to building failure. Cracks on building plaster in the tropical climate can be as a result of tradesmen overworking the trowel, causing the mix water to be drawn towards the surface. The resulting effect is that rapid evaporation of the water in the hot climate causes shrinkage cracks to appear due to the lack of moisture available for cement hydration.

Cracks that happen after the building has been finished can be attributed to either settlement of the building or, of a more serious nature, ground movement. Cracks can also appear due to contraction and expansion of materials that have not been properly taken into account by the building designers. Building materials will continually move during their life cycle, responding to daily or seasonal contraction and expansion. It is important that the building materials and detailing are specified to allow these small movements without resultant cracks.

Tile popping

Tiles can pop or 'tent' in a spectacular fashion in hot climates with a sudden and loud bang that happens seemingly out of nowhere.

The cause of the tile popping or tenting may be attributed to differential thermal expansion between the tiles and the backing surface. Where the tiles used start to expand, the build-up of pressure, especially in a large installation area, will need to be relaxed and the weakest point in the link would give way. Akin to what happens in an earthquake, this results in abrupt and sudden popping to the tiles. This breakage/relaxation of the pressure either happens at the tile-screed interface, or at the screed-slab interface.

This is often caused by tradesmen not using suitable adhesives to bind the tiles to the surface, or the screed backing not being prepared properly, causing a lot of hollow gaps between tiles and the backing screed. Another factor is the use of an inappropriate tile joint filler that is more rigid than the tiles, thereby not allowing the gap fill to become a sacrificial layer where its failure is not as catastrophic and can easily be made good (Figure 6.7).

This is obviously a workmanship issue which should be solved through better control of construction practice.

Flaking paints

Flaking paintwork occurs naturally due to old age and is usually a non-issue as the standard advice is to repaint surfaces every five to ten years, depending on the

Figure 6.7 Tile popping.

paintwork warranty. However, where the failure occurs prematurely, investigation often points to problems with regard to the paint-wall interface. This is particularly prevalent in tropical zones due to high moisture levels and the possibility of not providing a sufficiently prepared sub-base.

Surface preparation to receive paintwork is key to making sure that paintwork will last beyond its intentional warranty date.

Problems of flaking can be the result of leaching of soluble salts left at the wall surface, behind the paint film. Water in the plastered brick wall evaporates slowly bringing the soluble salts from within the wall outwards. As pure water evaporates, the soluble salts accumulate and push through from behind the paint work, causing flakes to fall off. This problem can normally be seen at the lower section of the wall, above the concrete beams. Water/moisture that cannot move gravitationally further down, as the concrete beams are denser, collects at the lower section of the wall, where it moves outwards to evaporate in order to achieve moisture equilibrium with the surrounding. See Figure 6.8, which illustrates both of these effects.

The source of water can be natural, such as rainwater splashes from the external wall, or artificial, such as from leaking pipes within the wall. Since the wall is normally single skinned in the tropics, the direction of water movement will become a factor where the flaking appears. On wall surfaces that are continually air conditioned, there is a big risk that the flaking will appear internally as moisture tends to evaporate inwards to compensate for the low humidity of the air-conditioned area, although the wall is an external wall on the other side.

Ficus and other vegetation

Ficus is a generic label for a group of shrubs and trees that extends to over 850 varieties that are widespread and native to tropical areas. They are often fast growing evergreens and vines that propagate easily and grow rapidly if unchecked. As members of the fig family they can be an important food resource for humans and wildlife. However, they present great challenges for owners and operators of buildings in tropical regions due to their propensity to grow on and in buildings and building elements. Seeds carried either by the wind or present in bird droppings can

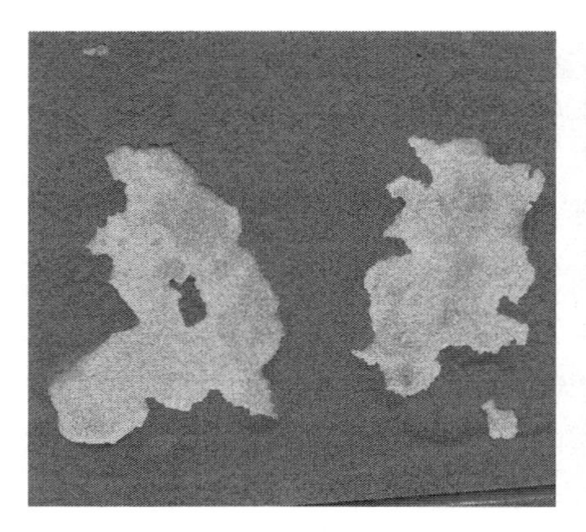

Figure 6.8 Flaking paint and moisture damage.

grow into small shrubs. These plants bury their roots into cracks and crevices of buildings, especially in corners where drainage and falls were not laid properly. The impact is that the growing roots force cracks wider due to the pressure exerted on the structure. Growing roots can also clog up rainwater downpipes, causing ponding of the rain gutters.

In tropical regions buildings that are afflicted with ficus or other vegetation growth can suffer significant damage to structure and fabric within a relatively short time frame. If left untended the building can effectively be consumed by the plant growth and disappear from view.

Figure 6.9 Ficus growth within building gutters.

Corrosion

Metallic corrosion is a worldwide problem. Nevertheless, due to the high humidity in the tropical regions, the susceptibility is higher and more prevalent. Metals corrode because in nature, the natural state of a lot of the common metal ores is in oxide form. Human intervention transforms the state of the metals to pure form and, as a result, metals are in a constant state of wanting to go back to their natural state.

Concrete structures near coastal areas are at risk of chloride-induced corrosion, a condition where chloride ions diffuse into bulk concrete and react electrochemically with the reinforcement bars, resulting in pitting corrosion. Once pitting corrosion has started, due to the large available corrosion potential, the rust propagation will occur rapidly causing volumetric expansion of the reinforcement bar, spalling the concrete from the resulting tensile stress, as shown in Figure 6.10.

Carbonation-induced corrosion can happen everywhere and is governed by the rate of CO_2 ingress and the resistance offered by the concrete cover and the mix composition of the bulk concrete. Carbonation mechanism works by the reaction between the carbonic acid formed as CO_2 diffuses into the pore solutions with CaOH (calcium hydroxide) and Si forming $CaCOH_3$. This reaction causes a drop in the alkalinity of the passivating layer that protects the reinforcement. Corrosion starts when the pH value drops to below 9. Because of the high atmospheric moisture content in the tropical region, the risk of carbonation is elevated significantly.

However, in general, concrete is still specified by strength characteristic and not based on the exposure classes. This means that generally in the tropics concrete framed buildings are not optimised for durability in withstanding carbonation risk where the risk is high. Consequently this is a common issue in such areas.

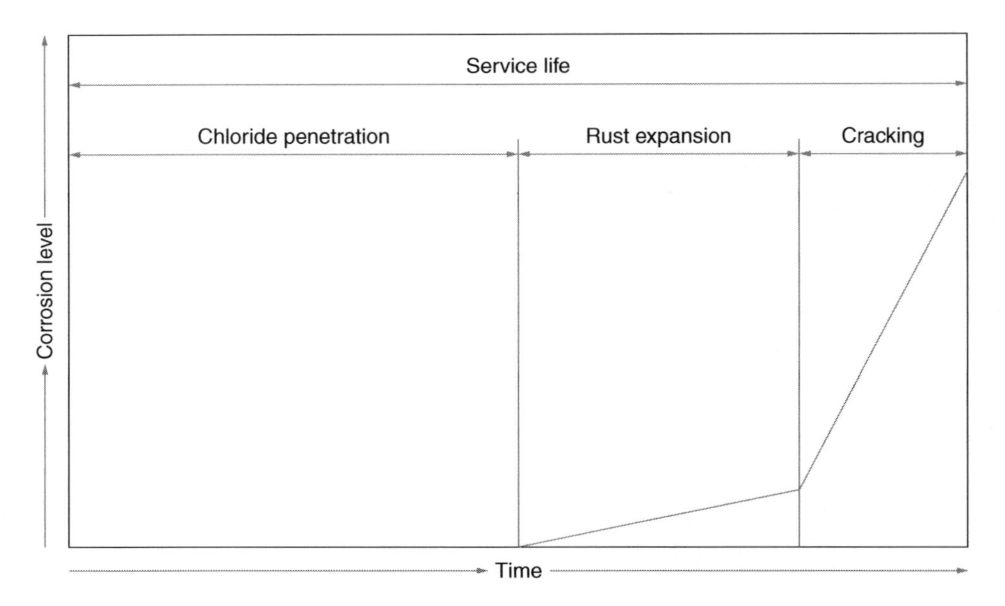

Figure 6.10 Chloride-induced reinforcement corrosion process.

Source: Chen and Mahadevan (2008).

Remediation of building defects

Patent vs latent defects

Whenever a defect occurs on a building, the classification that can be put in place is to determine whether the defect is either a patent defect or a latent defect. A patent defect is any defect usually found in a new build that becomes apparent when an inspection is carried out. An example might be a crack on the floor or wall tile, wrong window opening orientation, or crazing of plasterwork.

Latent defects are those defects that are not easily recognised initially, but become apparent well into the building life. A large proportion of latent defects is due to deficiency in design either architecturally or sometimes due to structural or geotechnical design deficiency.

Architectural design problems in the selection of building materials can cause latent defects, for example, galvanic corrosion due to the use of two dissimilar metals, tiles popping due to unsuitable substrate specification and premature flat roof failure due to wrong choice of waterproofing membranes.

Structural or geotechnical deficiency is normally evident when subsidence or structural movement cracks appear on buildings. This may be due to inadequate design of foundation, or geotechnical problems such as soil liquefaction. Whereas a patent defect is easy to spot and remedies can be straight forward, latent defects are harder to tackle or possibly be eliminated. This is because designers base their design on materials and specification that they have good experience with and rarely do they have time to research systematically construction works and materials, especially novel materials that clients want them to use in their building works. This leads to simplified assumptions of buildability.

Maintenance consideration and techniques

Assets, be they buildings, plants, machinery, equipment, etc., are prone to deterioration. This could happen over a short period of time, sometimes immediately after construction, or could happen over a long period of time.

In the tropics, developing countries in particular, building maintenance technology is currently vastly underrated and ignored by property owners, managers and professionals. Issues reported suggest that complaints about building defects have gone up in recent years, with common problems such as leaky roof and walls, floor defect and improper outlet pipes.

To retain whatever purpose a property is built to satisfy, the building must be constantly managed and maintained. The subsequent sections will look at the understanding of the property itself, obsolescence in buildings, consideration for maintenance, how to plan the work and a brief outlook on maintenance management.

Understanding the building elements

A building is a structure with walls and a roof. A building could be a residential building, a commercial building or an industrial building. Several elements make up the building structure. A completed and functioning building generally consists of the substructure, the superstructure, finishing, fittings and fixtures, and services (internal and external). The building owner must understand what structural, functional and aesthetic functions these elements bring to the building.

The *substructure* is the point of the building structure at and below ground level. The substructure's basic function could be seen as that of 'Atlas' shouldering the entire weight of all the other elements of the building structure above it. It consists of the foundation and all its components – walls, ground floors, damp-proofing and other membranes, etc. Sub-soil drainages and basement floors are also considered to be the substructure of a building.

The *superstructure* is made up of all the building elements above ground level. It consists of walls, staircases, upper floors, windows, doors, ceiling, roof, etc.

Finishings are the elements that add the final cosmetic outlook to a building structure. They could include: floor screeds; wood, metal, glass or PVC panelling; plastering and rendering; painting and other decorations, etc., and are meant to be visually appealing.

Fittings and fixtures are basically permanent attachments considered a part of the building structure such that removing them would affect either the fabric or make the building envelope incomplete, such as doors and window frames, built-in wardrobes/cupboards/shelf units, lighting fittings such as sockets, etc. There is a wide range of fixtures and fittings in a building. Some are included in the next element as services.

Services are installations for the general functionality of the building. They are basically fixtures and fittings. While some are removable and not considered part of the building envelope, some are considered permanent fixtures and fittings to the building. Services in a building could include water supply systems, HVAC systems, fire protection systems, lighting and power supply systems, communication systems, elevators, conveyors, escalators, etc. In automated buildings, monitoring and control systems such as building management systems (BMS) are also considered to be services. External services could include external sewer drainage channels, stormwater collection channels, etc. There is a wide range of services within and outside a building.

What is obsolescence?

Building elements, whether maintained or not, would gradually deteriorate over time and/or lose usefulness. While proper maintenance limits the level of this deterioration, lack of or poor maintenance aggravates the rate of physical wear and tear, functional incapacitation and inability to command value – be it income or occupancy satisfaction. Obsolescence, be it of any form, results basically from two things: deficiency (inadequacy) or over- or super-adequacy.

Obsolescence in general terms is a state in which an item or component, structure or building is no longer in a state of usefulness. Emphasis should be placed on 'usefulness' here, given the many factors that could adversely affect the usefulness of a property beyond just bad working condition, as a component, structure or building could be obsolete despite in being in good condition. That brings us to the question: *When is a building considered obsolete?* To answer this question, the three types of obsolescence in buildings will be discussed below.

Physical obsolescence: This type of obsolescence is caused by factors within the property. It is the deterioration of a property due to physical wear and tear or abuse.

This could be as a result of inappropriate use to which the property was constructed, excessive strain on facilities and structure, gradual wear and tear as a result of old age, etc. Physical obsolescence could be both curable and incurable. Curable means the rate of deterioration could be mitigated and the physical integrity of the property returned to normal. It could be incurable when the structural integrity of the property could no longer withstand any form of rectification and repair. In this case, the structure could be due for demolition and redevelopment.

Functional obsolescence: When a property can no longer perform its functional requirements, such a property is considered functionally obsolete. Note that a functionally obsolete property could still be in good physical condition. Functional obsolescence is a vital factor to loss of value in property. It is also caused by factors within the property such as inadequacy, inability to perform designed functions or deficiencies in technology. It could result from improper design and improper installations. Functional obsolescence could be both curable and incurable. Curing it might require slight to large modifications to address either the inadequacy or over-adequacy. At its worst stage, the property might require a complete redevelopment or change of use.

Economic obsolescence: This form of obsolescence is caused by factors beyond the property such as external and environmental factors. A building could suffer loss in desirability and subsequently loss in value as a result of external activities not directly within the property's influence and control. This could be as a result of change in construction fashion, change in demand for type of accommodations, regulations, location, and so many other external activities. A building located by the highway where there is a tremendous amount of traffic could suffer loss in value compared to its counterparts located away from the highway traffic but within the same neighbourhood. The traffic is an external activity affecting the value of the property in question. This is also referred to as external or environmental obsolescence. Economic obsolescence is generally considered to be incurable because the factors causing it are from external sources and are beyond the property owner's influence or control.

What is maintenance in buildings?

Maintenance is simply any action or work carried out to preserve or sustain the required functionality of a building, structure or component. British Standard 3811 of 1993 defined maintenance as 'the combination of all technical and administrative actions, including supervision actions, intended to retain an item in, or restore it to, a state in which it can perform a required function'.

From an organisational viewpoint, on the other hand, ISO 9001:2008 is vastly accepted as a standardised approach to quality management for maintenance operations, given the international reference for quality management model. This standard turns into a generic guide for a maintenance process operation in which fulfilment with requirements should be demonstrated. The ISO 9001 embraces the following elements which maintenance organisations need to achieve:

1 quality management (process approach, sequence and interaction of the processes, description of the elements of each process, generation of documents or records);

2 management responsibility (entailment with strategic aims and objectives of the organisation, working definition, top management champion, clear definition of responsibilities and authorities, suitable communication);
3 resource management (humans beings, material resources and infrastructure);
4 measurement, analysis and continuous improvement approach (audits, evaluation of user post occupancy, information analysis, corrective and preventive actions, improvement).

Need for maintenance

Maintenance is a continuous process once a building is constructed and its life cycle revolves around it. How long a building lasts and functions depends greatly on how well it is maintained. To the building owner, the scope of maintaining a building has transcended day-to-day operations to a strategic (the thinking) level. The intention is conducting maintenance with the needs and preferences of the organisation and building occupants in mind as employee and occupancy work productivity is becoming increasingly tied to the workplace. This has called for serious understanding of how the design stage of a building's construction affects its maintenance during its life cycle. The need to accommodate the maintenance requirements of a building right at the design stage of its construction so as to reduce costs and maintenance needs is gradually gaining popularity among professionals in the construction process. Maintenance of buildings is carried out to achieve the following: retain value of investment, ensure functionality, good aesthetics, increase productivity (work-play-live), offer health comfort and safety.

Nature of maintenance strategy

When planning for maintenance, factors such as costs of the maintenance works, availability of resources to carry out the maintenance works, the urgency of the nature of work, future use of the component or structure or building, age of the property, and social considerations (effects on occupants' sensibilities and the environment) must be considered.

The nature of maintenance works could take the following forms:

Servicing or day-to-day maintenance

These are basically routine cleaning or servicing operations carried out at stipulated schedules that could range from daily routines to years' intervals to keep components and the building in an acceptable working condition. Such works could include sweeping and mopping floors daily, washing or cleaning windows, repainting of walls and decorations, lawn mowing, clipping flowers, HVAC servicing, etc.

Rectification

This involves actions or work carried out to correct faults. It could occur early in the life of the building as a result of faulty design, inherent faults or simply unsuitable components, incorrect assembly or installation, items damaged while in transit, etc., or could occur when some on maintenance works have been carried out arising from some of the same reasons stated above. Rectification work is done to ensure that the right components are installed for the right purposes.

Replacement

This is work done to completely replace components, as the name implies. Components could be replaced either because they have broken down completely or have reached the end of their lives, in which case they could still be functional, but, say, safety or operating conditions require that they be replaced.

Maintenance approaches

Maintenance falls generally into two categories: Planned and Unplanned Maintenance. Under these two categories fall the various types or approaches to maintenance activities. Planned maintenance is thought out, documented, monitored and executed in an organised manner. Unplanned maintenance is the direct opposite. There is no plan or forethought whatsoever to this type of maintenance. Unplanned maintenance is basically of a corrective nature and could or could not involve emergency maintenance. All other approaches of maintenance, including corrective and emergency maintenance, fall under planned maintenance.

Preventive maintenance

As the name implies, this approach involves carrying out measures that tend to avoid failures in components and the entire building. Measures are taken to replace components or keep them in good condition before failure could occur. This type of approach is further broken commonly into two forms: *scheduled preventive maintenance* and *condition-based preventive maintenance*. In the former, maintenance work is carried out on the item or component based on the scheduled period, which could be usage or time based, despite its actual condition. Most scheduled maintenance actions are recommended by the suppliers based on either limited knowledge of the actual use conditions of the component or from past experiences. In the latter, a preventive measure is initiated based on the actual condition of the item or component. Routine and continuous monitoring of the component informs its condition and the need to carry out a preventive action to avoid its eventual failure. A preventive maintenance checklist should be developed and adhered to.

Corrective maintenance

This is an action carried out to restore an item or component that had already failed. It is intended to return the component or item in a condition in which it can again satisfactorily perform its required function.

Emergency maintenance

This approach is employed to rectify issues that might result in disastrous failures or serious consequences. It could be day-to-day actions resulting from incidents as they occur. For example, a plumbing pipe leak would require an emergency fix that if left unattended could lead to serious water loss or a gas pipe leak within a house that could lead to a fire outbreak if no immediate action is carried out to fix it.

Run-to-failure maintenance

This is also referred to as *just in time maintenance*. In this approach only routine servicing is carried out on the item or component until the item or component is broken or ceases to function. This type of approach can only be employed when the cost of preventive measures to failure outweighs the impact the actual failure would cause. It is also justified only when the impact of failure of the item or component is negligible to the overall functioning of the building. Thereby, investing heavily in preventive measures would be an unwise decision.

Building refurbishment

This consists of restoration, overhaul or modification from original design and specification of the building with the aim of improving the original design. This aspect is treated in more detail further below. On a one-time basis, renovations are more capital intensive and should be planned very well in advance; lack of proper budgeting and costing, and knowledge of the true extent of renovation works to be done, could result in abandoned renovation projects.

In tropical countries, there are large numbers of buildings which have stood for more than a hundred years, some of which are listed as heritage buildings. Often the external building enclosure needs to be sustained to preserve the significant financial, functional or cultural value of the building. In many cases these historic structures need to be refurbished, upgraded and revitalised for modern uses that differ from their original purpose. Refurbishment is one of the main alternatives to renovate or upgrade the building internally to fit the needs of the evolving user, rather than demolishing and rebuilding.

The refurbishment sector, globally, is growing rapidly as much of the vacant land has been fully developed and the lifespan of buildings aims to be prolonged to ensure cost effectiveness and full optimisation. Increasingly, it is claimed that demolition and rebuilding existing buildings is slow, costly and unpopular which directly impacts on refurbishment gaining popularity.

This section attempts to highlight some of the important elements and features of refurbishment in tropical countries. This links, also, to sustainability, which is one of the most prominent elements influencing the construction and use of buildings in recent times. The existing building stock worldwide often falls below optimal standards of energy efficiency, sustainability, building condition and standards. Therefore, it is important to note that the issue of sustainability will be strongly related to refurbishment of buildings in tropical countries as it is elsewhere. However, sustainability is a very subjective matter. The definition of the term 'sustainable refurbishment' must be well understood in order to determine what the key features of such refurbishment are in tropical countries.

Refurbishment is defined as a process of upgrading and repairing the building to extend the lifespan and transform the usage of the building to reflect the contemporary demands of the market. Sustainable refurbishment is a much more subjective matter. As such, different authorities may view this term in different ways.

Some may think the process of refurbishment conducted in a sustainable manner is known as sustainable refurbishment. On the other hand, others posit the view that sustainable refurbishment is the process of refurbishment to create a sustainable building (Cotgrave and Riley 2013). In reality it is a combination of both as the aim of a refurbishment project is to upgrade an existing building to become a better building; both the process of refurbishment conducted in a sustainable manner and the outcome of producing a more sustainable building are equally important. The decision for refurbishment is, itself, a potentially more sustainable way forward than new build as the process of construction is reduced by maintaining the structures and facilities of the existing building. This approach is, of course, only tenable if the resulting building is fit for purpose.

Ineffective building maintenance and management in many tropical countries has been cited as causing the reduction of building life cycle and there is increased recognition of the growing importance of the refurbishment industry. Refurbishment or demolition works become an alternative when buildings fail to perform optimally and to satisfy the required functional needs. The most challenging decision is whether to refurbish or to undertake new build. The purpose of this section is to highlight the changing trend of refurbishment work in tropical countries and the available refurbishment options, complexities and uncertainties faced by the players in executing refurbishment tasks in a sustainable manner. The decision-makers face difficulties in deciding whether to demolish or refurbish. As the issue of sustainability is gaining more importance worldwide, this section will also discuss the potential of sustainable refurbishment in tropical climates.

The Malaysian example

The Malaysian refurbishment sector has changed rapidly through the years due to several factors. These include the application of sophisticated technology, globalisation and changes in the economy. The construction industry plays an important role in the economy of Malaysia in generating wealth and improving quality of life through the translation of the government's socio-economic policies into social and economic infrastructure and buildings.

The total number of projects awarded for refurbishment and new projects in Malaysia, for example, increased from 2010 to 2013 but slightly decreased in 2014 and 2015 (CIDB 2016). However, the total project value has increased every year from 2010 to 2015. This suggests that refurbishment projects of higher individual value are increasing. The development of new build drastically increased also. This indicates that refurbishment projects in Malaysia, as well as other countries in South-East Asia, are gaining popularity. However, many refurbishment projects carried out are unreported, especially those undertaken by house owners and small businesses or community groups. Therefore, the actual value of refurbishment works in this area is, in reality, rather higher than officially recorded.

Decision to refurbish: consideration and assessment

Refurbishment works become an alternative when a building has reached the end of its service life or fails to perform its required function. However, the decision to refurbish is influenced by several factors, such as the building's physical deterioration and obsolescence which may be associated with change in use, economic change, investment decisions, historical value and change in condition. Moreover, building refurbishment is also influenced by other factors which include changes in technology, social or societal factors and sustainability. The main drive of refurbishment is to fulfil the evolving needs of the building users. Some buildings, which have a strategic location, will be considered for conversion and refurbishment to fit current demands due to the lack of possible sites for new build. For example, SS2 Kuala Lumpur is a well-known area for bridal shops which has become a famous spot for couples that plan to get married. However, the available shop lots are limited, causing most of the bridal shops to be created through conversion from terraced houses. As shown in Figure 6.11, in this example the house condition within the terrace is poor but they form the basis of successful refurbishment and conversion to become shops. This illustrates, to some degree, a situation associated with perception of the relative merits of new-build and refurbishment. In many tropical countries, particularly those that might be described as 'developing', there is sufficient affordable, undeveloped land available to facilitate the relatively cost effective construction of new buildings. This does generate a degree of sprawl in urban development but it does not yet present the sort of policy issues that are present in more heavily developed urban areas internationally. There is, in some areas of the world, including many tropical countries, a perception that refurbishment is a poor alternative to new build, which demonstrates progress, wealth and growth. The 'disposability' of relatively new buildings in places such as Hong Kong and Dubai illustrates the cultural favouring of new buildings over old in some areas.

Figure 6.11 Refurbished terrace houses.

The needs of the building users may vary from time to time. As time passes, more advanced technologies evolve and the building needs to be upgraded to fulfil the changing demands of the user. Buildings may also need to be refurbished to achieve accordance with updated regulations and standards, better and safer structure as well as providing improved building services to be more cost and energy effective. This often occurs when there is a change of use within the building or in building forms that are subject to certification or external control such as schools and hospitals.

The time taken for demolition and new build is often longer than refurbishment and upgrading. However, in addition to the obvious issues around project duration, refurbishment can bring a lot of benefits compared to demolition. As refurbishment upgrades the existing building, the basic structures of the building are preserved. In addition, refurbishment might be beneficial to the neighbouring land as the appearance of the whole neighbourhood will be upgraded and give an impression that it is worth investing in. For example, the Central Market Kuala Lumpur, Malaysia, has been transformed from a wet market into the centre for Malaysian Culture, Heritage, Art and Craft. In order to preserve the significant values of the building in showcasing the nation's arts, culture and heritage, the building has undergone upgrading and alteration which has attracted tourists from all over the world.

The approach to building refurbishment is totally different in tropical countries. The refurbishment of buildings in tropical climates will often be focused on improving the air circulation and lowering the relative humidity as the temperature will be largely the same throughout the year. Effective usage of natural ventilation will be one of the considerations in renovating or upgrading buildings in a tropical climate. There is an irony in the situation regarding some buildings that are facing refurbishment as they fall between distinct periods of awareness around technology and sustainability. Buildings constructed prior to the advent of air conditioning and advanced environmental controls generally include vernacular features associated with natural ventilation, solar shading, thermal mass and so forth. Later buildings, particularly those of the 1960s, 1970s and 1980s, will often be based on more homogenised international construction forms, with air conditioning and environmental controls combating the environmental conditions of the tropical climate. Paradoxically, this may make older buildings more suitable for refurbishment using a sustainable approach than some newer buildings in which it is difficult to incorporate passive, sustainable design features.

Building deterioration is generally associated with one or more agencies or causes such as dampness, bio-decay and movement. Unfortunately, all buildings are exposed to these effects to some degree and they can be extreme in tropical areas. High humidity causes condensation. Fungal attack causes bio-decay and subsidence causes movement. Exposure to the environment causes carbonation and corrosion of reinforcement bars. Moisture and high rainfall could cause dampness in the building. Underground water penetrating through the ground floor slab can cause damage within the building. In addition, vegetation and mosses grow in damp floors, especially on the building aprons. Perimeter drains tend to become clogged after only a few months due to high water levels and fast-growing vegetation. Movement, due to vegetation growth, such as ficus, could also cause cracks in walls. These processes normally affect the exterior facade of the building, followed by the interior. Cracks in walls could lead to the occurrence of dampness inside the building. All of these issues can be noted in tropical regions where building defects are often more extreme and faster in developing than they are in temperate regions.

However, building deterioration can be controlled by implementing planned maintenance. This type of maintenance consists of a routine, preventive and corrective approach. Properly planned maintenance for a building is required to ensure that the building could function as needed by its users. By taking a combination of all technical and administrative actions, including supervision actions, the performance of the building can be maintained or restored and the building can perform to a required function. The physical condition of a building generally deteriorates with age, and lack of regular maintenance may cause the building to deteriorate faster. In many tropical areas the cost of planned maintenance is almost prohibitive due to the pace of developing conditions and defects. This is exemplified by the extent to which severe weathering is visible on relatively new buildings in tropical areas.

Obsolescence is the process of an asset going out of use, which indicates the tendency for the objects and operations to become out-of-date or old-fashioned. Old buildings that were completed in the 1960s and 1970s generally have no extra space for additional new data, and technological change makes the existing building system become obsolete faster. The need for buildings to accommodate the latest automation and electronic systems require the building owners to refurbish their buildings. The complexity of modern building automation systems requires sensitive design, particularly on their services layout. This is important for business organisations to provide better building equipment, quality workspace for their staff and a high standard of building appearance to enhance the building's position in the city. Refurbishment is an option to meet the change of demand of a building necessitated by the installation of modern facilities. In addition, information technology (IT) has changed the demand for new premises and the working environment of many people.

Condition and assessment

Obviously, there are many factors that need to be considered when dealing with the decision to refurbish in tropical climates. Some of the important factors are discussed in this section.

- *Different types of building*

Different types of buildings differ in their degree of uncertainty and the problems faced in refurbishment projects. Building types are linked to the concept of complexity, which influences the project's expenditure and resources. The uncertainty of refurbishment design concerns their statutory requirements, existing usage and complexity of the building layout. The extent to which buildings have unknown issues that are concealed is a major issue, particularly in tropical climates where fungal attack and other such defects develop very quickly if unobserved.

- *Public vs private ownership*

If a building is in public ownership, the decision to refurbish should generally provide some benefit or value to the public. Some governments provide incentives to encourage private resources to finance refurbishment projects in order for the public to enjoy the facilities. There is a very large degree of variation across tropical regions in this approach, and the wealth of the country, its degree of development and cultural issues all impact upon the approach taken.

On the other hand, private refurbishment projects put more focus on the return of investment and profit that will be gained from a particular project. The owner will consider whether the refurbishment project is worth investment. Some of the characteristics of the building, such as strategic location, condition of building structure and market demand, will influence decisions.

- *Physical condition*

The decision to refurbish likewise relies on the state of the current building structure. Frequently, there are situations where data for the structures, as-fabricated drawings and manuals are not legitimately recorded, deficient or missing. This is true in all areas of the world and is by no means unique to tropical countries. In addition, a portion of the accessible information is inaccurate as the changes made, which include alterations and modifications throughout the duration of the life cycle of the building were not updated in the documents causing the actual design to be different from the final as-built configuration.

Effects of building refurbishment on tropical climate change

Efforts towards providing scientific perspective on climate change and environmental impact resulting from human activities have increased with the formation of the Committee by the World Meteorological Organization (WMO) and the United Nations (UN). The Committee is responsible for reviewing the actual cause of the problems associated with the climate through data collection as well as scientific research information produced by many countries around the world. The 'climate change 2013' report contains interesting evidence about climate change including the increase in the average temperature of the world in the twentieth century. Most of the effect was linked to human activities that could be controlled and reduced. Climate models have been developed and improved by taking into account more components and variables that affect temperature changes, such as environment, rate of departure of carbon, the quality of soil, and vegetation dynamics.

There is an indication that the world's climate is changing. Thus, control and monitoring of activities affecting climate change needs to be carried out. Among the activities that have been cited as influencing climate change in tropical areas are high-energy consumption, mainly due to building heating and cooling (in tropical areas air conditioning systems account for a high proportion of this).

Thus, in the refurbishment of buildings, most of the emphasis is on strategies for reducing greenhouse gas emissions by increasing the efficiency of use of electricity within a building.

Refurbishment case study in Malaysia

Since the building boom of the 1970s, many of Malaysia's historic buildings have been demolished. Recent large scale urban development continues to threaten pre-war buildings, while other historic buildings are simply deteriorating due to age, neglect and the high cost of maintenance. In order to preserve the significant values of buildings, laws for historic building conservation are established through legislation whereby a national inventory of historic buildings includes lists and schedules of old buildings for protection. Refurbishment becomes an alternative to preserve the identity, values and structure of the buildings.

One well-known building in Kuala Lumpur was refurbished to become an assembly hall and place to organise activities. The original function of the building was as a gathering place for the Chinese societies and other races to run activities. This hall was used as shelter for those Chinese societies who came from South China and did not have any place to go. When China was colonised by Japan, this building was the place for the Chinese community in Malaysia to obtain the latest news from China and here was the place to collect funds for China in 'Anti-Japan' operations. The building was listed as a national heritage building and therefore refurbished due to its unique architectural style and values.

The main purpose was to preserve the building to provide sufficient information and knowledge to the public. The building has a library for Chinese Studies Research and most of the sources are in Chinese language. The history of the building gives awareness to the young generation about the importance of unity. The building has undergone several renovations in order to restore its original appearance, as shown in Figure 6.13. Thus, refurbishment and upgrading work play an essential role in tropical countries whereby a lot of historical buildings need to be upgraded to preserve their significant values.

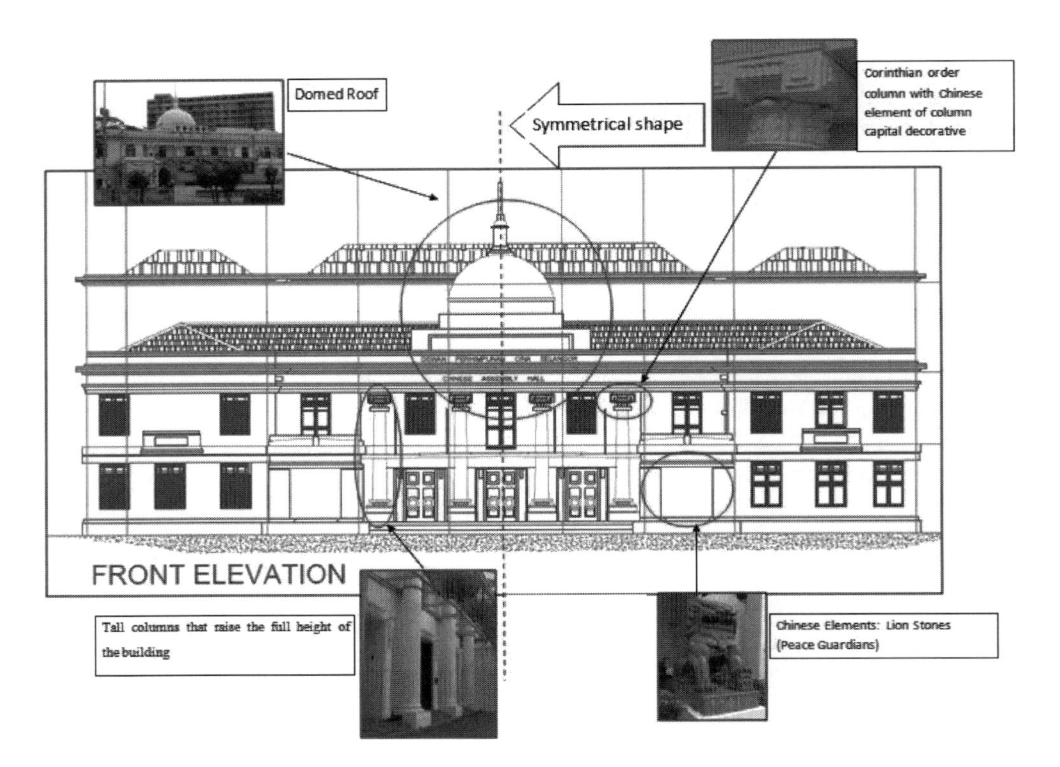

Figure 6.12 Illustration of Case Study.

Figure 6.13 Renovation of Case Study.

Bibliography

Abbas, M. Y., Marhani, M. A., Adnan, H. and Ismail, F. (2013). 'AcE-Bs 2013 Hanoi (ASEAN Conference on Environment-Behaviour Studies), Hanoi Architectural University, Hanoi, Vietnam, 18–21 March 2013. OHSAS 18001: a pilot study towards sustainable construction in Malaysia', *Procedia – Social and Behavioral Sciences*, 85, 51–60.

Abbas, P. D. M. Y., Marhani, M. A., Jaapar, A. and Bari, N. A. A. (2012). 'AicE-Bs 2012 Cairo (Asia Pacific International Conference on Environment-Behaviour Studies), Mercure Le Sphinx Cairo Hotel, Giza, Egypt, 31 October–2 November 2012. Lean construction: towards enhancing sustainable construction in Malaysia', *Procedia – Social and Behavioral Sciences*, 68, 87–98.

Aikivuori, A. (1996). 'Periods and demand for private sector housing refurbishment', *Construction Management and Economics*, 14(1), 3–12.

Ali, A. S. (2009). 'Cost decision making in building maintenance practice in Malaysia', *Journal of Facilities Management*, 7(4), 298–306.
Ali, A. S., Rahmat, I. and Hassan, H. (2008). 'Involvement of key design participants in refurbishment design process', *Facilities*, 26(9), 389–400.
Alias, A., Sin, T. K. and Aziz, W. N. A. W. A. (2010). 'The green home concept: acceptability and development problems', *Journal of Building Performance*, 1(1), 130–139.
Almås, A.-J., Bjørberg, S., Haugbølle, K., Vogelius, P., Huovila, P., Nieminen, J. and Marteinsson, B. (2011). 'A Nordic guideline on sustainable refurbishment of buildings', World Sustainable Building Conference, Helsinki, RIL – Finnish Association of Civil Engineers, 146–147.
Appleby, P. (2013). *Sustainable Retrofit and Facilities Management*, Routledge, New York.
Board of Architects Malaysia (2004). *The Malaysian Architect Act 1967 and Rule: International Law Book Services*, Board of Architects Malaysia.
Burton, S. (2012). *Handbook of Sustainable Refurbishment: Housing*, Routledge, London.
Chau, K. W., Wong, S. K., Leung, A. Y. T. and Yiu, C. Y. (2003). 'Estimating the value enhancement effects of refurbishment', *Facilities*, 21(1/2), 13–19.
Chen, D. and Mahadevan, S. (2008). 'Chloride-induced reinforcement corrosion and concrete cracking simulation', *Cement and Concrete Composites*, 30, 227–238.
Construction Industry Development Board (CIDB) (2007). *Construction Industry Master Plan Malaysia 2006–2015*, CIDB, Malaysia.
Construction Industry Development Board (CIDB) (2016). *Construction Quarterly Statistical Bulletin 2016*, CIDB, Malaysia.
Construction Industry Research and Information Association (CIRIA) (1994). *A Guide to Management of Building Refurbishment*, CIRIA report, UK.
Cotgrave, A. and Riley, M. (2013). *Total Sustainability in the Built Environment*, Palgrave Macmillan, Basingstoke.
Cox, I. D., Morris, J. P., Rogerson, J. H. and Jared, G. E. (1999). 'A quantitative study of post contract award design changes in construction', *Construction Management and Economics*, 17(4), 427–439.
Department of Energy and Climate Change (2007). *Estimated Impacts of Energy and Climate Change Policies on Energy Prices and Bills Report*, Department of Energy and Climate Change, London.
DETR (1998). *Sustainable Construction Report*, The Stationery Office, London.
Douglas, J. (2004). *Building Adaptation*, Butterworth-Heinemann, Oxford.
Egan, J. (1998). *Rethinking Construction (The Report of the Construction Task Force)*, The Construction Task Force, UK.
Egbu, C. O., Young, B. A. and Torrance, V. B. (1996). 'Refurbishment management practices in the shipping and construction industries – lessons to be learned', *Building Research & Information*, 24(6), 329–338.
Elmualim, A., Shockley, D., Valle, R., Ludlow, G. and Shah, S. (2010). 'Barriers and commitment of facilities management profession to the sustainability agenda', *Building and Environment*, 45(1), 58–64.
Flanagan, R. N. G. (1993). *Risk Management and Construction*, Blackwell, Oxford.
Gaterell, M. R. and McEvoy, M. E. (2005). 'The impact of climate change uncertainties on the performance of energy efficiency measures applied to dwellings', *Energy and Buildings*, 37(9), 982–995.
Gohardani, N. and Björk, F. (2012). 'Sustainable refurbishment in building technology', *Smart and Sustainable Built Environment*, 1(3), 241–252.
Government of Malaysia (2011). *Tenth Malaysia Plan [2011–2015]*, The Economic Planning Unit – Prime Minister's Department Putrajaya.
Government of Malaysia (2016). *Eleventh Malaysia Plan [2016–2020]: Anchoring Growth on People*, The Economic Planning Unit – Prime Minister's Department Putrajaya.

Hodges, C. P. (2005). 'A facility manager's approach to sustainability', *Journal of Facilities Management*, 3(4), 312–324.

International Specialised Skills Institute (2011). *Preservation and Restoration of Timber Heritage Structures*, International Specialised Skills Institute, Melbourne.

Kagioglou, M., Cooper, R., Aouad, G. and Sexton, M. (2000). 'Rethinking construction: the Generic Design and Construction Process Protocol', *Engineering, Construction and Architectural Management*, 7(2), 141–153.

Kamaruzzaman, S. N., Lou, E. C. W., Zainon, N., Mohamed Zaid, N. S. and Wong, P. F. (2016). 'Environmental assessment schemes for non-domestic building refurbishment in the Malaysian context', *Ecological Indicators*, 69, 548–558.

Laufer, A., Shapira, A., Cohenca-Zall, D. and Howell, G. A. (1993). 'Prebid and preconstruction planning process', *Journal of Construction Engineering and Management*, 119(3), 426–444.

Locke, E. A. and Schweiger, D. M. (1979). 'Participation in decision-making: one more look', in B. M. Staw (ed.). In Straw, B. M. (ed.) *Research in Organizational Behaviour*, JAI Press, Greenwich, CT.

McKim, R., Hegazy, T. and Attalla, M. (2000). 'Project performance control in reconstruction projects', *Journal of Construction Engineering and Management*, 126(2), 137–141.

MDC Legal Advisers (1984). *Uniform Building Bye Laws*, MDC Publisher Sdn Bhd Malaysia.

Meyer, V. W. (1999). 'Distribution and density of termite mounds in the northern Kruger National Park, with specific reference to those constructed by Macrotermes Holmgren (Isoptera: Termitidae)', *African Entomology*, 7(1), 123–130.

Mitchell, T., Tanner, T. and Wilkinson, E. (2006). 'Overcoming the barriers: mainstreaming climate change adaptation in developing countries', *Tearfund Climate Change Briefing Paper 1*, Tearfund, Middlesex.

Mitropoulos, P. and Howell, G. A. (2002). 'Renovation projects: design process problems and improvement mechanisms', *Journal of Management in Engineering*, 18(4), 179–185.

Miyatake, Y. (1996). 'Technology development and sustainable construction', *Journal of Management in Engineering*, 12(4), 23–27.

Organisation for Economic Co-operation and Development (OECD) (2013). *Environmentally Sustainable Buildings: Challenges and Policies*, OECD, Paris.

Pitt, M., Tucker, M., Riley, M. and Longden, J. (2009). 'Towards sustainable construction: promotion and best practices', *Construction Innovation*, 9(2), 201–224.

Price, S., Pitt, M. and Tucker, M. (2011). 'Implications of a sustainability policy for facilities management organisations', *Facilities*, 29(9/10), 391–410.

Rahmat, I. B. (1997). *The Planning and Control Process of Refurbishment Projects*, Degree of Doctor of Philosophy, University of London.

Shah, S. S. (2007). *Sustainable Practice for the Facilities Manager*, Blackwell Publishers, Oxford.

Staufer, L. A., Ullman, D. G. and Dietterich, T. G. (2000). 'Protocol analysis of mechanical engineering design', International Conference on Engineering Design, 74–85.

Suruhanjaya Tanaga Malaysia (2003). *Energy Efficiency: How to Take Advantage of it*, Suruhanjaya Tanaga Malaysia, PWTC, Kuala Lumpur.

The Commissioner of Law Revision Malaysia (1976). *Town and Country Planning Act 1976*, Malayan Law Journal Sdn Bhd and Percetakan Nasional Malaysia Bhd Malaysia.

The Communities and Local Government Statistic Release (2006). *House Price Index 2006.*

Tucker, M. (2013). *Sustainable Facilities Management*, Palgrave Macmillan, Basingstoke.

United Nations (UN) (1992). *Agenda 21.*

United Nations Environment Programme (UNEP) (2009). *Buildings and Climate Change.*

Watkins, E. (1996). 'Why renovate?', *Lodging Hospitality*, 52(12), 16–18.

Zhang, X. (2015). 'Green real estate development in China: state of art and prospect agenda – a review', *Renewable and Sustainable Energy Reviews*, 47, 1–13.

7 Operational building performance in tropical climates

Nik Elyna Myeda, Syahrul Nizam Kamaruzzaman and Cheong Peng Au-Yong

After studying this chapter you should be able to:

- apply appropriate tools and technologies to assure sustainable performance in use;
- discuss links between issues and challenges and the application of sustainable performance in tropical zones;
- appreciate the application of appropriate tools and technologies to accommodate performance planning and control, resolve performance barriers and optimise building performance; and
- recognise the strategic way to move forward in adapting to challenges and strategies within tropical regions.

Introduction

The growing evidence of climate change and global warming has grabbed the attention of many nations which are now implementing strategies to enhance sustainability of the built environment. This is as true in tropical regions as it is in the rest of the World. Buildings have been the subject of much research and regulatory attention as they have a significant impact on the environment in terms of energy consumption and carbon emission. It is anticipated that the contribution from buildings towards energy consumption and greenhouse gas emission will increase. Therefore, sustainable development has become a main concern in many nations to reduce the impact of buildings on the environment. One of the strategies is the adoption of building assessment tools. Various assessment tools have been developed and adopted by different countries with the aim of attaining sustainability for buildings. Environmental assessment tools have emerged as a yardstick to measure and, hence, promote sustainability in the built environment (Crawley and Aho, 1999). Their adoption serves an important role in promoting awareness of sustainable building practice.

Overview

There are many assessment tools that have been developed in different countries. The most widely used assessment tools are BREEAM (Building Research Establishment Environmental Assessment Method), LEED (Leadership in Energy and Environment) in the USA, CASBEE (Comprehensive Assessment System for Building

Environmental Efficiency) in Japan, BEAM Plus (Building Environmental Assessment Method) in Hong Kong, the Green Building Labelling System (GBLS) in Taiwan, HQE (High Environmental Quality) in France, Green Star in Australia and Green Mark in Singapore. In the Malaysian context, GBI (Green Building Index) and MyCrest (Malaysian Carbon Reduction and Environmental Sustainability Tool) are the two assessment tools widely adopted for building assessment.

BREEAM was developed by the Building Research Establishment (BRE) to evaluate the performance of new and existing buildings. BREEAM also has a separate tool for refurbishment and fit-out for buildings developed in 2015 (BREEAM, 2015). Globally, almost 2.2 million registered buildings and over 500,000 buildings have been certified via BREEAM (BREEAM, 2015). BREEAM assessments have been used as a template and reference model for the creation of assessment tools in Canada, New Zealand, Hong Kong, China, Norway and Singapore (Ding, 2008; Lee, 2013).

LEED, developed by the US Green Building Council (USGBC), applies to new and major renovation projects (LEED-BD + C), existing buildings (LEED-O+M), interior projects (LEED-ID + C), homes (LEED-homes) and neighbourhood development (LEED-ND). The LEED assessment has been used in forty-one countries, including Canada, Brazil, Mexico, India and China (Lee, 2013).

The CASBEE tool was developed by the Japan Green Building Council (JaGBC)/ Japan Sustainable Building Consortium (JSBC) and their sub-committees (Lee, 2013). CASBEE is applied to pre-design, new construction, existing building and renovation, corresponding to the building life cycle. Similar to BREEAM, CASBEE includes a tool for refurbishing existing buildings to produce more environmentally efficient building stock. BEAM Plus is a voluntary tool first launched in 1996 (formerly known as HK-BEAM). It was based largely on the BREEAM assessment tool but was modified for Hong Kong's densely populated infrastructure. It applies to new and existing buildings and covers a wide range of issues related to the impact of buildings on the environment in terms of global, local and indoor scales (Lee, 2013).

In Taiwan, GBLS (also known as EEWH: ecology, energy saving, waste reduction and health) was developed based on the country's subtropical climate, with high temperature and humidity. It consists of five individual tools, including Basic for general green building practices, Residential Building, Factory, Renovation for existing buildings and Community (Construction Specifications Institute (CSI), 2013). By the end of December 2013, GBLS had certified a total of 4,300 buildings as green buildings. The Green Star assessment tool was developed in 2003 by a non-government organisation, the Green Building Council of Australia (GBCA). It is a comprehensive, national, voluntary environmental assessment tool and has been adopted by other regions; for instance, Green Star New Zealand and Green Star South Africa were developed by the New Zealand Green Building Council and Green Building Council of South Africa.

HQE was developed by the non-governmental organisation HQE based in Paris, France. It originated in 1996 at the initiative of the French Ministry of Equipment to set environmental and health criteria for buildings (CSI, 2013). This tool applies to new buildings, existing buildings and urban development, with defined performance criteria which are then implemented through a series of management requirements.

In Singapore, Green Mark was introduced by the Building and Construction Authority (BCA) in 2005. It aims to evaluate buildings for environmental impact and performance, and promote sustainable design, construction and operational practices throughout the City-State Republic of Singapore (CSI, 2013). It can be applied to new buildings, existing buildings, office interiors, landed houses, infrastructure and districts.

In 2010, the Malaysian Institute of Architects (PAM) and the Association of Consulting Engineers Malaysia (ACEM) developed GBI. It is derived from Singapore's Green Mark and the Australian Green Star, but developed within the contextual, cultural and social needs of Malaysia's tropical climate, environment and development (CSI, 2013). GBI is applicable to new and existing residential and commercial buildings in Malaysia. On the other hand, the Public Work Department (PWD) Malaysia and Construction Industry Development Board (CIDB) have recently developed MyCrest to guide industry players in designing, constructing and operating buildings that integrate low carbon and sustainable practices. It was developed by taking into account the whole building life cycle beginning from pre-design until the demolition stage.

Assessment trends

Environmental assessment tools share similarities in broad criteria. After holistic reviewing, among the fourteen tools, assessment criteria can be divided into:

- management
- sustainable site
- transport
- indoor environmental quality (IEQ)
- water
- waste
- material
- energy
- pollution
- innovation
- economic
- social
- culture
- quality of service.

Management

This criterion deals with how buildings can be sufficiently operated and maintained throughout the building life cycle. It aims to encourage the adoption of sustainable management practices throughout the building life cycle (planning, design, construction, commissioning, handover, aftercare and maintenance).

Sustainable site

This is generally divided into two aspects which are construction site and ecological value. It is concerned with site protection by avoiding development of inappropriate sites and encouraging habitat protection and improving biodiversity. The use of

Contaminated land Brownfield sites is encouraged for development selection due to low ecological value. However, proper investigation should be conducted, such as site contamination assessment, before sites are chosen. For Greenfield site building development, adequate mitigation should be taken to minimise the impact of building development on existing site ecology. Existing features should be protected prior to and during site operation to enhance site ecology. Moreover, biodiversity consideration should be included to improve or maintain biodiversity conditions. It is to minimise disturbance to flora and fauna that could be caused by site activities.

Transport

In order to reduce congestion and air pollution due to private vehicles, better access to sustainable means of transport is encouraged in most of the assessment tools. The aim of this criterion is to deliver a good level of communication, through easy access to public facilities and services and adequate provision for pedestrians, cyclists and drivers.

Public transport accessibility is assessed on distance from main building entrance to public transport node and frequency of the services. It aims to encourage people to use public transport to work or leisure. Proximity to amenities means community connectivity and access to local services and amenities. Cyclist facilities should be provided to promote exercise and help to reduce congestion. There should be a limit of car parking capacity to reduce private car usage. Travel plans refer to site specific travel assessments, travel patterns and strategy for managing all travel and transport in the building, together with an overall consideration of the neighbourhood. Use of green vehicles is highly encouraged. This can be done by providing preferred parking spots for low emission vehicles.

Indoor environmental quality (IEQ)

IEQ is one of the most popular criteria in all the assessment tools. The most popular sub-criteria that are considered in most of the assessment tools are noise and acoustics, lighting and illumination, thermal comfort, ventilation and contamination level, whereas odour is least considered in most of the tools. Noise and acoustic criteria are assessed to ensure a building's acoustic performance (noise level, sound insulation, absorption, background noise) meet the appropriate standards and comply with design ranges. Visual comfort and performance include glare control strategy, adequate daylight factor and views from windows, appropriate illuminance (lux) level to the undertaken required tasks; lighting should be zoned to allow for occupant control, with use of high efficient lighting fixtures.

Appropriate thermal comfort levels can be achieved through design and control to maintain a thermally comfortable environment. Ventilation is important for healthy indoor air quality through an appropriate design of the balance of natural ventilation and artificial ventilation systems. Indoor air quality (IAQ) plans refer to pre-occupancy flush-out or other strategies in order to minimise sources of air pollution and ensure building ventilation systems are not contaminated. An adequate supply of fresh air is essential for air change effectiveness (ACE). The ventilation system must allow incoming fresh air to be correctly diffused throughout all of the rooms. Air quality sensors for CO_2 monitoring should be installed to alert the building manager when CO_2 levels exceed the set point. Contamination levels should be monitored to limit the sources and effects of indoor air pollutants. Air quality measurement must be undertaken by taking into account these pollutants. It includes VOC level, formaldehyde level, smoke control, mould prevention, electromagnetic pollution and biological contamination. Odour criteria consist of determining appropriate odour levels, a strategy for controlling the source of odours and the installation of de-odourising devices.

Water

Water is considered a limited and hugely valuable resource, and as such all of the assessment tools include water efficiency and recycling in the quest for sustainable water use and management. Water consumption levels for major building services must be monitored. Main or sub-metering should be installed on the main water supply for monitoring. Leak detection devices can be installed to detect leaking to avoid water wastage. Efficient use of water should be provided using more efficient type of water efficient fittings. Water consumption can be minimised through water recycling such as rainwater harvesting, grey water recycling and efficient irrigation systems. Ground water can be recharged through permeable paving or landscaping. The building should also demonstrate a reduction in annual sewerage volumes. Cooling towers can be used for cooling purposes, to reduce potable water use.

Waste

Waste criterion are covered comprehensively in all assessment tools to ensure best practice in the management of construction and operational waste. Waste management plans are essential during construction in order to schedule in advance how the waste will be collected and sorted on construction sites. Different waste requires different treatment such as recycling, disposing or landfill. A proper waste treatment plan will encourage reuse or recycling materials to optimise material efficiency. Dedicated space for waste storage will help to facilitate waste collection, with waste facilities such as recycling bins for collecting different type of waste, to ease recycling purposes.

Material

Building materials are one of the most important criteria in the majority of assessment tools due to their complicated life cycle process from extracting raw materials until the disposal stage. Consideration can be divided into material selection, material disclosure information, efficient use of material and use of green products. In terms of material selection, materials with low environmental impact are considered in the majority of assessments too. The purpose of this concern is to raise awareness of the impact of different materials by taking account the life cycle assessment (LCA) of the material. Most of the tools encourage use of LCA tools such as building information modelling (BIM) to measure the environmental impact of the material over the building life cycle.

Besides, use of renewable material, recycled material and reuse of existing frame materials are encouraged to reduce the production of new materials. The material efficiency over a life cycle is assessed to integrate building design and buildability with selection of reused building materials. This is done by taking into account embodied energy, durability, carbon content and life cycle costs. Regional material refers to use of materials manufactured locally. Responsible sourcing of materials is emphasised in most of the assessment tools. It generally refers to timber or timber based products to ensure it is legally harvested and from sustainable sources. Material ingredients refer to products used in the building for which the chemical ingredients in the product are inventoried using an accepted methodology or select products are verified to minimise the use and generation of harmful substances.

Prefabrication and the use of modular and standardised designs are encouraged to reduce wastage of materials and quantities of on-site waste. Use of green products, such as environmentally friendly refrigerants and clean agents, are encouraged by all the standards.

Energy

Assessment tools place great importance on the energy criterion due to its significant impact on the environment. With increasing concerns about global warming and the greenhouse effect, energy assessment is concerned with energy performance for major building systems, their efficient operation and strategy, energy management and the use of natural resources.

The energy performance and consumption for HVAC systems, refrigerators, lifts/escalators, external lighting, car parks, roofs and building envelopes should be calculated to determine how much potential for improvement will be achieved. Energy saving for these building systems and equipment is essential for efficient operation to achieve optimum performance. CO_2 mitigation strategies such as low carbon design (passive design) or low carbon technologies should be considered. Energy efficient fittings or equipment should be used to achieve better energy saving and performance. To enhance the building energy efficiency, the built form and building orientation should be considered to enhance energy conservation. Energy metering systems can be installed to facilitate energy consumption monitoring. In order to reduce reliance on energy, the use of renewable energy should be encouraged.

Pollution

This criterion deals with outdoor sources of air pollution in order to reduce overall pollution and assesses whether adequate provisions are taken to limit the effect of the pollution. The impact of refrigerants should be assessed to reduce the impact through specification and leak prevention. Night light pollution refers to waste light from lighting that produces glare and obscures the night sky which reduces and avoids nuisance to the neighbourhood. Noise pollution should be assessed to reduce the likelihood of disturbance, arising as a result of noise, to noise-sensitive areas (residential, hospital, school).

In order to reduce watercourse pollution, construction site discharges should be properly managed through drainage design or decreasing the impermeable area. Heat island effects can be reduced by using shading, or planting trees through properly planned landscaping. NO_2 and CO_2 emissions should comply with appropriate criteria to control negative emissions. Construction activities generate dust and emissions during construction; hence, it is necessary to reduce the pollution through controlled soil erosion, waterway sedimentation and dust. It is noted that CASBEE evaluates wind pollution and earthquake resistance within the local conditions of Japan because this is a particular problem in that country.

Innovation

The majority of the assessment tools support innovation in their evaluation frameworks which can provide environmental benefits. Any new methods that can be shown to improve sustainability performance of a building are highly encouraged.

Economic

Most of the assessment tools lack the capacity to evaluate financial aspects, except CASBEE. CASBEE is concerned with construction cost, life cycle cost, operating and maintenance cost, investment risk, affordability of residential rental and commercial viability. The impact of the project on land value of adjacent properties and local economy also falls into this criterion.

Social

Social sustainability is essential in taking care of welfare (privacy, security, amenities) and equitably distributed among social classes and gender. It accesses privacy and participation of building users such as conduct user satisfactory assessment. Public open space can be provided such as shelter from rain or when waiting for people. Besides, disabled accessibility should not be neglected in the assessment tool. When designing buildings, the well-being of disabled persons should be catered for to enhance social integration. Building amenity features such as kiosks, stores and meeting rooms can be provided for the benefit of building occupants.

Culture

The culture aspect is neglected in most of the assessment tools except CASBEE. This tool proposes that building development should integrate local cultural value to enhance and promote cultural value in design. The presence of culture and heritage can be enhanced through its integration with recent designs and also through the use of locally significant materials. Historic interior and exterior spaces in buildings should remain preserved and maintained which can contribute to local culture. Improved streetscapes can also be considered to enhance the culture aspect.

Quality of service

CASBEE comprehensively evaluates the quality of service in the assessment tool, which is partly neglected in other tools. Assessment of service functions is crucial to keep the building in good condition in the long term.

Safety and security refer to safe access to/from the building or construction site. It can be provided by safety burglar alarm systems and CCTV. Functional and efficient refer to appropriateness of space provided for the required functions.

It means adequate floor area is allocated within the effective floor area or spaciousness. Flexibility and adaptability encourage consideration and implementation of measures to accommodate future changes to the use of the buildings and its system over its lifespan. Hence, a functional adaptation strategy can be conducted in design aspects to facilitate the replacement of all major plans within the life of a building.

Controllability of systems and maintenance of performance is considered broadly in most of the assessment tools. The effectiveness and design of control for lighting or technical systems is important for building occupants. In addition, maintenance of performance is important to consider by ensuring that maintenance of the building can be carried out with good access conditions to prolong the building life cycle. First, durability and reliability of materials, building services and systems is classified under this criterion to enhance durability and minimise frequency of replacement.

Way forward

Sustainability has attained worldwide attention due to increasing global warming, energy consumption and GHG emission. Various tools have been developed and adopted by different countries with the aim of attaining sustainability for buildings. However, it was found that there is a lack of assessment tools for building refurbishments. Existing buildings appear to have a big potential in reducing greenhouse gas emissions and energy consumption which eventually could achieve sustainability in the built environment. Building refurbishment is a way forward for every country to achieve carbon reduction commitments. Thus, a refurbishment assessment tool is crucial and requires further research.

Building information modelling

The construction industry is fragmented in nature which inhibits improvement in its performance. This fragmentation is the root cause of many problems in a construction project such as project delays, cost overrun, poor quality and dispute. It also creates adversarial culture, poor communication and coordination among the project parties. Much research has been conducted all over the world to try to solve the problems of fragmentation in construction. Many researchers and construction industry practitioners dealing with fragmentation of construction would agree that integrated practices could solve many problems within construction.

It is believed that the implementation of technology enhancements is one of the feasible ways to improve construction performance by reducing industry fragmentation. Building information modelling (BIM) is viewed as one of the solutions to eliminate industry fragmentation and inefficiencies. The application of BIM brings tremendous benefits and advantages to the construction industry. The concept of BIM is to virtually construct a building in a computer model prior to building it on site. Hence, it is possible to simulate and to analyse potential impacts, identify possible mistakes and errors and, most importantly, make adjustments before the building is constructed. This approach can help to avoid serious impacts on the project as most of the problems and issues have been identified and resolved earlier. Traditionally, errors and omissions in paper documents often cause unanticipated costs, delays and eventual lawsuits between various parties in a project team. One of the most common problems associated with paper based communication during the design phase is the considerable time and expense required to generate critical assessment information about a proposed design, including cost estimates, energy-use analysis, structural details, etc. These analyses are normally done last, when it is already too late to make important changes.

BIM is viewed as the solution for improving and rectifying the inefficiencies in the traditional business processes of the construction industry. Instead of sharing information through paper based documents, BIM utilises a single shared repository that contains all project information that can be accessed by all project participants. Thus, BIM has the capability to save cost, reduce time and improve the quality of work or, in other words, significant improvements can be attained in terms of time, cost, quality and efficiency in construction projects. Therefore, it is highly desirable for the construction industry to get immersed in BIM application.

Features of BIM

Eastman *et al.* (2008) defined BIM as "tools, processes, and technologies that are facilitated by digital and machine-readable documentation about a building, its performance, its planning, its construction, and later, its operation". BIM is an associated set of processes of using modelling technology to produce, manage and share information in a model with the use of BIM related tools. It is a process of project

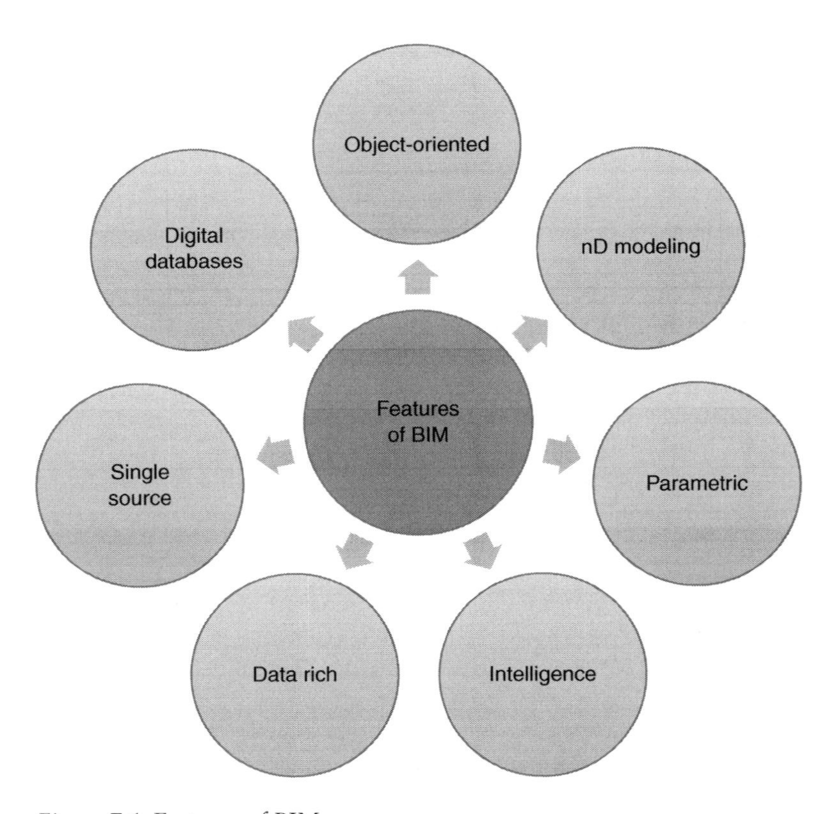

Figure 7.1 Features of BIM.

simulation through a 3D model and link information of project life cycle associated with it. BIM is an alternative to the traditional, paper based, 2D or 3D CAD based. It is known as a data-rich, object-oriented, intelligent and parametric digital representation of a facility. Thus, project participants are allowed to insert, extract, update and modify data and information that can be used to make decisions and to improve the process of delivering the building (Associated General Contractors of America (AGC), 2005). There are certain features and capabilities promised in BIM, which are different from CAD.

Object-oriented

BIM is object-oriented in nature as it contains specific characteristics, properties and rules. The rules and characteristics embedded in objects are allowed for adjustment to the objects automatically when a change is made to the model as the information in the BIM is interconnected. Traditionally, a designer would draw a line to depict the position of a wall or door which can only be interpreted by certain people. However, by using BIM, it creates an object-oriented database that is made up of intelligent objects, for example, walls, doors and windows with attributes and relationships between the building elements.

nD modelling

Multi-dimensionality is one of the well-known features of BIM. The objects of the model can be in different states in different phases of the life cycle in order to represent the 'N' dimensional information about the building (Isikdag *et al.*, 2007). Thus, BIM is not just limited to 3D geometric, but further dimensions include 'time' (4D), 'cost' (5D) factors, 'sustainability' (6D) and facility management (7D).

Parametric

One of the main and important features of BIM is parametric, which makes building information more reliable and coordinated. It is a process by which adjacent elements or assembly are automatically adjusted when one element is modified in order to maintain a previously established relationship (Stine, 2011). Parametric modelling uses parameters (numbers or characteristics) to determine the behaviour of a graphical entity and to define relationships between model components (Autodesk, 2007). Parametric modelling combines a data model (geometry and data) with a behavioural model (change management) that gives meaning to the data through relationships (Autodesk, 2003). Due to the relationships between model components, parametric change engines will determine which other related elements need to be updated and changed when the user modifies an element in the model.

Intelligence

BIM is 'intelligent' due to the relationships that are built into the model. When changes are made, components within the model know how to interact with one another to accommodate the changes. All information related to the building, including its physical and functional characteristics, project life cycle information, graphical and non-graphical information, appears in a series of 'smart' objects (Azhar *et al.*, 2008). Hence, the model is 'smart' by managing the attributes and the relationships between the building components.

Data rich

Graphical and non-graphical data such as drawings, specifications and schedules are included in the model. Information such as geometry, spatial relationships, geographic information, quantities and properties of building elements, cost estimates, material inventories and project schedule (Azhar *et al.*, 2008) are all available in the model. Hence, BIM is called a rich model because all objects in it have properties and relationships.

Single source

BIM provides a single and consistent source for all project information. All the required data and information throughout design, procurement, construction, operation and maintenance of a building will be stored and available in one accessible location. It makes the information available for use and reuse at every point in the project. Hence, it is a single repository of information base that can be accessed by all project participants.

Digital databases

All the information is created and stored in a BIM database instead of a 2D drawing file or spreadsheet. It will be available for use and reuse at a later stage of the project. The building information is presented on a presentation format that is suitable for a particular user to edit and to review. Although the type of presentation format is distinct for different project participants, all views are in a same information model. Once changes are made by one of the project participants in their presentation format, BIM assures that changes made in any of these views are reflected in all other presentations through digital databases (Autodesk, 2002).

Benefits of BIM implementation

BIM application has the ability to:

- i minimise project management;
- ii foster communication and coordination;
- iii identify errors early;
- iv reduce rework;
- v reduce costs; and
- vi improve quality.

Therefore, productivity gain is one of the major benefits which has made it widely accepted as a best practice approach for delivering major building projects. It enables project parties to make better decisions at very early stages. It can streamline and aid clear communication between client, consultant and contractor in construction projects by providing a single respiratory system for exchanging digital information in one or more agreed formats.

Another benefit generated by implementing BIM is that design errors and omissions can be discovered before construction. According to Eastman *et al.* (2008), design errors caused by inconsistent 2D drawings are eliminated in BIM because the virtual 3D building model is the source for all 2D and 3D drawings. In addition, because systems from all disciplines can be brought together and compared, multi-system interfaces are easily checked both systematically (for hard and soft clashes) and visually (for other kinds of errors). Conflicts are identified before they are detected in the field. Coordination among participating designers and contractors is enhanced and errors of omission are significantly reduced. This results in the acceleration of construction processes, reduction of costs, reduced likelihood of legal disputes and provision of a smoother process for the entire project team.

The next benefit of implementing BIM is the ability to synchronise design and construction planning. Based on the study conducted by Huang *et al.* (2007), the study showed that 4D modelling enables the users to visualise the constructability of the proposed construction approach. The system also assists the project team to design a precise construction schedule so as to remove any potential unproductive activities. Effectively, as opposed to traditional bar chart techniques, 4D models can identify and eliminate construction related problems which might not be foreseen during the planning stage. This has directly contributed towards saving costs and time in the long term. The project scheduling process can be enhanced with the aid of a preliminary schedule, where a detailed construction sequence can be stimulated. This benefit would allow the project team to consider alternative approaches pertaining to process flow, equipments and resources towards achieving the best value of the project.

BIM also contributes largely in terms of cost estimating, planning and scheduling with the incorporation of time and cost elements. In terms of cost estimating, BIM can facilitate the quantity surveyor quantifying the cost and the material of the project in a shorter time which can be reduced by up to 80 per cent compared to traditional methods. BIM also ensures costs can be saved through ensuring that minimum changes are required throughout the whole process.

The most significant value of BIM is through 3D visualisation where it enhances the understanding of the buildings, where previously the information was more reliant on 2D drawings. This would typically reduce the number of Total Request Information (TRI) during implementation. The emergence of improved 3D visualisation also supports the decision-making process and aids the evaluation process by both technical and non-technical parties.

BIM technology facilitates simultaneous work by multiple design disciplines (Eastman *et al.*, 2008). While collaboration with drawing is also possible, it is inherently more difficult and time consuming than working with one or more coordinated 3D models in which change control can be well managed. This shortens the design time and significantly reduces design errors and omissions. It also gives earlier insight into design problems and presents opportunities for a design to be continuously improved. This is much more cost effective than waiting until a design is nearly complete and then applying value engineering only after the major design decisions have been made. Early collaboration facilitates the project planning towards reducing the potential issues and risks.

BIM deliverables throughout the project life cycle

BIM is a standardised machine-readable information model to improve design, construction, operation and maintenance processes throughout a project life cycle. The model can manage building design and project data throughout all phases of the project life cycle from conception, through design and construction, to operations and maintenance.

Design phase

Visualisation is the main feature of BIM during the design phase. It enables architects, engineers and contractors to have visualisation what is to be built in a simulated environment. Any potential issues and uncertainties related to design, construction or operations can be identified and detected earlier in the virtual environment. In this context, all the relevant aspects of the project can be considered and planned before the actual construction takes place to avoid abortive work and to reduce construction waste.

BIM is able to improve the accuracy of cost estimation. Material quantities are automatically quantified and extracted from the model which could reduce human errors and save time. Changes are inevitable in construction projects, which requires costing to be updated regularly. BIM allows drawing sections, perspectives, plan views and quantity take-offs to be updated effectively to ascertain potential costs. Changes are detected and updated automatically when they are made in the model, which ensures accuracy and consistency. Furthermore, one of the popular uses of BIM is for clash and conflict detection. All major systems can be visually checked in BIM for interferences to detect clashes. Any design conflict and clashes can be identified and resolved early to avoid change orders which often cause delay and extra costs.

Construction phase

BIM plays an important role in planning, scheduling and sequencing of the construction phase. Traditionally, it is difficult to view and coordinate the work planning at construction sites, especially for large and complex projects. However, by using BIM, the construction schedule can be developed in the model, which facilitates planning, monitoring and visualisation of construction progress.

BIM allows for the consideration of alternative approaches to sequencing, site logistics, site access, site planning and layout, and crane and material placement (Haron *et al.*, 2009). Hence, contractors can coordinate the site, more efficiently to develop traffic layouts and identify potential hazards at the construction site, which can aid in preparing a more realistic site safety plan. Moreover, constructability is crucial to ensure synchronisation of design and construction planning. Based on a study conducted by Huang *et al.* (2007), 4D modelling enables its users to visualise the constructability of the proposed construction approach. Any construction issue or problem can be identified and avoided before going to the construction site.

Management stage

Building information models contain complete information about a building from planning to construction. In the management phase, information such as building spaces, systems and components on the use or performance of the building; its occupants and contents; the life of the building over time; and the financial aspects of the building can be leveraged for use by facility managers.

BIM provides two major benefits to facility managers which are: (1) the same critical information is present in a single electronic file and (2) facility managers do not have to sift through piles of information to gather data. For example, facility managers can click on any equipment or fixture to obtain information on the product, warranties, life cycle of the product, maintenance checks, replacement cost, installation and repair procedures, and even place an order for a replacement online (Jordani, 2008). Facilities managers are able to access, track, update and maintain this information to improve the effectiveness of operations throughout the life cycle of the building.

BIM provides owners and facilities managers with digital records of information in relation to architectural, structural, mechanical, engineering and plumbing elements of the building. By using information in a BIM record model for asset maintenance, facilities managers can evaluate the cost implications of changing or upgrading building assets. Facilities managers can produce accurate quantity takeoffs of current company assets for financial reporting and estimating the future costs of upgrades or replacements. Furthermore, using BIM for space management will provide area information for space and occupancy, which enables the facilities team to analyse the existing use of space, evaluate proposed changes and effectively plan for future needs.

BIM adoption has been gaining increased support from industrial bodies and regulators in many countries. It is essential to consider the application of BIM in practice as the application will be beneficial for the construction industry in future. The implementation of BIM is certain to become increasingly important as it offers abundant benefits to the construction industry. Hence, more research on the area of BIM application is needed as it will be beneficial for the construction industry.

FM practice in tropical zones

The practical relevance of FM to organisations in all sectors of the economy is now growingly recognised. The attraction of FM is becoming increasingly common, as forward-looking organisations are beginning to realise FM as a function with clearly defined objectives, and as a strategic and commercially oriented discipline. Organisations are increasingly realising that they must seek some form of competitive advantage from every part of their organisation, which must include the costs of running the working environment. Accordingly, organisations seek to improve their competitiveness by introducing a core business philosophy and restructuring to release senior management time to improve effectiveness. In these organisations, although the focus is on the core activities, facilities are no longer of marginal significance. The strategic role of FM is recognised and the opportunities provided through effective management are better understood. Managements have begun to realise that for organisations to benefit from the enormous investment in facilities, they have to begin managing them actively and creatively with commitment and a broader vision (Pathirage *et al.*, 2008).

In tropical zones the mean annual temperature is not as important as the annual temperature range which gives an indication of the variation throughout the year.

Even more important is the daily range. For example, temperatures of stone surfaces of the Borobodur temple complex in Indonesia were recorded as having risen from 25°C to 45°C in four hours to a depth of approximately 5 cm. The same applies to relative humidity. The distribution of rain is an important factor and it is the daily fluctuation that counts for conservators. During the day the maximum levels of relative humidity occur a little before sunrise and the minimum levels in the early evening, approximately the reverse of temperature (Davison, 1981). The main climatic factors affecting human comfort and relevant to construction are: air temperature, its extremes and the difference between day and night, and between summer and winter; humidity and precipitation – incoming and outgoing radiation, the influence of the sky condition, air movements and winds (Gut and Ackerknecht, 1993).

When an economy faces an increase in market value of goods and services produced over time, it positively affects consumption including business investments to improve facilities management. Forecasting for future business planning means understanding that global economic power will continue to shift eastwards. Perhaps the strongest empirical relationship in the wealth and poverty of nations is one between ecological zones and per capita income. Economies in tropical ecozones are usually poor everywhere, while those in temperate ecozones are often rich. And when temperate economies are not rich there is typically a straightforward explanation, such as decades of communism or extreme geographical isolation. Of the thirty economies classified as high-income by the World Bank, only two large economies – Hong Kong and Singapore – are in the geographical tropics, and these two constitute just 1 per cent of the population of the rich economies. Since sea navigable regions are generally richer than landlocked regions, regions that are both temperate and easily accessible to sea based trade almost everywhere have achieved a very high measure of economic development. Tropical and landlocked regions, by contrast – such as Bolivia, Chad, Niger, Mali, Burkina, Faso, Uganda, Rwanda, Burundi, Central African Republic, Zimbabwe, Zambia, Lesotho and Laos – are among the very poorest in the world.

The pattern of temperate-zone development and tropical-zone underdevelopment can also be seen within large nations that straddle ecozones and within highly integrated, multi-national regions such as Western Europe. Thus, the sub-tropical US South lagged behind the temperate US North in industrialisation; Brazil's tropical North-east, though the original site of Europe's colonisation of Brazil, has lagged the temperate South-east for one-and-a-half centuries; temperate North-east China has long had higher per capita income than the sub-tropical South-east China; and of course Northern Europe industrialised roughly a half-century or more ahead of Southern Europe (Sachs, 2001).

FM in tropical fully humid or tropical wet

In the tropical wet region, very few designers take into account the climate of the surroundings and the variety of building construction methodologies usually employed in Europe and America, which are becoming increasingly more prevalent in Africa. As a result, thermal discomfort dominates these types of buildings and necessitates the occupants using artificial heating, ventilation and air conditioning systems (HVAC). Owing to the high outdoor relative humidity in this region, some indoor ambience problems have been identified in the past few years. In particular, in newly renovated buildings located in central and eastern Cameroon, the high indoor moisture-laden relative humidity is a persistent problem. Consequently, the indoor air almost always contains a higher percentage of water vapour than the outdoor air which can cling to the walls with the risk of condensation and wall damage. Furthermore, another side effect of this higher level of relative humidity is poor indoor air quality and its related effects on the occupants. The key to controlling this drawback is related to building construction materials. Studies have shown how a structure made of solid wood can control changes in humidity even if the ventilation rate is relatively high. Despite this, it can be concluded that it is difficult to precisely quantify the phenomenon and the effect in terms of comfort for people and their health. A few studies have concluded that indoor temperature is affected by moisture and that the increase in relative humidity depends on the rate of air exchanges, outside air and moisture transfer between structures and indoor air (Nematchoua and Orosa, 2016).

FM in tropical monsoonal

For a significant proportion of the tropical zones there is a unique seasonal phenomenon each summer – the monsoon. This meteorological maelstrom of events encompasses more than just rain; it includes extreme heat, flooding and severe storms which are all hazards associated with this time of year.

According to MonsoonSafety.org, monsoon season in the USA runs from 15 June to 30 September. During this time, regions in the country's south-west can expect any number of extreme weather events, including heat, high winds and heavy rain.

The monsoon in the USA brings more than just lots of rain. The season usually begins with an extreme heat wave that can cause significant fire hazards in areas that are already particularly arid. The heat is often followed by a series of extreme thunderstorms, high winds and even flash-flooding. Together these phenomena have accounted for an average of ten deaths and sixty injuries every year since 1995.

There is a lot that the monsoon can throw at buildings that can result in costly damage. The thunderstorms alone can often result in winds exceeding 60 mph. This makes any building a very appealing stationary target for flying debris that could damage the facility's exterior. The monsoon can throw everything from dust storms to tornadoes.

Poor and improper building maintenance will definitely cause more damage and costly repair work if left unattended. In Malaysia, for example, buildings are built in accordance with British Standard and under strict supervision. Unfortunately, the maintenance aspects of the buildings are still weak. Making it worse, sometimes building maintenance is perceived as merely about the mechanical and electrical systems in the buildings, without much consideration given to structural elements. Some common problems include waterproofing systems, cracks and soil settlement. For instance, with a tropical climate and average rainfall of 250 cm in a year, a good waterproofing system is crucial for buildings with flat roofs. However, the performance of waterproofing systems depends on many factors, i.e. quality of the materials, the skill of workers, application methods, substrates condition, weather, maintenance, etc. (Sulaiman, 2013), and the quality assurance processes required to guarantee water highness can sometimes be below par.

FM in tropical wet and dry

Tropical wet-dry climates, a major climate type of the Köppen classification, are characterised by distinct wet and dry seasons, with most of the precipitation occurring in the high-sun ('summer') season. The dry season is longer than in tropical monsoon and trade-wind littoral (Am) climates and becomes progressively longer as one moves poleward through the region. The tropical wet-dry climate is abbreviated as Aw in the Köppen-Geiger-Pohl system. Temperatures in tropical wet-dry climate regions are high throughout the year but show a greater range than wet equatorial (Af) and Am climates (19–20°C in winter and 24–27°C in summer).

In the past few years, various research works have shown that people spend about 90 per cent of their time in an indoor environment. In view of this, the major concern of building designers is to ensure that buildings are built so as to provide comfortable and healthy conditions for their occupants. Nowadays in new cities, lifestyles have changed dramatically, especially in sub-Saharan Africa. In these regions, very few designers take into account the climate of the surroundings and the variety of building construction methodologies usually employed in Europe and America, which are becoming increasingly more prevalent in Africa. As a result, thermal discomfort dominates these types of buildings and necessitates the occupants using artificial heating, ventilation and air conditioning systems (HVAC). Owing to the high outdoor relative humidity in this region, some indoor ambience problems have been identified in the past few years. In particular, in newly renovated buildings located in central and eastern Cameroon, the high indoor moisture-laden relative humidity is a persistent problem. Consequently, the indoor air almost always contains a higher percentage of water vapour than the outdoor air which can cling to the walls with the risk of condensation and wall damage. Furthermore, another side effect of this higher level of relative humidity is poor indoor air quality and its related effects on the occupants. The key to controlling this drawback is related to building construction materials. Hameury's studies have shown how a structure made of solid wood can control changes in humidity even if the ventilation rate is relatively high (Hameury, 2005). Despite this, from these same studies, it can be concluded that it is difficult to precisely quantify the phenomenon and the effect in terms of comfort for people and their health. A few studies have also concluded that indoor temperature is affected by moisture and that the increase in relative humidity depends on the rate of air exchanges, outside air and moisture transfer between structures and indoor air (Nematchoua and Orosa, 2016).

Performance planning and control

The global atmosphere is undergoing a period of rapid human-driven change, with no historical precedent in either its rate of change or its potential absolute magnitude (IPCC, 2002). There is considerable concern at how this change may affect the Earth's ecosystems, and in turn how these system responses may feedback to accelerate or decelerate global change (Malhi and Wright, 2004).

Impact of climate change on tropics

The climate is changing in the tropics, as it is in the rest of the world (IPCC, 2013). The effects of steadily rising concentrations of greenhouse gases on the climate may be less obvious to tropical residents, however, because they are overlain by considerable natural variability. The tropics have warmed by 0.7–0.8°C over the last century – only slightly less than the global average – but a strong El Nino made 1998 the warmest year in most areas, with no significant warming since.

More than anything else, fear of economic fallout has blocked significant steps to cut US carbon dioxide emissions. That fear has led opponents of emissions reductions to disregard both scientific evidence and probable consequences when it comes to global warming.

Consider that the European Union has cut greenhouse gas emissions by 5 per cent since 1990. Some steps can be taken for little or no cost. Others involve proven, energy-efficient technologies that are cost-effective today, though they may require longer payback periods than many organisations now accept. And alternative energy sources are moving closer to being cost-competitive with utility power – power that is not priced to reflect the future costs of undoing the harm caused by the burning of fossil fuels to produce electricity.

In addition serious effort to drive down energy use in buildings would certainly spur technology development, just as the CFC phase-out led to more efficient chillers.

Reducing greenhouse gas emissions won't be painless. The scientific evidence demands action. It is irresponsible to argue that economic risk is a reason not to take significant action.

Optimising building performance

The development of performance measurement in building management aims to optimise the quality and service of buildings, as well as meeting energy and cost parameters. Measurement of building performance is an assessment that helps to identify the strengths and weaknesses of operation and maintenance activities. In addition, the result of performance measurement indicates the effectiveness of existing operation and maintenance strategies. Consequently, management teams are able to plan and make appropriate decisions for future planning of strategy.

In general, building performance assessments involve all stakeholders, but typically those who design, manage, operate, maintain, reside or work in a building as its cores. In some circumstances, stakeholders can also be the people who live in the communities next to, or living systems that are influenced by the environmental impacts of buildings. Their experience, comment, feedback, complaint, suggestion and satisfaction are reviewed thoroughly through the assessments. Subsequently, their responses can be utilised to optimise the design, operation and maintenance outcomes.

The strength of the assessments is the extraordinary interdisciplinarity, which incorporates the expertise of architects, planners, facilities managers, psychologists, sociologists, anthropologists, geographers, and many others. This diversity helps to address the inherent complexities of human-environment relationships. Meanwhile, numerous benefits are attributed to the evaluations and assessments, including:

- improved communication among the design teams, clients and stakeholders through integrative and collaborative/participative processes;
- clear and measurable performance criteria;

- informed briefing/programming from lessons learned from previous buildings;
- support for predicting the effectiveness of emerging or alternate designs;
- promotion of better functioning and facilities management of buildings in-use;
- a means to monitor and diagnose potential problems;
- support for programme review when planning alterations and adaptive reuse; and
- a platform for learning through the development of design guidelines, standards and best practices.

Commonly, the assessment of building performance can be obtained through the level of success or failure in terms of scheduling, cost and functionality. It should not only focus on quantifiable aspects, but should also emphasise the quality of operation and maintenance activities. Nowadays, in tropical regions, the most emphasised performance aspects are energy efficiencies and cost parameters, as well as health, safety and environmental issues. Nevertheless, they are not limited to these and many other aspects are taken into account in building performance measurement, including:

- health, safety and environmental issues
- operational and maintenance costs
- energy efficiencies
- property and asset values
- organisational achievements
- customer or user satisfactions
- operational and maintenance downtime
- building functionality and quality.

Building is one of the largest energy consumers as it consumes about 30–40 per cent of the world's primary energy consumption. Undeniably, energy consumption of air conditioning systems covers the substantial proportion of energy consumption of a building. The climate in tropical regions is generally hot and humid throughout the year. The use of air conditioning systems is vast in order to enhance the indoor thermal comfort. In tropical countries, the air conditioning system can easily consume up to half of the electricity use of the building. Due to climate change and the growing demand for living quality, it is foreseeable that primary energy consumption might be growing continuously in the construction and building industry. In order to make the first move, an energy management programme is introduced. This programme does not only lead to energy efficiency, but it even produces significant financial benefits. Indeed, energy efficiency attained through a proper planned energy management programme inspires lower running costs, making an extensive cost saving over the building life span. Meantime, the programme encourages less consumption of natural resources and reduced emissions. Enhancing energy efficiency can be beneficial creating, better buildings with greater comfort, a conducive working environment, more satisfied users or customers and increased productivity (Haji-Sapar and Lee, 2005)

Energy management programmes closely link to the application of green technologies. Theoretically, the integration of energy management and green technology applications greatly encourages the optimisation of building performance in the aspects of energy and environmental performance. In fact, various technologies and applications have been introduced in the construction and building industry to minimise energy usage and environmental impact, such as the use of renewable energy to replace the fuel source. Some of the technologies and applications are suitable in tropical regions due to the climate conditions. Currently, appropriate applications of renewable energy in the construction and building industry in the tropics include:

- photovoltaic energy generation
- solar water heating
- water supply and recycling – rainwater harvesting system.

(Groenhout *et al.*, 2008)

Therefore, one of the sustainable performance objectives is to optimise the building energy cost. There are a number of researches focusing on the energy performance optimisation in tropical regions. For example, Kong *et al.* (2012) appraised that centralised air conditioning system consumes less energy compared to split units. However, the thermal imbalance occurs as a result of hydraulic imbalance in the centralised system, causing lower temperature in lower spaces and higher temperature in higher spaces in multi-storey buildings. The researchers suggested installing a pressure balancing valve in the air conditioning system for each floor to regulate the circulation of water flow.

Furthermore, Pillai *et al.* (2014) noted the building integrated photovoltaic (BIPV) system as a sustainable power generator and it tends to minimise greenhouse gas emissions. The BIPV system can also be easily installed in both new and existing buildings, and thereby save on land requirement and construction cost. The system is very suitable for tropical regions as the regions receive good solar insolation all year round. On average, a country which has 250 sunny days per year can translate to 5,000 trillion kWh per year, which generates significant amounts of power sustainably.

Meanwhile, Lei *et al.* (2016) studied the energy performance of building envelopes integrating with phase change materials (PCMs) for cooling load reduction in the tropical zone. They demonstrated that the PCMs can efficiently decrease the heat gains through building envelopes in a range of 21–32 per cent throughout the year in the tropics; while other regions with PCMs are only efficient in certain seasons. Subsequently, the use of energy to operate the air conditioning system can be minimised without jeopardising the occupants' comfort.

In addition, Wan Mohd Nazi *et al.* (2017) revealed that reducing cooling load and increasing the air conditioning system efficiency is a major component in the decarbonisation of buildings in tropical countries. They recommended several actions to optimise the energy performance of air conditioning systems by reducing the cooling load as follows:

- lighting system – includes automatic daylight dimmer in off-activity zones and replaces existing lamps with high-efficiency LEDs;
- air conditioning system setting – identify a suitable temperature setting and operation schedule while ensuring thermal comfort is achieved;
- renewable energy – installation of solar panels to support power supply, reduce energy cost and limit building envelope heat gain.

In common, the environmental conditions under tropical climates change frequently in the aspects of temperature and pressure. The frequent change of these aspects increases the rate of building deterioration. In order to optimise the building performance, it is essential to mitigate the increasing rate of deterioration. Thus, consideration on environmental conditions cannot be neglected in building operation and maintenance activities. Maintenance detailing, materials and components that provide technical feed for easy maintenance under tropical climate change are the approaches to optimise building performance by saving maintenance costs. For instance, use of materials that are able to stand the local exposure conditions and for long service life is critical to enhance system quality and performance (De Silva *et al.*, 2012).

The sustainable performance of a building relies on the characteristics and dynamic interaction of building elements, external weather conditions, internal climate settings, facilities, mechanical and electrical systems, and occupant behaviour. The interaction of these characteristics is very sophisticated and can substantially influence performance. In order to optimise the building performance, it is vital to take all characteristics and their interaction into account during the operational stage.

References

Associated General Contractors of America (AGC) 2005. *The Contractor's Guide to BIM*, Las Vegas, NV, AGC Research Foundation.

Autodesk. 2002. *Building Information Modeling: White Paper* [online]. Available: www.laiserin.com/features/bim/autodesk_bim.pdf [accessed 14 October 2012].

Autodesk. 2003. *Building Information Modeling in Practice: White Paper* [online]. Available: www.ddscad.com/BIM___In_Practice.pdf [accessed 15 October 2012].

Autodesk. 2007. *Parametric Building Modeling: BIM's Foundation* [online]. Available: www.consortech.com/bim2/documents/bim_parametric_building_modeling_EN.pdf [accessed 13 October 2012].

Azhar, S., Nadeem, A., Mok, J. Y. and Leung, B. H. 2008. Building Information Modeling (BIM): A new paradigm for visual interactive modeling and simulation for construction projects. First International Conference on Construction in Developing Countries, 4–5 August 2008. 435–446.

BREEAM. 2015. *BREEAM International Refurbishment and Fit-out* [online]. Available: www.breeam.com/refurbishment-and-fit-out [accessed 4 April 2017].

Construction Specifications Institute (CSI) 2013. *The CSI Sustainable Design and Construction Practice Guide*, Hoboken, NJ, Wiley.

Crawley, D. and Aho, I. 1999. Building environmental assessment methods: applications and development trends. *Building Research & Information*, 27, 300–308.

Davison, S. 1981. Conservation of museum objects in tropical conditions. *In:* S. McCredle *et al.* (eds) *Conservation of Museum Objects in Tropical Conditions.* Kuala Lumpur, International Council on Archives.

De Silva, N., Ranasinghe, M. and De Silva, C. R. 2012. Risk factors affecting building maintenance under tropical conditions. *Journal of Financial Management of Property and Construction*, 17, 235–252.

Ding, G. K. 2008. Sustainable construction: the role of environmental assessment tools. *Journal of Environmental Management*, 86, 451–464.

Eastman, C., Teicholz, P., Sacks, R. and Liston, K. 2008. *BIM Handbook: A Guide to Building Information Modelling for Owner, Managers, Designers, Engineers, and Contractors*, Hoboken, NJ, Wiley.

Groenhout, N., Hyde, R., Prasad, D., Chandra, S., Yoshinori, S. and Lim, C. H. 2008. Green technologies, performance and integration. *In:* Hyde, R. (ed.) *Bioclimatic Housing: Innovative Designs for Warm Climates*, London, Earthscan.

Gut, P. and Ackerknecht, D. 1993. *Climate Responsive Building: Appropriate Building Construction in Tropical and Subtropical Regions*, St Gallen, SKAT, Swiss Centre for Development Cooperation in Technology and Management.

Haji-Sapar, M. and Lee, S. E. 2005. Establishment of energy management tools for facilities managers in the tropical region. *Facilities*, 23, 416–425.

Hameury, S. 2005. Moisture buffering capacity of heavy timber structures directly exposed to an indoor climate: a numerical study. *Building and Environment*, 40, 1400–1412.

Haron, A. T., Marshall-Ponting, A. J. and Aouad, G. 2009. Building information modelling in integrated practice. 2nd Construction Industry Research Achievement International Conference (CIRIAC 2009), 3–5 November 2009, Kuala Lumpur.

Huang, T., Kong, C. W., Guo, H., Baldwin, A. and Li, H. 2007. A virtual prototyping system for simulating construction processes. *Automation in Construction*, 16, 576–585.

IPCC 2002. *Climate Change 2001: The Scientific Basis*, Cambridge, Cambridge University Press.

IPCC 2013. *Climate Change 2013: The Physical Science Basis*. Working Group 1 contribution to the Fifth Assessment Report of the Intergovernmental Panel on Climate Change.

Isikdag, U., Aouad, G., Underwood, J. and Wu, S. 2007. Building information models: a review on storage and exchange mechanisms. Bringing ITC Knowledge to Work. 24th W78 Conference Maribor, 2007. 135–144.

Jordani, D. 2008. BIM: a healthy disruption to a fragmented and broken process. *Journal of Building Information Modelling*, 2, 24–26.

Kong, X. F., Lu, S. L., Gao, P., Zhu, N., Wu, W. and Cao, X. M. 2012. Research on the energy performance and indoor environment quality of typical public buildings in the tropical areas of China. *Energy and Buildings*, 48, 155–167.

Lee, W. L. 2013. A comprehensive review of metrics of building environmental assessment schemes. *Energy and Buildings*, 62, 403–413.

Lei, J. W., Yang, J. L. and Yang, E. H. 2016. Energy performance of building envelopes integrated with phase change materials for cooling load reduction in tropical Singapore. *Applied Energy*, 162, 207–217.

Malhi, Y. and Wright, J. 2004. Spatial patterns and recent trends in the climate of tropical rainforest regions. *Phil. Trans. R. Soc. Lond. B*, 359, 311–329.

Nematchoua, M. K. and Orosa, J. A. 2016. Building construction materials effect in tropical wet and cold climates: a case study of office buildings in Cameroon. *Case Studies in Thermal Engineering*, 7, 55–65.

Pathirage, C., Haigh, R., Amaratunga, D. and Baldry, D. 2008. Knowledge management practices in facilities organisations: a case study. *The Journal of Facilities Management*, 6, 5–22.

Pillai, R., Aaditya, G., Mani, M. and Ramamurthy, P. 2014. Cell (module) temperature regulated performance of a building integrated photovoltaic system in tropical conditions. *Renewable Energy*, 72, 140–148.

Sachs, J. D. 2001. Tropical underdevelopment. *NBER Working Paper (Vol. 8119)*.

Stine, J. D. 2011. *Design Integration Using Autodesk Revit 2012*, Mission, KS, SDC Publications.

Sulaiman, A. S. 2013. Some common maintenance problems and building defects: our experiences. *Procedia Engineering*, 54, 101–108.

Wan Mohd Nazi, W. I., Royapoor, M., Wang, Y. and Roskilly, A. P. 2017. Office building cooling load reduction using thermal analysis method: a case study. *Applied Energy*, 185, 1574–1584.

8 Case studies

Michael Farragher and Mike Riley

This chapter brings together examples of projects undertaken in a range of tropical areas. Projects from Uganda, Brazil and Australia illustrate the range of issues that must be dealt with in designing and constructing within a tropical setting. Increasing focus on the passive approaches to environmental control reflect a growing movement across developed and less developed countries to address sustainability as a core issue in the development of the built environment. Imaginative architecture and robust science combine in many of these projects to maximise the ability to respond in a sustainable fashion to the design challenges of residential, commercial and institutional buildings in these locations.

The authors are grateful to the contributors who have allowed the use of their projects within this text.

Africa: Uganda

Introduction

The urban core of sub-Saharan Africa, despite being relatively poor, still has a thirst for survival. Uganda as an example is positioned 19th on the Global Poor List based on GDP and PPP (purchasing-power-parity) published in 2016 by *Global Finance Magazine*. Yet Uganda survives by a cacophony of micro-businesses. As an example, Ndeebe district in Kampala hosts a hot bed of small and large building supply companies and is relatively affluent compared to the tropical slums of Mutundwe close by. Moving into the centre of Kampala, one can find Entebbe Road which is endowed with printing/office and graphics shops. It seems like this tropical city has businesses of the same typology that cling to each other.

In this melee of concrete buildings interspersed with informal shack-like developments, it is hard to find real examples of architecture that are intrinsically African. All types of buildings in Kampala, whether informal settlements, government or commercial in character, have to deal with the same environmental issues. Heat and water are the biggest problems to overcome in designing sustainable tropical architecture in this region. Lake Victoria and the Rwenzori mountains to the west produce great swathes of rain clouds at different times throughout the year which cause a chequerboard of hot spells and torrential rain that have to be dealt with by varying degrees.

In dealing with these climatic extremes, there are some of the worst examples of human habitation in terms of open-drain sanitation and construction. At the same time there are some real gemstones in the green and burnt-umber earth of this city.

FBW Architects has been established in Uganda for 18 years and is run by a Liverpool John Moores University graduate of architecture, Paul Moores. The company produces a blend of European and African architecture and the following three examples show excellently the solutions to tropical climatic problems that Uganda faces.

Case Study 1 Ndere Cultural Centre Facility, Kampala, Uganda

The Ndere Foundation is arranged as a composition of distinct but linked elements housing the main exhibition area along with a theatre, a cafe, an external theatre, a traditional Ugandan building section and a journalism school. Beyond the main entrance with its large over-sailing canopy, a courtyard provides a sheltered environment and focus for the social activity within the centre. The canopy is designed to allow trees to grow through its cover which allows the building a special integration with nature.

Figure 8.1 Ndere Cultural Centre – entrance view.

Source: courtesy of FBW Architects.

In response to the climate, the building has the capacity to be completely naturally ventilated. Wherever possible, high level openings in the split faced local natural stone walls provide cross ventilation and where there are windows, these are large, openable and shaded. The different elements of the building are situated in a lush, landscaped garden which provides shade and a cool microclimate. Materials have been chosen for solidity, longevity and low maintenance, drawn wherever possible from local materials and skills. The steel-framed buildings will be finished in locally sourced split stone with clay pan-tile roofs and hand crushed brick render. From the start the building has been designed to be low energy and to make minimal environmental impact. The building was completed on budget and on programme.

The Ndere Cultural Centre in Kampala, Uganda, required the new building to be conceived, designed and constructed as an African building, using local resources and materials. As well as being in a challenging location, the building also had to meet the standards of a modern international cultural centre.

Figure 8.2 Ndere Cultural Centre – internal courtyard view.

Source: courtesy of FBW Architects.

The contractor embraced the objectives of this challenging project from the outset. The requirement to use local resources while achieving the required environmental standards presented one of the team's main challenges. For example, the contractor was faced with a choice of either sending compliant European timber all the way to the equator and causing significant CO_2 emissions during transportation, or attempting to manage the local hardwood supply chain.

The project team opted for tackling the latter option and, by diligent research and continuous monitoring, were able to source appropriate local resources. This included working with the National Forestry Authority in Uganda to identify suitable tropical hardwoods based on a detailed, tree-by-tree survey, and then implementing a replanting regime.

Figure 8.3 Ndere Cultural Centre – theatre.

Source: courtesy of FBW Architects.

Another challenge was the local climate. The weather is extremely hot in Kampala year-round and temperatures vary little between day and night. As a result, natural ventilation is more difficult to achieve because it is not possible to introduce a night-time purging strategy. Thermal modelling was therefore used to prove the proposed solution, which was based on natural ventilation and passive solar control using a permanent sunscreen from locally produced steel and clay tiles.

The added problem in Kampala comes from the use of galvanised steel sheet roofing materials which cause an acoustic problem in buildings that require good sound reduction such as the entrance foyer and theatre. These areas were clad in steel sheet and the steel sheet was covered internally with a dark hessian which was glued onto the underside of the sheet to reduce noise.

Additional capital was invested to achieve the natural ventilation solution and the strategy has paid off by minimising the need for air conditioning and hence reducing energy use. Local room control was also added to further increase energy efficiency. In addition, the building was designed to make the maximum use of thermal mass to keep out heat and minimise cooling requirements.

Figure 8.4 Ndere Cultural Centre – high level ventilation detail.

Figure 8.5 Ndere Cultural Centre – external amphitheatre.

Case Study 2 Nissan Motorcare new showroom and offices, Kampala, Uganda

The motor showroom was commissioned by Nissan in 1999 and was the first project that was administered by Paul Moores in Uganda. The aesthetic is one of hi-tech modernity interspersed with African materials such as mahogany for the entrance ramp and curtain wall cladding. The building is an extravagant essay in how to deal with hot air passively. As can be seen from the cross-sectional drawing, the rising hot air is cleverly dealt with by the split roof. There is a plenum void above the two curved ceilings in which hot air can be gathered and directed to ventilation grilles in the split zone between the two roofs.

Figure 8.6 Nissan Motorcare Centre, Entebbe Road, Kampala – internal mezzanine showing cool ceiling.

The building is remarkably cool considering the large glass frontage, which is important for the viewing and sampling of the vehicles on display and the thermal comfort of the staff.

The steel frame of the building is elegantly extended to produce flagpoles which are braced by steel rods.

The use of the slope in the site is clever. A huge storm drain has been cut at the front of the showroom which allows water to pass to the rear of the property. A large mahogany clad ramp allows clientele and vehicles to access the showrooms for viewing.

Figure 8.7 Nissan Motorcare Centre, Entebbe Road, Kampala – view from mezzanine.

Figure 8.8 Nissan Motorcare Centre, Entebbe Road, Kampala – entrance facade.

Figure 8.9 Nissan Motorcare Centre, Entebbe Road, Kampala – stormdrain.

Case Study 3 Virika Hospital Reconstruction, Fort Portal, Uganda

The Virika Hospital was built in the rural location of Fort Portal in the North West of Uganda. The area is an earthquake zone so the building had to be built to withstand the resultant forces of such a natural disaster. It was determined at an early stage that steelwork would have to be used. A simple 50×50mm SHS was used for the framework so that the sections could be transported on the back of a van rather than an articulated lorry that is used for large scale steel sections.

Figure 8.10 Virika Hospital, Fort Portal – view of steel framed loggia.

Source: courtesy of FBW Architects.

Figure 8.11 Virika Hospital, Fort Portal – entrance view.

Source: courtesy of FBW Architects.

Locally engineered materials were used throughout such as steel, handmade clay-bricks and timber ceilings. The Mobati metal sheet roofs are ventilated at the ridge so that hot air can escape and passively cooled internal spaces can be created. Solar shading walkways and devices are used throughout to provide protection from the searing heat in this region.

Windows are all covered with mosquito netting as the building is situated in a serious malaria area.

Windows are also fully openable in horizontal glass panels. High level cladding is carried out in sustainably sourced timber which is out of the reach of termites and also provides a cool material at the hottest part of the building.

Figure 8.12 Virika Hospital, Fort Portal – roof detail.

Source: courtesy of FBW Architects.

Figure 8.13 Virika Hospital, Fort Portal – various views.

Source: courtesy of FBW Architects.

South America: Brazil

Case Study 4 Experiences of thermal performance analysis in wooden dwellings

Courtesy of: Thalita Gorban Ferreira Giglio (1) and Ercilia Hitomi Hirota (2)

1 Architecture and Urbanist, MSc, Dr, Reader at Construction Department, State University of Londrina – email: thalita@uel.br
2 Civil engineer, MSc, Dr, Professor at Construction Department, State University of Londrina – email: ercilia@uel.br

In spite of its continental territory and huge amount of forests compared to many countries, Brazil is in a poor position concerning the use of wood in the building industry. Throughout its history, the use of inadequate technology in Brazil for building houses in particular, has led to a misleading concept of wooden buildings that conjures up poverty, short-term building and low performance.

In southern Brazil, the first wooden dwellings had vertical fences formed by massive boards with only 2.2 cm thickness. Another alternative was the use of roundwood, thicker wood, usually used in cold regions such as Curitiba, Paraná (Giglio, 2005).

Zani (1997) is an architect who registered the most typical wooden houses, built from 1930 to 1970, in the northern region of Paraná State, mainly focusing on architectural design and technology. According to him, the most used wood was the native aspidosperma polyneuron, called *Peroba Rosa*, with 22 cm by 2.2 cm thick boards, perpendicularly nailed in a lower and upper frame, with 1 cm joints, and side-splices 6 cm by 1.2 cm thick, externally and internally fixed (Figure 8.14). The buildings were usually painted with latex or oil paint, with strong colours such as blue, yellow or grey.

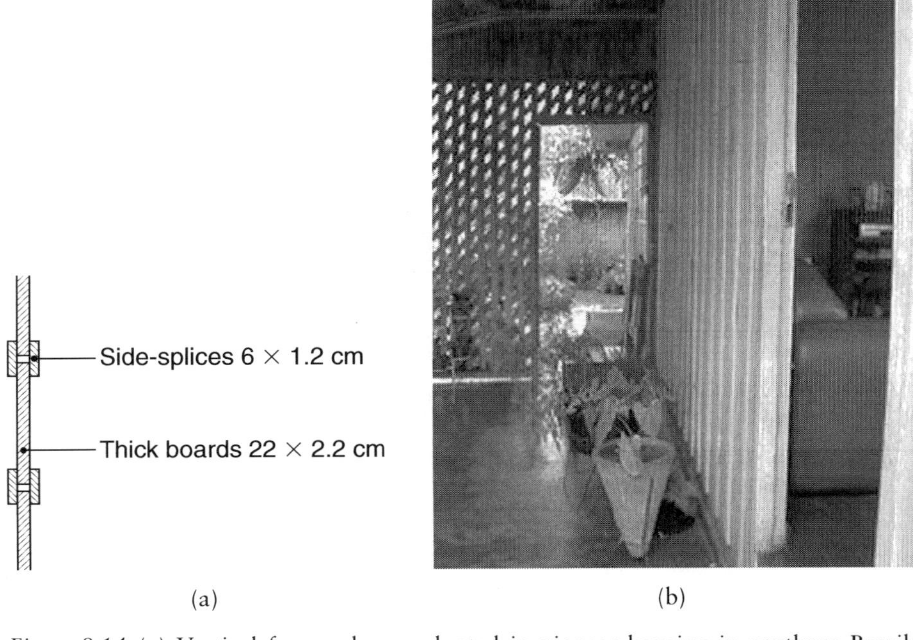

Figure 8.14 (a) Vertical fence scheme adopted in pioneer housing in southern Brazil; (b) exemplar of pioneer housing in Londrina, Paraná.

Over the years, the thermal performance of these pioneer dwellings was considered inadequate for the Brazilian tropical climate, both in summer and winter. In research developed by Bogo (2003), in which the thermal performance of thin wooden panels was evaluated, the author concluded that these panels are inadequate for most regions of Brazil. In Donadello *et al.* (2013), the evaluation, by simulation, of the thermal performance of a dwelling built with a prefabricated solid wood system resulted in a low level of energy efficiency level D (Giglio, 2005), demonstrating that the composition of the panel in wooden technology is still weakly approached in the Brazilian buildings. Levels from A to E are defined according to the Technical Quality Regulation for the Level of Energy Efficiency of Residential Buildings (RTQ-R – Brazilian Regulation).

Such past experiences have presented as barriers for implementing light fence systems in the building sector. However, in recent years, through the Brazilian government's housing programme 'Minha Casa, Minha Vida' (My Home, My Life), there has been an increase in the number of social houses built in wood frame or in low apparent density concrete (monolithic system for walls). This fact indicates a paradigm shift regarding the introduction of light systems in social housing projects, especially with the use of wood as a wall element.

Few studies focus on the evaluation of the thermal and energy performance of light building systems, mainly for the understanding of the composition of materials that best meets the climatic conditions of a region. In Silva and Basso (2001), Atem (2002), Giglio (2005), Invidiata (2013) and Galindo (2014), it was observed that the choice of constructive components of light system wall panels is a fundamental step to ensure good performance of the system.

Computational simulations developed by Giglio (2005) for the climate of Londrina, Paraná State (23° 18' 37" N and 51° 09' 46" W), the wood frame system with double panels and air chamber without thermal insulation, does not reach a satisfactory thermal performance in a 44 square meter house, without the shadowing of the openings. Such conditions resulted in more than 1,000 hours of discomfort in one year quantified based on hours of temperature above 29°C and below 18°C. However, it has also been observed that the wall panels do not need high thermal capacity to soften the external peaks of heat and cold. Better results were observed for panels with thermal insulation or with a double air chamber, achieving hours of thermal comfort close or superior to a masonry house.

In Galindo (2014), new computational simulations were carried out for the Londrina climate, aiming to carefully investigate the thermo energetic performance of an innovative energy efficient social house designed by researchers from the State University of Londrina, using mass customisation and target costing strategies in the design process, and wood frame technology for the panels. The analysis considered the parameters from the Brazilian regulation for energy efficiency labelling of residential buildings (Brasil, 2012). The results showed a strong synergy between the decisions made in the architectural design and the panel composition adopted. Double panels with air chambers filled with a thermal insulation guarantee good thermal performance when the roof is isolated and painted in white. However, maximum energy efficiency levels (Level A) were only reached with the addition of more oriented strand board (OSB) panel layers, both externally and internally. Thus, the panels gained high thermal resistance to compensate for the low thermal capacity of the construction system.

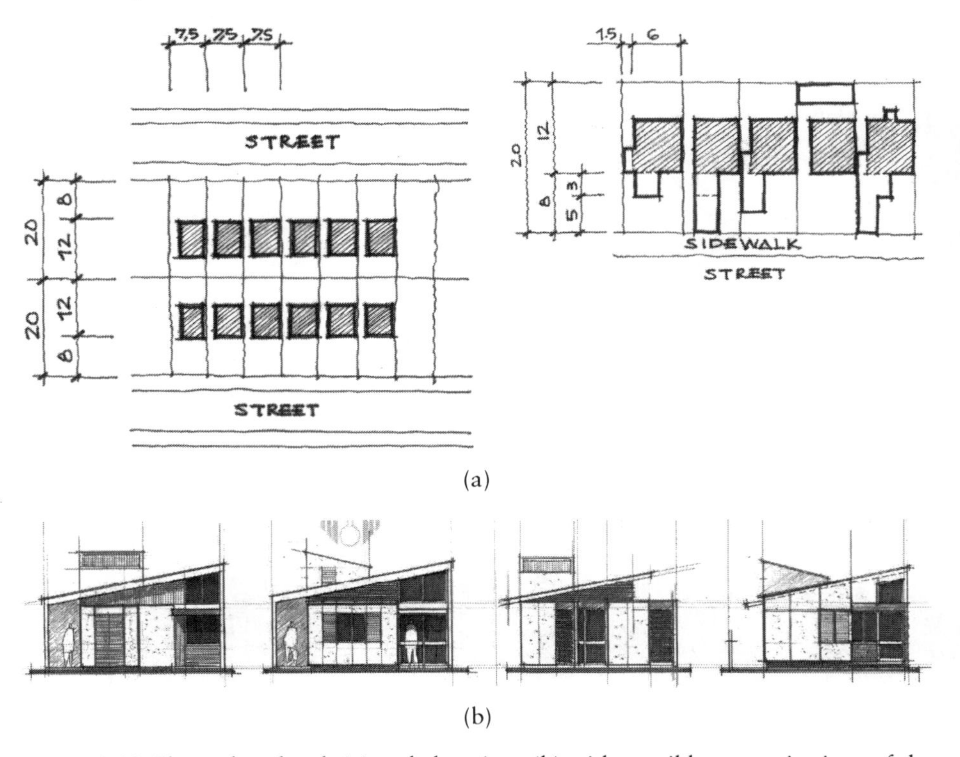

(a)

(b)

Figure 8.15 Floor plan sketch (a) and elevations (b) with possible customisations of the housing units developed by the State University of Londrina.

In view of the climatic diversity of the Brazilian territory, we highlight the study of Invidiata (2013). Researchers from the University of São Paulo (USP) and Federal University of Santa Catarina (UFSC) developed the Ekó House, for the international Solar Decathlon competition held in Madrid. Invidiata (2013) adapted the wooden house designed for the cold climate of Madrid to the eight different Brazilian bioclimatic conditions. Figure 8.16 shows the division of eight Brazilian bioclimatic zones, applied to the thermal performance Brazilian normative.

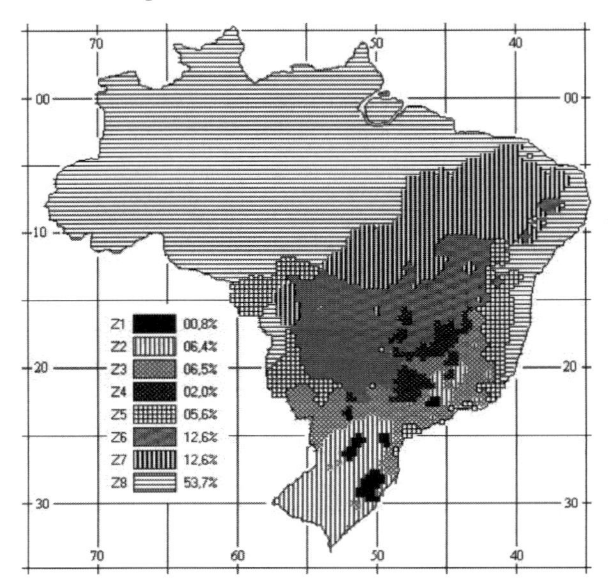

Figure 8.16 Division of Brazilian territory by different bioclimatic zones.

Source: Associação Brasileira de Normas Técnicas (2005).

The results pointed out the need for different adjustments of the wood panels for the different bioclimatic conditions concurrently with the characteristics of natural ventilation and protection of the house openings. Thus, maximum efficiency levels could be achieved for most bioclimatic zones, except for bioclimatic zone 1 (Curitiba) and bioclimatic zone 6 (Campo Grande), which reached efficiency level B.

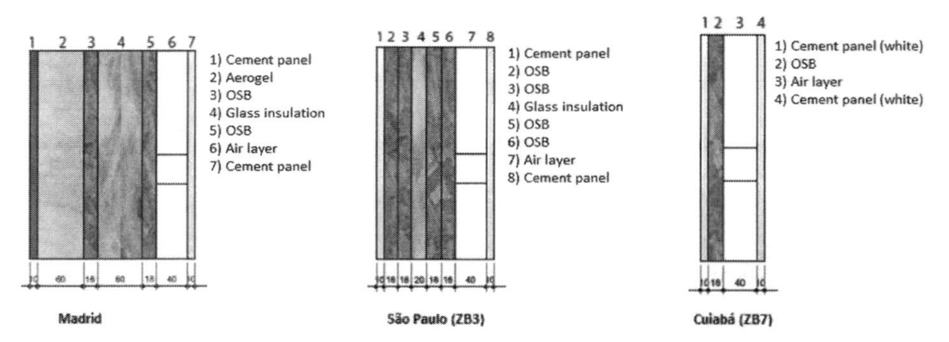

Figure 8.17 Adjustments suggested for the panel composition according to Brazil's different bioclimatic conditions.

Source: Invidiata (2013).

Discussions on the low thermal inertia of light building systems were observed in Brito *et al.* (2013) and Krüger and Zannin (2006). In the latter, the authors evaluated, by measurement, the thermal performance of buildings with a light building systems envelope. The authors concluded that light panels with components with higher thermal capacity provided higher thermal inertia and better results of thermal performance. In the case of buildings situated in climatic conditions of high daily thermal amplitude, it is important to use high thermal inertia walls as it eases the effects of the daily temperature peaks. The daily thermal amplitudes in some Brazilian regions such as Londrina, São Paulo and Curitiba, among others, is above 10°C. So, it is necessary to have a deeper investigation about the lack of thermal inertia in light panels in the thermo energetic performance of the building.

Another characteristic of thermal performance in tropical countries is the need for maintaining contact between the building and the soil, as discussed in Souza *et al.* (2011), and in Brito *et al.* (2013). In Souza *et al.* (2011) the study with simulation showed that envelopes composed of light building systems have a satisfactory thermal performance in the cases where the building was in contact with the ground. In those cases, the cooling of the building was provided and the operating temperature got closer to the ideal conditions. In Brito *et al.* (2013) the simulation study for the São Paulo climate indicated that the contact with the ground caused a lower range of the daily air temperature, especially in buildings with a lightweight construction system. The authors mention that the reduction in the daily thermal amplitude refers to the reduction of the maximum internal temperature, "which can be a distinctive aspect so that the system can meet the minimum criteria established in the ABNT NBR 15575 code". It should be noted that in spite of the significant influence of the soil contact with the building in the thermal performance results, NBR 15575–1 (Associação Brasileira de Normas Técnicas, 2013a) does not consider such influence on the analysis criteria.

In order to secure a careful evaluation of technological innovations, the Brazilian code for performance of buildings NBR 15575–4 (Associação Brasileira de Normas Técnicas, 2013b) defines minimum standards of performance that eventually fulfilled the lack of Brazilian parameters related to the quality of buildings. However, minimum thermal performance procedures and parameters specified in that code are guided by the conventional, higher density construction systems, which entails the need for simulation studies for a deeper analysis. Furthermore, those parameters lead to very high indoor air temperature in the summer, and very low indoor temperatures in the winter. This fact may undermine the code credibility concerning users' satisfaction with the thermal comfort conditions. In a context of cultural barriers to technological innovations like the construction industry, reports of a few unsuccessful cases can connect the technology to an image of bad practice.

Further studies on the improvement of the wood frame building technology is necessary to ensure that better thermal quality housing is provided. The user's dissatisfaction with the indoor environment not only discourages the use of the technology, but also leads to the use of artificial air conditioning systems, increasing the energy consumption in the building. In the case of low-income housing, the consequence of high energy consumption will impact both the Brazilian electric generation system and, mainly, the already critical family budget.

Australasia: Australia

Case Study 5 Elements Resort of Byron Bay

Architect: Shane Thompson Architects

The project was won by means of a design competition. The thing that struck a chord with the client was the fact that the true essence of the site was conveyed and how that could be exploited and interpreted through the experience of the built form.

The new resort, recently completed on a spectacular beachfront site, is intended to be distinctly Byron Bay and Australian in character. Byron Bay is the easterly most point of Australia, located about two hours south of Brisbane. The region is very well known for its rich agricultural history and the abundant fauna and flora, which flourish in and around the ancient Mt Warning volcanic caldera. It is designed to be sensitive and respectful of its site, which includes rare littoral rainforest, ponds, a diverse wildlife, native landscape, ocean and wetlands.

The central facilities and accommodation refer to both the traditional fibro beach shack and local naturally occurring forms and colour of the sand dunes. Leisure and conference facilities are housed under three separate pavilions, arranged to embrace the north-eastern aspect and views to the dunal landscape. The organic form and loose planning of these central facilities is informed by the studies of place making in the indigenous South Pacific villages and in the windswept shape of the sand dunes. A large lagoon style pool at the heart of this area weaves through the buildings and landscape, providing discreet places for sun lounges, a fire pit and water play.

The resort is intended to be an exemplar of sustainable design, with a minimal physical and ecological footprint, the re-establishment of the original dunal and wallum landscape and the lightweight construction based on the best passive solar design principles. There are 94 keys of accommodation provided in 75 small villas spread throughout the site, each sited to preserve existing vegetation and with regard for the north-east aspect, views and privacy. The timber framed and fibre cement sheeted villas are raised above the natural ground, allowing natural overland flows to be maintained, and are designed to provide shelter, retreat and a quiet place to contemplate the sublime natural environment. The landscape design saw the introduction of over 65,000 new native trees and plants.

The overall character and design has sought consciously to create a sense of community and a place of luxurious escape where there is a heightened appreciation of the natural world. The resort offers an identifiable, laidback and Australian experience.

The concept design was done in a collaborative manner between Shane Thompson, Bill Ellyett and Larissa Fouche, with Shane steering the overall design intent. Initial sketch design iterations were all done by hand as it kept the design process fluent and organic, and it was in itself a representation of the organic matter in which the understanding of the site and the design came together.

Moving into design development, Shane Thompson Architects collaborated with ARUPS in particular to establish an economical means of construction methodology. Shane Thompson used Revit for Design Drawing documentation, which was then translated by means of Rhino and Grasshopper to produce structural calculations, data and documentation.

As for ESD (Environmentally Sustainable Development), the team collaborated with a multi-disciplinary service provider, EMF Griffiths, who delivered design advice that was integral with other service designs and specifications. EMF Griffiths also provided service consultancy for mechanical, electrical and hydraulic. This resulted in an ESD analysis that was truly integrated.

EMF Griffiths were engaged to report on the principles of ESD that would be applied to the proposed development at North Byron Bay.

The development presented a fantastic opportunity to create a sustainable development which sits lightly in the landscape.

The North Byron Beach Resort Ecologically Sustainable Design philosophy uses passive design principles and eco-friendly technologies to provide a natural and open experience. The development:

- minimises energy and water use;
- enhances site ecology; and
- provides a sustainable service to the community.

Through the Passive Design Ethos, natural and passive design principles minimise ecological impacts as follows:

- selective orientation of the buildings and a light footprint scheme to enhance the connection to the site habitat;
- use of insulated lightweight construction to minimise embodied energy in the building fabric;
- revegetation areas with native plant species;
- site infrastructure configuration with integrated water sensitive design principles and to minimise ground disturbance; and
- natural ventilation during favourable climatic conditions, shading devices and comprehensive roof insulation minimise the electrical consumption due to operation of air conditioning.

The project provides a high quality eco-resort experience with:

- informative and educational features explaining the ecological design;
- enhanced landscaping in a sub-tropical design setting;
- sensitive selection of materials finishes for the building using recognised third party eco-certification;
- low noise building services and design for acoustic privacy;
- selection of finishes with low solvent content to improve indoor air quality; and
- daylighting and vision corridors to the site flora for maintaining a continuous connection with the subtropical environment.

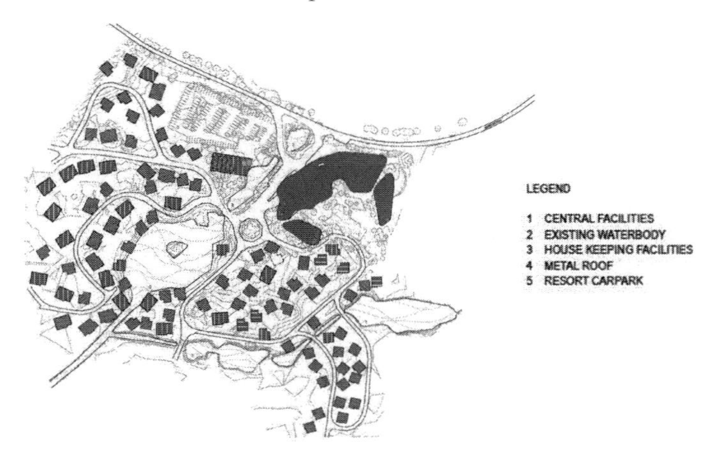

Figure 8.18 Masterplan of the Elements Resort, Byron Bay, NSW.

Source: courtesy of Shane Thompson Architects.

It was intended that the Ecological Performance should be maintained throughout the operational phase of the project by:

- ongoing management of energy, water and waste use during the operational phase;
- minimising resource use through integration of robust and efficient building services into the design; and
- establishment and maturity of site landscaping.

AIMS

The aims of the project were to make the development:

- as self sufficient as possible, working as an integrated system;
- as environmentally responsible as possible without adversely effecting amenity for the occupants;
- as cost effective as possible to build and operate; and
- comfortable and enjoyable for the people who live and work there.

STRATEGY

- To use any natural resources available on site.
- To work with, rather than against, the local climate and prevailing weather conditions.
- To reduce the long-term effect on the site to a minimum: 'touch the earth lightly'.
- To integrate systems so they operate in unison for maximum efficiency.
- To become a focal point for the local community.
- To minimise energy usage and generate as much on site as practical.
- To minimise water usage and recycle water used on site.

Byron Bay has a humid subtropical climate (*Cfa* in the Köppen climate classification) with hot summers and mild winters. Winters are not cold, with daily maximums usually reaching 19.4°C and a minimum of 11°C. Summer can be hot, with a daily average of 27°C. Summer evenings can be wet, cooling the day down.

The overriding factors in designing the building envelope were to reduce solar gain through solar absorptance (given as a factor between 0.0 and 1.0), achieve a fairly low thermal resistance factor (R value) in comparison to UK values in order to improve passive cooling, and to improve insulation factors with regard to orientation. The Building Code for Australia (BCA) JV3 verification has been used in order to ensure the thermal compliance of the building. In this, a Reference Building is the process of modelling the Deemed-to-Satisfy provisions in the BCA to create a 'target' for annual energy consumption. Table 8.1 shows in the first row that the reference model requires a total thermal resistance of 4.2 m^2K/W whereas the recommendations by the engineers are to use a thermal resistance of 4.2 with insulation resistance of 3.66 m^2K/W or 2.7 m^2K/W with foil face. The second option would provide excellent reflectance of the rays of the sun plus a medium level of thermal resistance that could be efficient in the summer and the winter. Note that the solar absorptance is lower than the reference model at 0.5 by using a reflective material for the roof.

As an example of how this can affect the technological design of building wall elements, we can see that the Reference Building in Table 8.1 has a higher thermal resistance for north facing walls which makes sense in that the sun in Byron Bay, which is south of the Tropic of Capricorn, would be north or overhead at midday. Therefore, higher resistance to heat would be required at these positions so that the facades to the north repel heat. A lower thermal resistance to the south elevations would produce a greater heat loss through the 'cooler' facade. Insulation in the building elements is generally foil faced to reduce the heat gain capabilities of the material, so that the SHGC (Solar Heat Gain Coefficient) can be at a reasonable level to allow the flow of heat from the interior to the exterior. Despite the requirements of the Reference Building row for concrete walls, the engineers have adopted a low thermal resistance so that heat can escape from the walls on north and south facades and within the floor slab. This is stated as $0.27\,m^2K/W$.

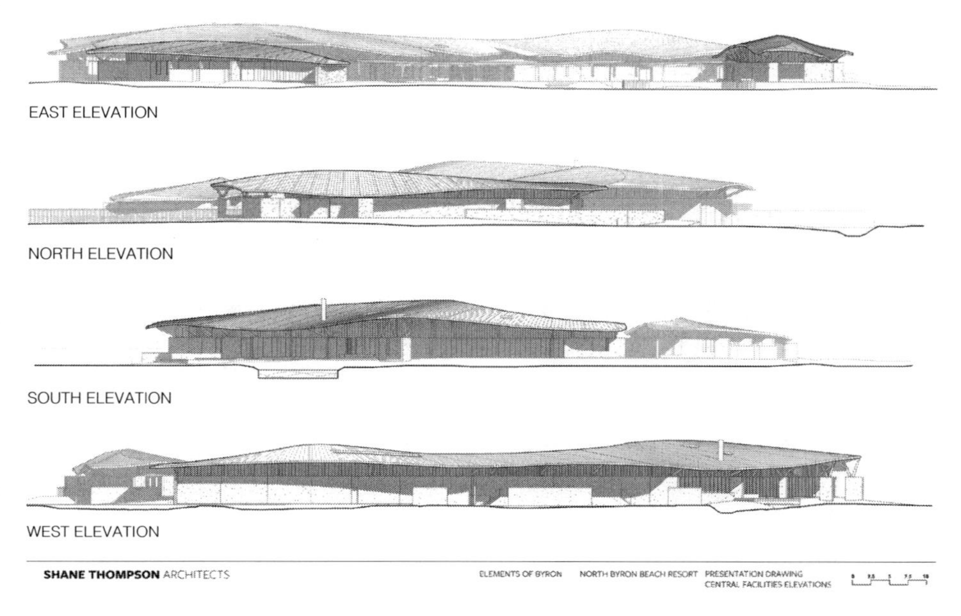

Figure 8.19 Elevations of buildings 1 and 2.

Source: courtesy of Shane Thompson Architects.

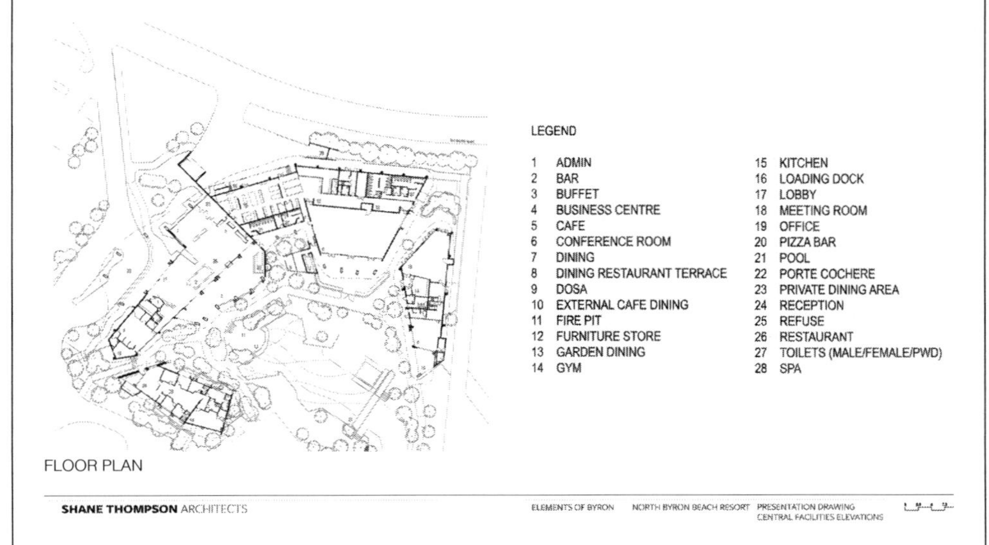

Figure 8.20 Plan of buildings 1 and 2.

Source: courtesy of Shane Thompson Architects.

Table 8.1 Synopsis results of thermal resistance of building elements for the JV3 method of compliance with BCA

Description/current construction	*Reference model value (DTS requirement)*	*Proposed construction value*	*Notes/recommendations*
Sheet Metal Roofing Sheet Metal Roofing Colour Assumed Medium (Solar Absorptance = 0.5)	$R_{TOTAL} = 4.2$ Solar Absorptance = 0.7	$R_{TOTAL} = 4.2$ Insulation Options (roof construction TBC) 1. $R_{INSULATION} = 3.66$ or 2. $R_{INSULATION} = 2.7$ + reflective foil (0.9 outer, 0.05 inner emittance) Solar Absorptance = 0.5	Roof Colour TBA. Note that light roof colour recommended (low solar absorptance) to reduce insulation requirements. Example colorbond colours: • Shale grey 0.43 • Dune 0.47 • Windspray 0.58
External Stud Walls	Generally $R_{TOTAL} = 3.3$ Solar Absorptance = 0.6 South or shaded $R_{TOTAL} = 2.8$ Solar Absorptance = 0.6 South (shaded) $R_{TOTAL} = 1.8$ Solar Absorptance = 0.6	$R_{TOTAL} = 2.3$ $R_{INSULATION} = 1.88$ (Bradford Gold Wall Batts R2.0 90mm)	Insulation reduced based on JV3 method
External Walls Concrete	Generally $R_{TOTAL} = 2.8$ Solar Absorptance = 0.6 South or shaded $R_{TOTAL} = 2.3$ Solar Absorptance = 0.6	$R_{TOTAL} = 2.7$ No insulation required	Insulation reduced based on JV3 method
Internal Walls Concrete (between conditioned/ unconditioned spaces)	$R_{TOTAL} = 2.3$	$R_{TOTAL} = 2.7$ No insulation required	Insulation reduced based on JV3 method
Slab on Ground	$R_{TOTAL} = 2.7$	$R_{TOTAL} = 2.7$	No insulation required
Glazing	As per glazing calculator	Building 1: Single glazing U = 6.4; SHGC = 0.7 Building 2: Single glazing U = 6.4; SHGC = 0.6	Glazing requirements reduced to some windows, based on JV3 method (example glazing: Gjames 425 Series fixed window 6.38 mm Comfortplus Clear Low E Laminate)

Source: courtesy of Shane Thompson Architects.

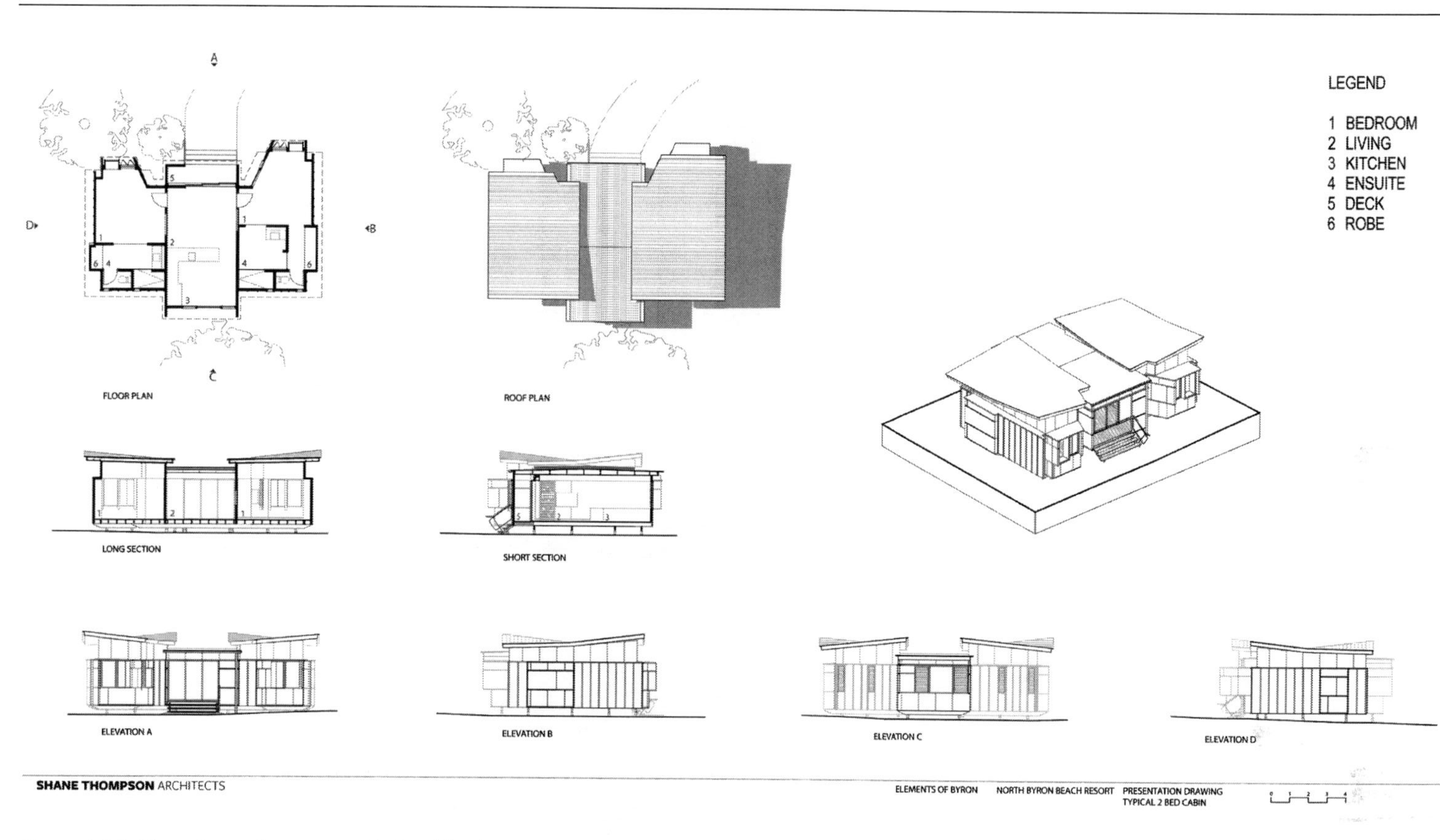

Figure 8.21 Elevations of accommodation blocks.

Source: courtesy of Shane Thompson Architects.

Together with the thermal resistance measures of the roof, walls and ground floor, buildings 1 and 2 have large overhangs which shade the facades from the heat of the summer sun, which show that not only the technological make-up of building components is important in tropical regions but also the form. The accommodation units are laid out around ponds and small lagoons surrounded by fauna and foliage. Figure 8.18 shows how the site is arranged.

The smaller accommodation units have a different approach to compliance in that they are designed to the BASIX standard, which is an energy use toolkit similar to the now outdated UK Code for Sustainable Homes.

The below summary outlines the requirements to achieve compliance with Energy Efficiency sections for Class 1b or Class 3 buildings under the National Construction Code 2013. As shown in Figure 8.21, the configuration of the accommodation units comes in three layout types. A typical double unit is shown in this figure, which indicates a roof abutment detail that will allow heat to escape from high level. The ground floors are also elevated from the ground to allow ventilation at a low level.

BASIX compliance for Class 1b buildings targets higher levels of sustainability, including water and energy efficiency targets. To achieve a Class 1b or a BASIX equivalent it is shown that increased scope for energy and water efficiency should be included in each accommodation unit design, such as:

- rainwater tanks or waterless toilets;
- PV or gas/solar hot water;
- tinted glass.

Figure 8.22 Example of roof-mounted photovoltaics.

Source: courtesy of EMF Griffiths Engineers.

Our basis for understanding of the applicability of the building code to the development is as follows:

- The prototypes are considered Class 1b. A Class 1b development under NCC 2013 would have to achieve BASIX compliance.

- The Development Application (DA) for the prototypes refers to 1987 Code compliance and that energy efficiency sections and BASIX are not applicable.
- The development will be Class 3. Preliminary assessment has been completed assuming the accommodation buildings are Class 3.

It is understood that the development would like to achieve the highest level of sustainability and therefore is exploring the option of complying with BASIX for the accommodation units.

Undergoing BASIX assessment provides higher sustainability aims such as decreases in energy and water use by 40 per cent as well as maintaining high levels of thermal comfort in the building. A BASIX assessment would be a way of ensuring good sustainability outcomes whether or not the buildings are classified 1b. It should be said that the IES VE software has been used for all the buildings to comply with the BCA as this is a software that is approved.

An example of a printout of the glazing elements is shown in Figure 8.23. It should be noted that the glazing has a height U value to allow heat to escape from the accommodation units (BASIX).

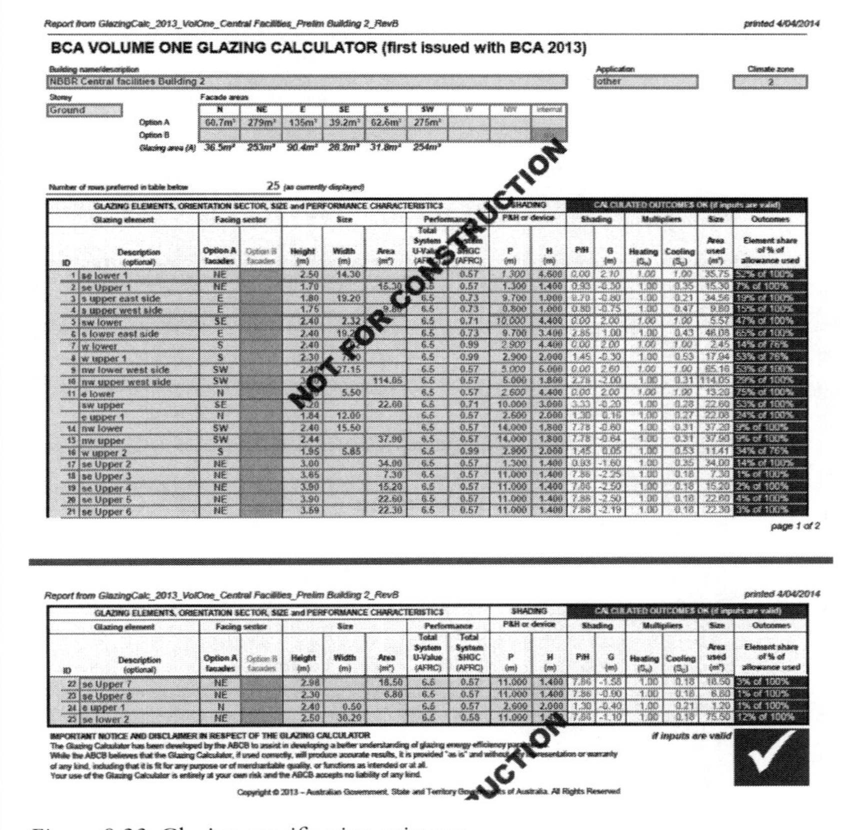

Report from GlazingCalc_2013_VolOne_Central Facilities_Prelim Building 2_RevB — printed 4/04/2014

BCA VOLUME ONE GLAZING CALCULATOR (first issued with BCA 2013)

Building name/description: NBBR Central facilities Building 2 — Application: other — Climate zone: 2

Storey: Ground

Facade areas

	N	NE	E	SE	S	SW	W	NW	internal
Option A	60.7m²	279m²	135m²	39.2m²	62.6m²	275m²			
Option B									
Glazing area (A)	36.5m²	253m²	90.4m²	28.2m²	31.8m²	254m²			

Number of rows preferred in table below 25 (as currently displayed)

GLAZING ELEMENTS, ORIENTATION SECTOR, SIZE and PERFORMANCE CHARACTERISTICS										SHADING		CALCULATED OUTCOMES OK (if inputs are valid)					
Glazing element		Facing sector		Size			Performance		P&H or device		Shading		Multipliers		Size	Outcomes
ID	Description (optional)	Option A facades	Option B facades	Height (m)	Width (m)	Area (m²)	Total System U-Value (AFRC)	Total System SHGC (AFRC)	P (m)	H (m)	P/H	G (m)	Heating (S_H)	Cooling (S_C)	Area used (m²)	Element share of % of allowance used
1	se lower 1	NE		2.50	14.30		[illegible]	0.57	1.300	4.600	0.00	2.10	1.00	1.00	35.75	52% of 100%
2	se Upper 1	NE		1.70		15.30	[illegible]	0.57	1.300	1.400	0.93	-0.30	1.00	0.35	15.30	7% of 100%
3	s upper east side	E		1.80	19.20		6.5	0.73	9.700	1.000	9.70	-0.80	1.00	0.21	34.56	19% of 100%
4	s upper west side	E		1.75		[illegible]	6.5	0.73	0.800	1.000	0.80	-0.75	1.00	0.47	9.80	15% of 100%
5	sw lower	SE		2.40	2.32		6.5	0.71	10.000	4.400	0.00	2.00	1.00	1.00	5.57	47% of 100%
6	s lower east side	E		2.40	[illegible]		6.5	0.73	9.700	3.400	2.85	1.00	1.00	0.43	46.08	65% of 100%
7	w lower	S		2.40			6.5	0.99	2.900	4.400	0.00	2.00	1.00	1.00	2.45	14% of 76%
8	w upper 1	S		2.30	[illegible]		6.5	0.99	2.900	2.000	1.45	-0.30	1.00	0.53	17.94	53% of 76%
9	nw lower west side	SW		2.40	27.15		6.5	0.57	5.000	6.000	0.00	2.60	1.00	1.00	65.16	53% of 100%
10	nw upper west side	SW		[illegible]		114.05	6.6	0.57	5.000	1.800	2.78	-2.00	1.00	0.31	114.05	29% of 100%
11	e lower	N		[illegible]	5.50		6.5	0.57	2.600	4.400	0.00	2.00	1.00	1.00	13.20	75% of 100%
	sw upper	SE		[illegible]		22.60	6.6	0.71	10.000	3.000	3.33	-0.20	1.00	0.28	22.60	53% of 100%
	e upper 1	N		1.84	12.00		6.6	0.57	2.600	2.000	1.30	0.16	1.00	0.27	22.08	24% of 100%
14	nw lower	SW		2.40	15.50		6.5	0.57	14.000	1.800	7.78	-0.60	1.00	0.31	37.20	9% of 100%
15	nw upper	SW		2.44		37.90	6.5	0.57	14.000	1.800	7.78	-0.64	1.00	0.31	37.90	9% of 100%
16	w upper 2	S		1.95	5.85		6.6	0.99	2.900	2.000	1.45	0.05	1.00	0.53	11.41	34% of 76%
17	se Upper 2	NE		3.00		34.00	6.5	0.57	1.300	1.400	0.93	-1.60	1.00	0.35	34.00	14% of 100%
18	se Upper 3	NE		3.65		7.30	6.5	0.57	11.000	1.400	7.86	-2.25	1.00	0.18	7.30	1% of 100%
19	se Upper 4	NE		3.90		15.20	6.5	0.57	11.000	1.400	7.86	-2.50	1.00	0.18	15.20	2% of 100%
20	se Upper 5	NE		3.90		22.60	6.5	0.57	11.000	1.400	7.86	-2.50	1.00	0.18	22.60	4% of 100%
21	se Upper 6	NE		3.69		22.30	6.5	0.57	11.000	1.400	7.86	-2.19	1.00	0.18	22.30	3% of 100%

page 1 of 2

Report from GlazingCalc_2013_VolOne_Central Facilities_Prelim Building 2_RevB — printed 4/04/2014

GLAZING ELEMENTS, ORIENTATION SECTOR, SIZE and PERFORMANCE CHARACTERISTICS									SHADING		CALCULATED OUTCOMES OK (if inputs are valid)					
Glazing element		Facing sector		Size			Performance		P&H or device		Shading		Multipliers		Size	Outcomes
ID	Description (optional)	Option A facades	Option B facades	Height (m)	Width (m)	Area (m²)	Total System U-Value (AFRC)	Total System SHGC (AFRC)	P (m)	H (m)	P/H	G (m)	Heating (S_H)	Cooling (S_C)	Area used (m²)	Element share of % of allowance used
22	se Upper 7	NE		2.98		18.50	6.6	0.57	11.000	1.400	7.86	-1.58	1.00	0.18	18.50	3% of 100%
23	se Upper 8	NE		2.30		6.80	6.5	0.57	11.000	1.400	7.86	-0.90	1.00	0.18	6.80	1% of 100%
24	e upper 1	N		2.40	0.50		6.6	0.57	2.600	2.000	1.30	-0.40	1.00	0.21	1.20	1% of 100%
25	se lower 2	NE		2.50	30.20		6.6	0.58	11.000	[illegible]	7.86	-1.10	1.00	0.18	75.50	12% of 100%

IMPORTANT NOTICE AND DISCLAIMER IN RESPECT OF THE GLAZING CALCULATOR
The Glazing Calculator has been developed by the ABCB to assist in developing a better understanding of glazing energy efficiency [illegible]
While the ABCB believes that the Glazing Calculator, if used correctly, will produce accurate results, it is provided "as is" and without [illegible] representation or warranty of any kind, including that it is fit for any purpose or of merchantable quality, or functions as intended or at all.
Your use of the Glazing Calculator is entirely at your own risk and the ABCB accepts no liability of any kind.

if inputs are valid ✓

Figure 8.23 Glazing specification printout.

Source: courtesy of EMF Griffiths Engineers.

The assessment process has used a scorecard scheme for the categories of technologies, design and materials to evaluate the concept sustainable initiatives. These scorecards and their elements are detailed in the following section.

Table 8.2 Synopsis of technologies scorecard used

	Principles	*Design*	*Operations*	*Ethos*	*Resources*	*Experience*	*Verdict*
Photovoltaic power generation	Recognisable ESD principle for reduction in energy use	Energy efficiency technology	Reduction in energy and emissions	Integrated and responsible low energy solution	Reduction of resources associated with grid energy generation	Recognisable ESD principle – symbol of green design	Inclusion of PV energy generation = energy efficiency. Recommended
Solar organic Rankine cycle combined electricity and domestic hot water generation	Sustainable energy hot water generation	Efficient energy and hot water production	Uses local climatic conditions	Integrated and environmental design	Reduction of energy resources and use of local site conditions	Pushing green building boundaries	Not recommended
Solar hot water	Proven technology to decrease energy costs associated with water heating	Energy efficient design	Reduces energy, emission and cost of water heating	Low impact design using available energy on site	Reduce the need for energy resources	Natural expected technology in sub-tropical design	Solar hot water systems = energy efficiency
Geothermal air conditioning	Decreases air conditioning energy use	Energy efficient system	Decrease emissions and energy	Low impact solution	Using local climate resources	Pushing green building boundaries	Feasible to determine cost implications
LED lighting	Low maintenance and reliable lighting	Energy efficient lighting	Minimises lighting energy use, and GHG emissions	Low impact, simple, responsible and efficient design	Low energy use minimises resource depletion	Quality of design	LED lighting = smart design. Recommended
Energy monitoring and management strategies	Proven and reliable way to monitor use and discover faults	Using technology to increase energy efficiency	Assist in maintenance and management of the building	Integrated design solution	Using technology to minimise energy resources	High quality, luxury design	Energy monitoring = smart design. Recommended
Greywater treatment	Proven sustainable technology	Water efficiency by reuse on site	Minimise water use and discharge	Low impact, water conservation	Recycling and reuse of water	Eco friendly solution	Explore feasibility/cost implications
Blackwater treatment	Proven sustainable technology	Water efficiency by reuse on site	Minimise water use and minimise sewage waste	Integrated design. Low impact, water conservation	Minimisation of water resources and discharge	Eco friendly solution	Not recommended
Biomass gasification	Sustainable solution to energy generation	Onsite energy generation	Reduction of waste and decreasing in energy use from the grid	Pushing green building boundaries	Minimising waste and resources	Pushing green building boundaries	Not recommended

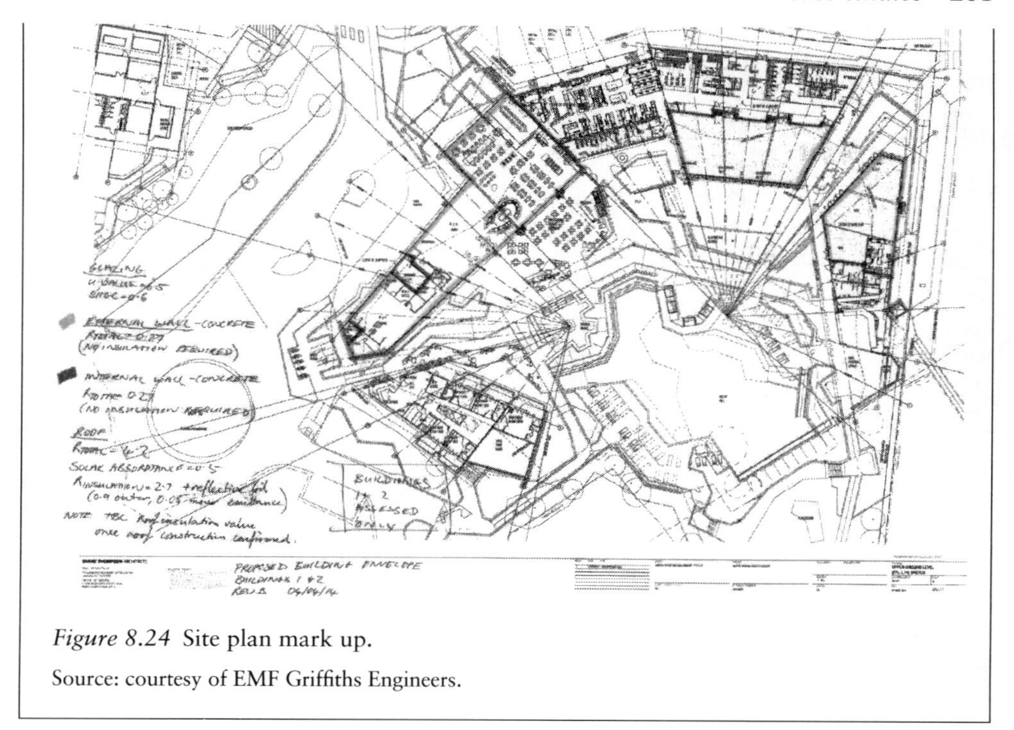

Figure 8.24 Site plan mark up.

Source: courtesy of EMF Griffiths Engineers.

Photovoltaic power generation

Description

Solar Photovoltaic Technology converts light energy into electrical energy. This electrical energy can be used on the site to reduce the energy required from the public grid. There are three basic solar systems types:

- Grid Connected Solar System without battery: this system must have an incoming grid connect.
- Stand Alone System: this system requires batteries to store energy to be released later. This system does not require a grid connect.
- Hybrid Solar System: this system is a combination of batteries and grid connection.

The simplest to implement and most cost effective is Grid Connected Solar. This type of system is quick to implement and involves no maintenance or knowledge to operate; this type of system is what you see on most homes.

This system can be designed as an open field system (solar farm) or installed on individual buildings. The Grid Connected Solar System will simply put energy into the grid supply up to the limit of the grid. When the solar grid system is operating, it will displace any electrical load up to and equal to the value that the solar PV is producing. Any excess energy produced will be sent to the grid network. Some energy providers will pay for this excess energy, currently Origin Energy is paying \$0.06–\$0.08 per kWh. The ideal system design would only produce enough solar PV energy to match the customer's load.

Table 8.3 Comparison of properties of different types of solar panel

Properties	*Crystalline solar*	*Thin film solar*	*Flexible thin film*
Wattage per m^2	140 Watts to 175 Watts per m^2	60 Watts to 90 Watts per m^2	60 Watts per m^2
Effect of heat on output	High	Low	Low
Cost per Watt Budget	$4.15 to $5.40	$3.25 to $4.00	$6
Cloudy day output	Low	High	High
Installation cost	$40 per m^2	$56 to $80 per m^2	Nil
Warranty	25 years to 80 per cent of rated output	20 years to 80 per cent of rated output	20 years to 80 per cent of rated output

Geothermal air conditioning

Description

This technology uses the constant ground temperature to act as a heat sink to allow for improved heat rejection for cooling systems. A ground source heat pump (GSHP) is a heating and cooling system that transfers heat to or from the ground, using the ground as a heat sink in the summer and heat source in the winter. This technology utilises the refrigerant cycle in combination with the constant ground temperature for heat rejection/capture.

It can also operate as a heat pump to provide efficient cooling. The system relies on drilling multiple bores of 100–150 m deep. In cooling mode, water is pumped in a closed loop arrangement by a heat exchanger where heat is transferred from the A/C system. The water circulates through the bores and the heat gained is rejected to the lower temperature sub-surface.

Site application

The site is relatively open and has a large surface area, and there is good access for pipe installation. Refrigerant and water ground source loops are both possible,

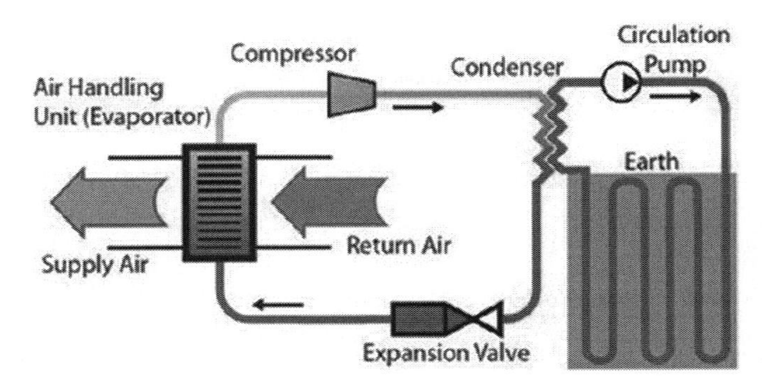

Figure 8.25 Ground source exchange schematic.

though water source loops are more cost effective and have reduced risk of refrigerant loss as key elements of the circuit are localised.

A temperature survey of soil at depth is recommended to determine heat rejection capacity. Heat rejection bores to 30 m depth for 3.5 kW heat pumps to the accommodation A/C fan coil units and 25 off bores at 60 m for 260 kW heat pumps for the central facilities A/C fan coil units (Griffiths, 2013).

Magney House, Bingi Point, New South Wales, 1982–1984

Architect: Glenn Murcutt

A chapter about tropical architecture in Australasia shouldn't pass by without mentioning the work of the architect Glenn Murcutt. Murcutt was born in London to Australian parents. He grew up in the Morobe Province of Papua New Guinea, where he developed an appreciation for simple, vernacular architecture. After moving to Sydney, he was educated at Manly Boys' High School and studied architecture at the Sydney Technical College, from which he graduated in 1961. After the Second World War, Australia witnessed a period of growth and became exposed to a variety of architectural styles and influences, mainly American and Scandinavian through the work of such architects as Mies Van Der Rohe, Bjorn Utzorn, Alvar Aalto and architects fleeing Nazi Germany such as Harry Sediler, who left his native Vienna for America and had studied under Walter Gropius and later worked with Oscar Niemeyer and Marcel Breuer before settling in Sydney. The work of Frank Lloyd Wright and Alvar Aalto had been disseminated through Australia by journals such as the *Architectural Forum*, *Architectural Record* and *Progressive Architecture*, with the culmination of the building of the Sydney Opera House by Bjorn Utzorn. It

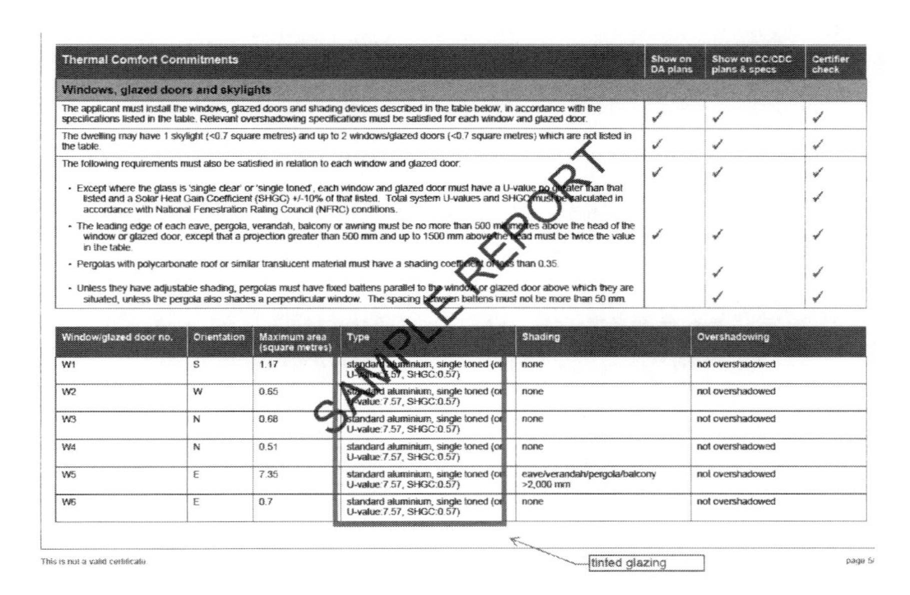

Thermal Comfort Commitments	Show on DA plans	Show on CC/CDC plans & specs	Certifier check
Windows, glazed doors and skylights			
The applicant must install the windows, glazed doors and shading devices described in the table below, in accordance with the specifications listed in the table. Relevant overshadowing specifications must be satisfied for each window and glazed door.	✓	✓	✓
The dwelling may have 1 skylight (<0.7 square metres) and up to 2 windows/glazed doors (<0.7 square metres) which are not listed in the table.	✓	✓	✓
The following requirements must also be satisfied in relation to each window and glazed door:	✓	✓	✓
• Except where the glass is 'single clear' or 'single toned', each window and glazed door must have a U-value no greater than that listed and a Solar Heat Gain Coefficient (SHGC) +/-10% of that listed. Total system U-values and SHGC must be calculated in accordance with National Fenestration Rating Council (NFRC) conditions.			✓
• The leading edge of each eave, pergola, verandah, balcony or awning must be no more than 500 millimetres above the head of the window or glazed door, except that a projection greater than 500 mm and up to 1500 mm above the head must be twice the value in the table.	✓	✓	✓
• Pergolas with polycarbonate roof or similar translucent material must have a shading coefficient of less than 0.35.		✓	✓
• Unless they have adjustable shading, pergolas must have fixed battens parallel to the window or glazed door above which they are situated, unless the pergola also shades a perpendicular window. The spacing between battens must not be more than 50 mm.		✓	✓

Window/glazed door no.	Orientation	Maximum area (square metres)	Type	Shading	Overshadowing
W1	S	1.17	standard aluminium, single toned (or U-value:7.57, SHGC:0.57)	none	not overshadowed
W2	W	0.65	standard aluminium, single toned (or U-value:7.57, SHGC:0.57)	none	not overshadowed
W3	N	0.68	standard aluminium, single toned (or U-value:7.57, SHGC:0.57)	none	not overshadowed
W4	N	0.51	standard aluminium, single toned (or U-value:7.57, SHGC:0.57)	none	not overshadowed
W5	E	7.35	standard aluminium, single toned (or U-value:7.57, SHGC:0.57)	eave/verandah/pergola/balcony >2,000 mm	not overshadowed
W6	E	0.7	standard aluminium, single toned (or U-value:7.57, SHGC:0.57)	none	not overshadowed

Figure 8.26 Example of BASIX assessment based on studio type accommodation.

Source: courtesy of EMF Griffiths Engineers.

is with this academic sustenance that Murcutt formed his early ideas concerning organic architecture and sought local influence from colonial and aboriginal architecture (Beck, 2002).

The history of the Australian house is essentially a succession of styles imported from Britain due to Australia being a colony. Australian houses, together with their awnings and verandas, were essentially a copy of the traditional American cottages of Virginia with all its British influences. Murcutt managed to blend this colonial style with traditional aboriginal features, keeping the elemental principles but delivering them with a modern organic idiom. In rural Australia, Murcutt found a source of reference in woodsheds, farm buildings, barns and greenhouses.

In these buildings he found an architecture that was specifically adapted to the Australian context and free of aesthetic preconceptions. Murcutt won the Alvar Aalto medal in 1992 which had been previously won by such illustrious architects as Aalto himself, Tadao Ando and Alvaro Siza. The singular iconic piece of architecture that typifies Murcutt's work is the Magney House in New South Wales. The house was constructed on 33 hectares on the pacific shores near the small town of Moruya, 350 miles south of Sydney. Murcutt's clients, who for years had camped on the site, wanted a lightweight shelter that was closer to a tent than a country house, in direct communication with nature (Fromonot, 1997).

Murcutt set the building out in an elongated rectangular plan at the high part of the sloping site with living rooms north facing for more sunlight in the daytime which is suitable for a site in the southern hemisphere. Living areas and services facilities were arrayed in two parallel bands along the elongated shape. The roof, an apparently free form, reflects the climatic consideration of the sun's varying penetration through the seasons, and its outline is adapted to the prevailing winds' force and direction to the interior ceiling heights. The smaller facade to the rear of the site which faces south is built of brick up to 2.1 m height and has a full length glazed sloping roof. The brick is consistent with the cold south winds from this direction.

Figure 8.27 Magney House, Bingi Point, New South Wales. 1982–1984. Architect: Glenn Murcutt. Roof detail.

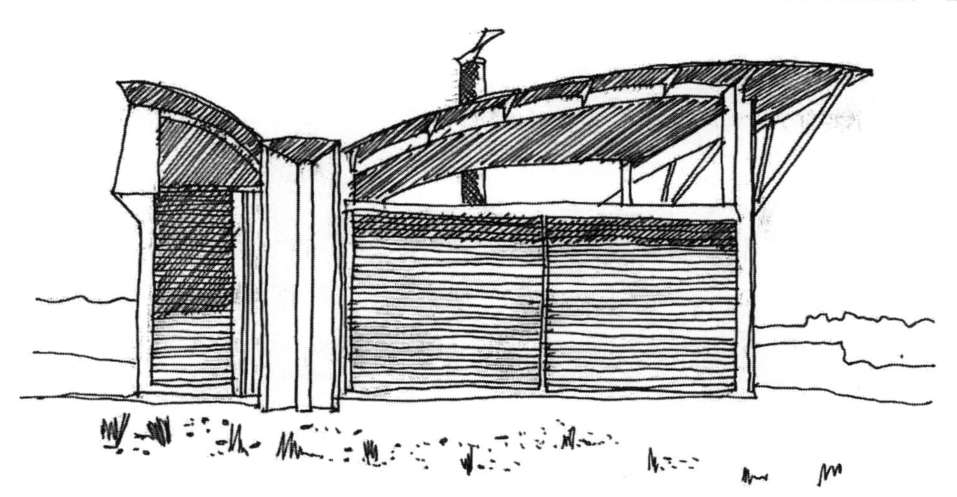

Figure 8.28 Magney House, Bingi Point, New South Wales. 1982–1984. Architect: Glenn Murcutt. Elevation.

There is a sloping shading device along this elevation which can also open up to provide cross ventilation in the hot summer months. The north facade rises to 3.4 m to maximise the sunlight entering the building and the views across the landscape. The large glazed doors, which replaced the louvres that were originally envisaged, are better suited to the climate and provide access to each room. The north facade is fitted with large aluminium venetian blinds, and the gable is partly glazed to frame the view of the headland.

The internal walls are brick up to 2.1 m and glazed above so that the whole building reads as a pavilion as the plasterboard ceiling that hides the roof structure floats and can be seen across the whole space. The gull-wing roof of galvanised corrugated iron can be seen as a thin profile on the landscape that has reminiscences of the agricultural architecture of the region and provides a lightweight structure that blends in perfectly with the surrounding nature.

Acknowledgements

The authors would like to thank Sidnei Junior Guadanhim for his support, and Tecverde Engenharia for its collaboration during the development of this research.

References

Associação Brasileira de Normas Técnicas (2005) – ABNT. NBR 15220–3 – Desempenho Térmico de Edificações. Parte 3 – Zoneamento Bioclimático Brasileiro e Diretrizes Construtivas para Habitações Unifamiliares de Interesse Social. Rio de Janeiro.

Associação Brasileira de Normas Técnicas (2013a) – ABNT. NBR 15575–1 – Edificações Residenciais – Desempenho. Parte 1 – Requisitos gerais. Rio de Janeiro.

Associação Brasileira de Normas Técnicas (2013b) – ABNT. NBR 15575–4 – Edificações Residenciais – Desempenho. Parte 4: Requisitos para os sistemas de vedações verticais internas e externas – SVVIE. Rio de Janeiro.

Atem, C. G. (2002). Desempenho Térmico de uma Habitação em Madeira (Thermal performance of a wooden house). In: Anais do Encontro Brasileiro em Madeira e em Estruturas em Madeira, Uberlândia.

BASIX. 'BASIX Certificates'. Retrieved 8 February 2017, from www.planningportal.nsw.gov.au/planning-tools/basix.

Beck, H. C. J. (2002). *Glen Murcutt: A Singular Architectural Practice*. Victoria, Australia, The Images Publishing Group Pty. Ltd.

Bogo, A. (2003). Avaliação de Desempenho Térmico de Sistemas Construtivos de Paredes em Madeira em Habitações (Thermal performance evaluation of building systems of wooden walls for houses). In: Anais do Encontro Nacional sobre Conforto no Ambiente Construído, Curitiba.

Brasil (2012). Instituto Nacional de Metrologia, Qualidade e Tecnologia – INMETRO. RTQ-R: Regulamento Técnico da Qualidade para o Nível de Eficiência Energética de Edifícios Residenciais. Rio de Janeiro, Brazil. Available at www.pbeedifica.org.br.

Brito, A., Akutsu, M., Vittorino, F., Aquilino, M. and Tribess, A. (2013). Efeito da Verticalização do Edifício e do Sistema Construtivo na Inércia Térmica de Habitações Unifamiliares (Effect of verticalization of the building and the construction system on the thermal inertia of single family houses). In: Encontro Nacional de Conforto no Ambiente Construído, Brasília. Anais do XII ENCAC.

Donadello, A., Nico-Rodrigues, E. and Alvarez, C. (2013). Análise do Desempenho Térmico de Dois Sistemas de Vedação Vertical Utilizados na Ilha da Trindade (Analysis of the thermal performance of two wall systems used in Trindade Island). In: Anais do Encontro Nacional de Conforto no Ambiente Construído, Brasília.

Fromonot, F. (1997). *Glenn Murcutt, Works and Projects*. London, Thames and Hudson, pp. 96–101.

Galindo, A. P. B. (2014). Análise da Eficiência Energética da Envoltória de Uma Habitação de Interesse Social em Wood Frame Aplicada ao Clima de Londrina – PR (Energy efficiency analysis of the envelope of a social house built in wood frame, for the climate in Londrina – PR). TCC. Engenharia Civil. Universidade Estadual de Londrina.

Giglio, T. G. F. (2005). Avaliação do Desempenho Térmico de Painéis de Vedação em Madeira Para o Clima de Londrina – PR (Evaluation of the thermal performance of wooden wall panels for the climate of Londrina – PR). Master dissertation. Programa de Pós-Graduação em Engenharia de Edificações e Saneamento. Universidade Estadual de Londrina.

Griffiths, E. E. (2013). *Energy Efficiency Provisions/Accommodation Units – Issue A*. E. Stewart: 8.

Invidiata, A. (2013). Solar Decathlon: Análise da Eficiência Energética da Casa Ekó House no Cenário Brasileiro (Solar decathlon: analysis of energy efficiency of the eco house in the Brazilian scenario). Master dissertation. Universidade Federal de Santa Catarina, Florianopolis.

Krüger, E. and Zannin, P. (2006). Avaliação Termoacústica de Habitações Populares da Vila Tecnológica de Curitiba (Evaluation of popular housing in Curitiba Technological Village). Ambiente Construído, Porto Alegre, v. 6, n. 2, pp. 33–44, abr./jun.

Silva, R. D. and Basso, A. (2001). Análise de Desempenho de Habitações de Interesse Social em Madeira: Estudo de Caso (Performance analysis of wooden social houses: a case study). In: Anais do II Encontro Nacional e I Encontro Latino Americano sobre Edificações e Comunidades Sustentáveis, Canela.

Souza, H. A., Amparo, L. R. and Gomes, A. P. (2011). Influência da inércia térmica do solo e da ventilação natural no desempenho térmico: um estudo de caso de um projeto residencial em light steel framing. Ambiente Construído. Porto Alegre, v. 11, n. 4, pp. 113–128, out./dez.

Zani, A. C. (1997). Arquitetura de Madeira: Reconhecimento de uma Cultura Arquitetônica Norte Paranaense – 1930/1970 (Wooden architecture: recognition of an architectural culture in the northern region of Parana State – 1930/1970). Doctoral thesis. Universidade de São Paulo.

Index

Page numbers in *italics* denote tables, those in **bold** denote figures.

acoustics **59**, 63, 66, 221
adaptability 172, 225
adaptation 9–10, 13, 16, 21–2, 34, 38, *52–3*, 67, 101, 189–90, 216–17, 225
agenda 3, 13–14, 23–5, 32–5, 177, 216, 217
air conditioning 3, 11, 40, *52–3*, 60, 78, 104, 176–7, 196, 211–13, 232, 234, 236–7, 243, 251, 253, *260*, 262
alpha 42
assessment tools 26, 33–4, 218–25, 238
atmosphere 15, 20–1, 40, *43*, 118, 143–4, 194, 234
awareness 13, 23, 25, 27–8, 36, 64, 211, 214, 218, 222

biocapacity 21, 32
bioclimatic 4–6, 12, 49–51, *53–5*, *57–9*, 62, 65–6, 68, 78, 239, 250
biodiversity 220–1; biodiversity loss 15, 19
Brundtland Commission 13
building/s 1–12, 21–35, 37–44, 49–55, 57–60, 62–70, 72–4, **77–85**, **88–91**, 96, 98–9, 101–4, 106–10, 114, 117–19, 121–3, 125–7, 129–37, 144–5, **148**, 150, *153*, 155, 158, 160–2, 170–7, 184–90, 192–5, 197–201, 203–14, 216–31, 233–41, 243–8, 251–5, *256*, 258–9, *260*, 26–6; Building Environmental Assessment (BEA) 26, *27*, 32, 34–5, 219, 238–9; Building Information Modelling (BIM) 114, 126, 222, 225–31, 238–9; building life cycle 27, 33, 209, 219–20, 222, 225; building pathology 2, 188–90, 193, 195, 197, 199, 201, 203, 205, 207, 209, 211, 213, 215, 217; building performance 11–12, 25–6, 33, 35, 37, 216, 218–19, 221, 223, 225, 227, 229, 231, 233, 235–9; building sector 3, 21–3, 25, 26, 248
built environment 3, 22–3, 25–7, 33, 35, 43, 50, 53, 65, 143, 216, 218, 225, 240

CAD 227
car parking 221
carbon 12–13, 19–21, 23, 25, *27*, 29, 30–5, 49, 213, 218–20, 223, 225, 235; carbon dioxide 15, 19, 21, 235; carbon emission/s 25, 218; carbon reduction 219, 225; carbon sequestration 19, 21
challenges 1, 2, 12, 17, 24, 200, 217–18, 240, 242
changes 4, 8, 10, 15, 24, 27, 39, 40–2, 54, 59, 78, 90, 139, 189, 197, 209–10, 213, 216, 225–30, 233–4
climate 1, 3, 5, 8–23, 25–6, 28–9, 33–8, 40–6, *48*, 50–5, 58–60, 62, 65, 66, 68–9, 78, 80, **83–4**, 89–90, 99, 101, 103–4, 131–2, 136–7, *138*, 139, 160, 167, 177–8, 186, 188–90, 195–6, 198, 211, 213, 216, 217–20, 232–4, 236–9, 241, 243, 248–51, 254, *260*, 265–6; climate change 3, 13–23, 25–6, 33–6, 51, 53, 177, 190, 213, 216–18, 234, 236, 238, 239
CO_2 3, 15–19, 21–3, 25–7, 30–3, 36, 202, 221–3, 242; CO_2 emissions 3, **17–19**, 21–3, 26–7, 30–1, 33, 36, 223, 242
co-efficient 39–40, 42–3; co-efficient of thermal expansion 42
cold-formed steel 42
comfort 9, 12, 44, 50–2, 56–60, 62–4, 66–7, 69, 78, 82, 104, 118, 153, 156, 178, 185, 206, 221, 232–4, 236–7, 244, 249, 251, 259
commercial 1, 3–7, 9, 11, 22, 25, 29, 34–5, 51, 80, 84, 97, 99, 101, 158, 160–3, 169, 170, 174–5, 177–8, 186, 189, 196, 203, 220, 224, 240; commercial buildings 9, 22, 29, 34, 80, 84, 101, 161, 163, 169, 170, 174, 178, 186, 189, 196, 220
compressive 39, 42, 167
construction 1–3, 5–14, 22–9, 31–43, 49, 51, 55, 64, 67, 69, 81–2, 90, 103, 107, 109, 113, 115, 117, 119, 121, 123, 125, 127,

construction *continued*
129, 131, 133, 135, 137, 139, 141, 143, 144–5, 149, 150, 152–3, 155, 157, 159, 161, 163, 165, 167, 169, **170–1**, 173, 175–7, 180, 183–90, 192, 195–7, 199, 203, 205, 206, 208, 209–11, 215–17, 219, 220, 222–6, 228–34, 236–40, 247, 249, 251–3, *256*, 258, 266; construction defects 188; construction industry 22–6, 35, 209, 216, 220, 225–6, 231, 239, 251
consumption 3, 16, 18, 22–4, 29–31, 35, 49–50, 60, 63, 131, 213, 218, 222–3, 225, 232, 236, 251, 253, 254
contaminated land brownfield 221
coordinated 227, 229
corrosion 141, 165–6, 190, 202–3, 211, 216
cost 12, 17, 20, 25, 29, 53, 55, 74, 137, 140, 149, 169, 172, 178, 180, 184, 208, 210–12, 214, 216, 224–31, 236–7, 254, *260–2*
cracks 81, 137, 145, 194, 197–8, 201, 203, 211, 233
cultural and quality of service 220, 224
cyclist facilities 221

dampness 8, 41, 80, 190, 211
database 29, 187, 226, 227, 228
decision/s 25, 27, 29, 31, 33, 35, 38, 51, 208, 209, 210, 212, 213, 216, 217, 227, 228, 229, 235, 249
deforestation 19, 20, 21, 32, 33, 36
deformation 39
demolition 23, 205, 208, 209, 211, 220
design 1–8, 12, 17, 22–9, 32, 35, 37–41, 43–4, 49–55, 57–60, 62–9, 71–2, 78, 80, 82, 84–5, **87**, 89–90, 99, 101–3, 108, 117, 119, 127, 129, 133, 153, 162, 170, 172, 177–8, 186–7, 195–6, 205–6, 208, 211–13, 216–17, 219, 220–1, 223–6, 228–30, 235–6, 238–40, 247, 249, 252–4, 258–61; design detailing 196; design phase 230
developing countries 3, 10–12, 14, 19, 25, 29–30, 32–4, 169, 189, 203, 217, 238
development 2, 3, 5, 7, 9, 11–14, 16–17, 20, 22–6, 28–35, *48*–51, 58, 64–5, 67–8, 90, 101, 104, 132, 162, 165, 194, 209–10, 213, 216–21, 224, 232, 235–6, 239–40, 252–4, 258–9, 265

early stages 28, 50, 228
earthquake 5, 143, 170, 172, 197–8, 224, 245
ecological 13–14, 16, 20–3, 26, 29, 32, 35, 66, 107, 193, 217, 220–1, 232, 252–4; ecological debt 16; ecological deficit 16, 21; ecological footprint **16**, 21, 252; ecological overshoot 16; ecological reserve 21
economic 2, 10, 16, 17, 22, 24–6, 28–30, 34–5, 58, 66, 82, 137, 166, 205, 209, 210, 216–17, 220, 224, 232, 235
effective 11–12, 32, 50–1, 53, 55, 58, 78, 82, 121, 135, 178, 190, 196, 210–11, 225, 229, 231, 235, 254, 261, 263
efficient 23–4, 27–8, 34, 39, 40, 42–3, 56, 63, 65, 120, 150, 196, 219, 222–3, 235, 237, 249, 254, *260*, 262
embodied carbon 29–31
embodied energy 3, 23, 29–31, 33, 183–5, 223, 253
energy 3, 12–13, 15, 19, 22–36, 40, *48*, 49–51, 53–5, 58, 60, 62–7, 82, 89, 101, 120, 131–2, 134–6, 153, 161–2, 176–8, 180, 183–7, 196–7, 208, 211, 213, 216–20, 223, 225–6, 235–7, 239, 241, 243, 248–9, 251, 253–4, 258–9, *260*–1, 266; energy consumption 3, 22–3, 30–1, 49–50, 60, 63, 131, 213, 218, 223, 225, 236, 251, 254; energy efficient 24, 28, 34, 63, 65, 120, 223, 235, 249, *260*; Energy Efficiency Index (EEI) 30, 32
environment 1–4, 6, 9, 12–13, 15–17, 19, 21–7, 29, 31–5, 37–8, 43–4, 49–51, 53–4, 59–60, **62**, 64–6, 69, 78, 80, 90, 101, 141, 143, 175, 186, 192, 196, 198, 206, 211–13, 215–21, 223, 225, 230–1, 234–6, 239–41, 251–3
environmental degradation 15–16, 19, 23 (*see also* climate change); environmental impact 13, 16, 26, 29, 33–5, 49, 213, 219, 222, 237, 241; environmental performance 26, 28, 186, 236
environmentally efficient building stock 219
errors 225, 228–30
evaluation 2, 27, 29, 41, 66, 186, 206, 224, 229, 248, 251, 266
existing buildings 2, 11, 12, 31, 89, 208, 219, 220, 225, 237

facade 51, 54–7, 63–4, 71, 108–9, 126, 129, 131–6, 153, 157–8, 162, 211, **245**, 254, 264–5
facility managers 231
features 1–2, 4–7, 17, 29, 37–9, 51, 65, 68–9, 89, 103, 145, 196, 208, 211, 221, 224, 226–7, 238, 253, 264
ficus 200–1, 211
flaking paint **200**
flexibility 27, 225
flora and fauna 11, 43, 69, 221
fragmentation 225
functional 2, 37–9, 41, 43, 49, 51, 53, 55, 57, 59, 63, 65, 67, 120, 204–5, 207–9, 224–5, 227

GHGs 15, 19; GHG emissions 3, 22–3, 25, *260*
global overshoot 16
global warming 15, 19, 218, 223, 225, 235; *see also* climate change
Green Building Index (GBI) 26, 49, 219, 220
Green Mark 27, 49, 219, 220
green products 222–3
green vehicles 221
greenhouse gas emission 218
greenhouse gasses 3

habitat protection 220
hazards 20, 230, 233
health 11, 15, 17, 28, 34, 37, 51, 56, 59, 69, 101, 186–7, 206, 219, 233–4, 236
high temperature 8, 39, 40, 168, 190, 196, 219
high-income 14, 232
human 14–15, 20, 22, 24, 35, 50–3, 55, 58–9, 63–4, 99, 104, 118, 121, 123, 202, 213, 230, 232, 234, 235, 240
humidity 4, 5, 8–11, 38, 40–1, 43–4, *47*, 51–2, 59, 60, 66, 85, 103–4, 108, 166, 178, 188–90, 194, 200, 202, 211, 219, 232–4
HVAC system 30, 178

indoor air quality 62, 67, 221, 233–4, 253
indoor environmental quality (IEQ) 59, 220–1
innovation 24, 32–3, 217, 220, 224
insect attack 11, 192
integration 59, 63, 224, 236, 239, 241, 254
interact 43, 108, 173, 227
interdisciplinarity 235

key 2, 8–9, 11, *15*, 19, 27, 32, 35, 39–40, 42, 57, 69, 78, 117, 131, 149, 178, 184, 190, 200, 208, 216, 233, 234, 263
Koppen-Geiger Climate Classification 17, 34

latent defects 203
Leadership in Energy and Environmental Design (LEED) 26, 27, 28, 33, 35, 49, 218–19
life cycle 13, 23, 26, 27, 29, 33, 35, 178, 186, 193–4, 196, 198, 206, 209, 219–20, 222–5, 227, 230–1; Life Cycle Assessment (LCA) 13, 26, 29, 222 (*see also* Life Cycle Approach); life-cycle approach (LCA) 23
lifespan 23, 25, 208, 225
lighting 23, 49, 50, 55, **59**, 62–3, 66, 71–2, 78–9, 82, 178, 204, 221, 223, 225, 237, *260*
loading 7, 8, 38, 39, 40
low carbon 23, 30, 32, 220, 223

macroclimate 43, 44, 145
maintenance 2, 11–12, 23, 25, 30–1, 33, 59, 78, 165, 186, 189, 193, 195, 197, 199, 201, 203–9, 211–17, 220, 224–6, 230–1, 233, 235–6, 238–9, 241, *260*, 261
management 2, 11–13, 33, *45*, 65, 180, 203–6, 209, 215–17, 219, 220, 222–3, 227–8, 230–2, 235–6, 239, 254, *260*
material/s 1–12, 19, 26, 28–32, 37–43, 49, 53–5, 57, 60, 64, 69, 74, 78, 80, 82, 91, 97, 103–4, 107–8, 137, 144–5, 149–50, 152–3, 157, 162, 164–5, 167–9, 172, 176, 180, 182–5, 188–90, 192, 194, 196–8, 203, 206, 220, 222–3, 225, 228–30, 233–4, 237–9, 241, 243–4, 246, 248, 253–5, 259; material ingredients 223
mesa climate 43–4
microclimate 43–4, 50, 57–8, 103, 132, 241
mitigation 13, 16, 19, 20–2, 25, 34–5, 58, 127, 143–4, 221, 223
model 66, 109–10, 114, 117, 133, 158, 162, 205, 213, 219, 222, 225–31, 238–9, 243, 254, 256
moisture 9, 39, 40–2, 60, 81, 104, 168
mould 9, 11, 166–7, 183, 190–1, 194–5, 222

National Property Information Centre 49
new and existing buildings 219, 237
new buildings 2, 3, 11, 24–5, 49, 74, 210, 212, 219

operation/s 3, 12–13, 17, 21–3, 25, 27–8, 32–5, 59, 196, 205–6, 212, 214, 217–23, 225–30, 235–8, 253
operational stage 25, 238
organisation 34, 186, 205–6, 217, 219, 231
owners 74, 200, 203, 209, 212, 231

participants 216, 226–8
passive 4, 5, 6, 111–12, 25, 49, 50–1, 53–4, 57, 64, 65, 68, 80, 84, 101, 103–4, 157–8, 161, 176, 186, 211, 223, 240, 243, 252–4; Passive and Low Energy Architecture 50, 101, 186
patent defect 203
performance 2, 5, 6, 8, 11–12, 25–8, 32–3, 35, 37–9, 41, 49, 59, 63–7, 96, 103, 150, 165, 186–8, 192, 196, 212, 216–21, 223–7, 229–31, 233–9, 247–51, 254, 266
planning stage 229
pollution 10, 23, 27, 56, 127, 129, 186–7, 220–4
populated infrastructure 219
power 18, 65, 133, 177, 204, 232, 235, 237, 240, *260*, 261
pre-design 219–20
precipitation 17, 38, 43, *47*, 66, 232, 234

prefabrication 223
preliminary schedule 229
productive 21, 51
productivity gain 228
projects 25, 29, 108, 144, 208, 209, 212, 213, 217, 219, 226, 228, 229, 230, 238, 240, 248, 266
provisions 223, 254, 266

quality 2, 3, 11, 14, 24–8, 58–9, 62, 64, 67, 78, 80–1, 149, 155, 166–7, 178, 205, 209, 212–13, 219–22, 224–6, 228, 233–6, 238–9, 248, 251, 253, *260*

rating tool 26
reduce time 226
refurbishment 2, 23, 188, 208, 209–19, 225, 238
reinforced cement concrete 42
relationships 227–8
relative humidity 5, 11, 51, 59, 60, 104, 178, 190, 194, 211, 232–4
reliable 227, *260*
renewable energy 23, 51, 65, 89, 186–7, 223, 237, 239
renovation *27*, 49, 208, 215, 217, 219
repair 12, 30–2, 188, 190, 205, 231, 233
replacements 231
repository of information base 228
research 2, 24, 26, 28–33, 63, 65–6, 85, 101, 120, 174, 185–7, 194, 203, 213–19, 225, 231, 234, 238–9, 242, 248, 265
residential 1, 3, 4, 6, 22, 29, 34, 49, 51, 63, 65, 67, 84, 97, 99, 101, 120, **121**, 195, 203, 219–20, 223–4, 240, 248, 249
responsive 38, 65, 71, 80, 169, 178, 239
restoration 11, 208, 217
restructuring 231

safety 206–7, 224, 230, 236
satisfactions 236
save cost 226
sea level rise 15, 20
series of management 219
shading 4, 12, 51, 53, 55–7, 63, 65, 67, 85, 104, 108, 109, 114, 118, 119, 120, 123, 139, 150, 152–5, 158, 160, 162, 174, 178, 186, 211, 223, 246, 253, 265
smart 216, 228, *260*
social 13, 17, 22, 24–8, 35, 67, 97, 101, 118, 206, 209, 210, 215, 220, 224, 241, 248, 249, 265, 266
social integration 224
spaces 4, 6, 49, 50, 51, 66, 69, 71, 84–5, **88**, 108, 117, 120–1, 123, 125, 127, 150, 155, 157, 165, 178, 224, 231, 237, 246, *256*
stakeholders 23, 25, 27, 69, 235
standards 24, 28, 50, 70, 101, 196, 208, 211, 221, 223, 236, 241, 242, 251
strategies 13, 16, 21, 30, 35, 49, 50, 51, 53, 57, 58, 60, 63, 65, 144, 172, 213, 218, 221, 223, 235, 249, *260*
structural 7, 8, 10–12, 31, 38–40, 42–3, 56–7, 117, 120, 162, 168, 169–72, 186, 197–8, 203–5, 226, 231, 233, 252
sustainable 2, 10, 12–14, 16–17, 22–9, 32–6, 49–50, 58, 65–7, 69, 74, 101, 104, 162, 180, 182, 185–8, 208–9, 211, 215–23, 237–8, 240, 252–3, 258–*60*
sustainability 2, 3, 13–17, 19, 21–7, 29, 31, 33–5, 37, 49, 50, 65, 183, 208–11, 216–19, 224–5, 227, 240, 258–9
sustainable construction 9–10, 13–14, 22–5, 27–8, 32–3, 36, 67, 180, 215–17, 238; sustainable development 13–14, 16–17, 24, 26, 28–9, 32–3, 35, 49, 58, 67, 101, 162, 185, 218, 252, 253; sustainable refurbishment 188, 208, 209, 216

temperature 4, 5, 8–9, 15–16, 38–42, *46*–7, 51–3, 55–9, 78, 85, 103–5, 108, 138–9, 140–5, 149–50, 152–3, 159, 168, 178, 188, 190, 193–4, 196–7, 211, 213, 219, 231–4, 237, 239, 249, 251, 262–3
tensile 38, 42, 202
thermal 1, 4, 6, 8, 11, 39–42, 50, 54–5, *57*, 59, 60, 62, 64, 66, 78, 101, 123, 133–6, 144, 150, 152–3, 155, 165, 168, 178, 185–7, 197–8, 211, 221, 232, 234, 236–7, 239, 243–4, 247–9, 250–1, 254–6, 258–9, 266
tile popping 198, **199**
Total Request Information 229
transport 30–1, 69, 187, 220–1
tropical 1–21, 28–34, 37–49, 51, 54, 59, 60, 62–6, 68–71, 74, 78, 80–4, 89, 90, 92, 97, 99, 101, 103–7, 109, 114, 117, 119, 126, 129, 132, 136–9, 142–5, 150–3, 155, 157–8, 162–4, 166–8, 170, 175–6, 178, 180, 183, 185–90, 192–8, 200–2, 208–14, 218, 220, 231–4, 236–40, 242, 248, 251, 253, 258, 263; tropical climate 1, 5, 9, 11, 12, 28–9, 38, *45*, *47*, *48*, 50, 55, 60, 68, **83**, 99, 101, 104, 189, 195, 196, 198, 211, 213, 220, 233, 238, 248; tropical region/s 17–19, 21, 40, 43, *45*, 74, 90, 145, 152–3, 190, 202, 239; tropical zones 5, 14, 72, 197, 200, 218, 231, 233
tsunami 5, 197, 198

upgrading 208, 211, 214, 231
urban 2, 7, 12–14, 22, 25, 32, 35, 44, 49, 50, 58, 64–6, 78, 99, 131–6, 171, 174, 178, 185–7, 210, 213, 219, 240; urban development 2, 12, 210, 213, 219; urban

heat island 58, 134; urban population 14, 49; urban sprawl 50
urbanization 3, 16, 30, 34, 78

value engineering 229
variance 4, 5, 7, 37, 41, 103, 104, **105**, 123, 172, 175
ventilation 3, 4, 5, 6, 11, 12, 28, 40, 49–52, 55–62, 64–5, 67, 71, 74, 78, **79**, 84–5, **86**, 99, 104, 106, 108, 117, 119–21, 131, 144–5, 149, 153, 155, 158–9, 167, 170, 174–6, 187, 211, 221, 232–4, 241, 243–4, 250, 253, 258, 265

waste 10, 16, 21, 23–5, 27, 178, 180, 184, 219–20, 222–3, 230, 254, *260*
water 9, 11, 16, 20, 27, 40–1, 43–5, 58, 65, 71–2, 82, **87**, 89, 98–9, 103, 137, 139, 140, 144–5, 150, 160, 165–6, 178, 180, 186, 188–9, 190, 192, 196–8, 200, 204, 208, 211, 220, 222, 233–4, 237, 240, 244, 252–4, 258–*60*, 262, 263
weather 9, 11, 15, 30, 33, 37, 44–6, *48*, 70, 78, 89, 103, 139, 186–8, 196, 233, 238, 243, 254
World Bank 14, 17–19, 24, 35
world footprint 16, 33

zone 20, 44, 51–2, 54, 59, 69, 131, 145, 171, 187, 232, 237, 244–5, 250